Re-Thinking Green

Alternatives to Environmental Bureaucracy

Re-Thinking Green

Alternatives to Environmental Bureaucracy

Edited by
Robert Higgs and
Carl P. Close

Oakland, California

The Independent Institute
100 Swan Way, Oakland, CA 94621-1428
Telephone: 510-632-1366 • Fax: 510-568-6040
Email: info@independent.org
Website: www.independent.org

Library of Congress Cataloging-in-Publication Data

Re-Thinking Green: Alternatives to Environmental Bureaucracy /
edited by Robert Higgs and Carl P. Close.
p. cm.
Includes bibliographical references and index.
ISBN 0-945999-97-6 (pbk. : alk. paper)
1. Environmentalism. 2. Green movement. 3. Environmental policy.
4. Environmental impact analysis. I. Title: Rethinking green. II. Higgs, Robert.
III. Close, Carl P.

GE195.R4 2004
333.72--dc22
2004029532

10 9 8 7 6 5 4 3 2 1

THE INDEPENDENT INSTITUTE is a non-profit, non-partisan, scholarly research and educational organization that sponsors comprehensive studies of the political economy of critical social and economic issues.

The politicization of decision-making in society has too often confined public debate to the narrow reconsideration of existing policies. Given the prevailing influence of partisan interests, little social innovation has occurred. In order to understand both the nature of and possible solutions to major public issues, The Independent Institute's program adheres to the highest standards of independent inquiry and is pursued regardless of political or social biases and conventions. The resulting studies are widely distributed as books and other publications, and are publicly debated through numerous conference and media programs. Through this uncommon independence, depth, and clarity, The Independent Institute expands the frontiers of our knowledge, redefines the debate over public issues, and fosters new and effective directions for government reform.

THE INDEPENDENT INSTITUTE
100 Swan Way, Oakland, California 94621-1428, U.S.A.
Telephone: 510-632-1366 • Facsimile: 510-568-6040
Email: info@independent.org • Website: www.independent.org

Contents

1

Introduction

ROBERT HIGGS AND CARL P. CLOSE

The title of this book prompts some obvious questions: What exactly do we mean by "environmental bureaucracy," and why are we asking readers to rethink it? What principles inform this book's critique of current environmental policies, and what evidence suggests that these principles would provide a better guide to environmental policy?

The answers to these questions will emerge as the following pages are read, pondered, and debated, but asking readers to hold their questions until they discover the answers for themselves would be more than coyly evasive—it would forfeit an opportunity to explain why more and more analysts and observers are seeking alternatives to the current approach to environmental protection. The story of that search begins with a brief look at the rise in environmental regulation and how it came to exemplify the dreariness of politics as usual.

ENVIRONMENTAL "AWAKENING" AND POLICY REALISM

Contrary to popular mythology, the environmental "awareness" that came of age publicly in the United States with the first Earth Day in 1970 did not spring fully formed from the brow of the Woodstock generation or even from Rachel Carson's *Silent Spring* eight years earlier. Rather, it was the culmination of earlier efforts by John Muir, Gifford Pinchot, Theodore Roosevelt, Aldo Leopold, and others. These thinkers and policymakers' legacy is hardly monolithic. Many writers today distinguish, for example, between the "conservationists" and the "environmentalists" among these seminal influences, thus revealing inconsistencies in the goals of environmental policy (Chase 2001).

Nor are historians' interpretations of these legacies uniform. Samuel P. Hays (1959) has argued that the conservationists of the late nineteenth and early twentieth centuries were more concerned with promoting wise environmental stewardship for human purposes than is suggested by the rhetoric of preserving wilderness for its own sake. More recently, however, a case has been made that the early conservationists were not the efficiency-driven, public-interest technocrats that Hays suggested and that their environmental evangelism was more spiritual than scientific (Rubin 2000). The point here, however, is not to trace the history of environmental thought, but merely to indicate that environmental history is a tangled thicket.

Similarly, significant differences exist in how best to interpret the history of environmental regulation. One source of such differences (one not touched on in this volume) is the inherent complexity of the scientific aspects of numerous environmental issues. Another source, which may often be more difficult to resolve, is the problem of understanding how particular regulations operate *in the real world.* The last four words are emphasized because for too long the goals and efficacy of particular regulations were taken at face value. The goals of a regulation, to the extent that they are influenced by electoral politics, were often assumed to coincide with "the public interest," and seldom were proposed regulations examined sufficiently to establish whether they would advance a reasonable conception of this goal in the presence of real-world political and institutional constraints.

Another type of "awakening" has also taken place in recent decades. Public choice, a young hybrid discipline of economics and political science, aims more or less explicitly to remedy the deficiencies that result from taking public-policy proposals at face value by prescribing a strong dose of policy realism. Rather than ask, idealistically, "What would we like governments to do?" public choice asks, realistically, "What are governments likely to do, given the incentives and constraints of politicians, voters, bureaucrats, and special-interest groups?" Like the consumer who discovers that sausages are less appetizing after he or she sees how they are made, students of public choice learn that political processes—even revered democratic ones—often result in outcomes far different from those described in high school civics classes.

The dethroning of idealized conceptions of political institutions has not been solely a "destructive" affair, however. Public choice has contributed a number of positive insights that can promote the improvement of our institutions. It recognizes, for example, that voters typically display "rational ignorance" in public-policy matters, which means that they often view the opportunity costs of becoming well informed about particular political candidates and proposed changes in the law as greater than the perceived benefits of possessing such information. Rational ignorance thus contributes both to voter "apathy" and to politicians' tendency to pose as "doing something" to fix a perceived problem rather than to deal with the problem's root causes. Recognizing the risks of rational ignorance is the first step toward reforming our institutions.

Public choice also emphasizes that government programs have disparate impacts on people and that such disparities affect the political process. In many cases, for example, a program's relatively few beneficiaries are subsidized by a greater number of taxpayers or others disproportionately burdened by the program's costs. The combination of concentrated benefits and dispersed costs encourages lobbying efforts by the potential beneficiaries. In fact, it often means that a formal lobbying organization, strategically located near political or media centers, will emerge to court favor with policymakers in order to maintain or expand the benefits and to capitalize on the rational ignorance of those who must pay the bill. Thus, the risk of disproportionate influence by special-interest groups is another hazard that must be considered when reforming political institutions.

The preceding public-choice insights also cast aspersions on the assumption (still taken seriously in many classrooms) that politicians and bureaucrats act as selfless representatives of the electorate. This assumption seems especially odd given the widespread recognition that when these government functionaries act as private consumers, their behavior corresponds predictably to their perceived material self-interest. Yet government functionaries do not undergo a profound personality transformation each time they don their "public-interest" vestments. Rather than clinging to the childlike assumption that they do, public choice holds that material self-interest helps to explain their behavior (for example, their quest to win reelection or their desire to detect new problems to fix in order to justify a larger budget).

In the real world, this incentive problem combines with the knowledge problem: even if politicians and bureaucrats place the "public interest" above their material self-interest, the problem remains of how to define that goal and how best to discover and aggregate the knowledge necessary to achieve its realization.

These points are but a few of the better-known concepts that inform public-choice analysis (see Mitchell and Simmons 1994 for a more thorough treatment), but they are sufficient to suggest an important piece of public-policy advice. Proposed reforms must take into account the cold reality that political institutions seldom perform in the real world as we would like, and government policy must be assessed in a real-world context, not in relation to a textbook list of ideal public-policy criteria (the "nirvana" approach). The risk of government failure cannot be ignored without cost.

THE FAILURES OF ENVIRONMENTAL BUREAUCRACY

Environmental policy in the United States is not entirely without success stories, but the contributors to *Re-Thinking Green* argue that, for the most part, it has been unexpectedly costly, corrosive to America's liberal political and legal traditions, and not very effective in enhancing environmental quality. These failures are rooted in the bureaucratic, top-down approach that has characterized U.S. environmental policy.

Ineffective Policies

If environmental policies worked as they were intended to work, there would be no need for this book (although other problems in environmental policy would still merit discussion). In too many cases, however, environmental policies have failed to advance policymakers' stated goal—namely, to improve environmental quality.

Policies intended to protect endangered species, for example, have rescued few species from their endangered or threatened status. The Endangered Species Act, in particular, has created disincentives for private landowners, who lose much of the economic value of their land when a member of an endangered species is discovered on it. In Africa, the global ban on trade in ivory has created a profitable black market in ivory, leading to widespread poaching of elephants. Protection of endangered species thus fails to live up to Hippocrates' directive that healers should "first, do no harm," as do many other environmental policies that create perverse incentives.

Policy fads have also caused their share of policy failures. In the United States, for example, cities whose governments have sought to improve the urban environment by implementing "smart-growth" policies have experienced increased crowding, traffic congestion, escalating housing prices, and deterioration in the quality of urban living.

In any event, the fact that environmental policies often fall short of their aim has been perhaps the most influential reason for seeking new approaches to environmental problems.

Costly Policies

The prevailing command-and-control approach to environmental policy has entailed heavy costs. For example, Clyde Wayne Crews Jr. has recently estimated that in 2002 the costs of enforcing and complying with U.S. environmental regulations amounted to $201 billion (2003, 5). Such costs spill over from the businesses, households, and governmental entities directly subjected to the regulations and affect everyone: the costs of production and hence the market prices of many goods and services have been elevated greatly; other goods have been restricted or eliminated entirely (e.g., asbestos, Freon), compelling consumers to make do with inferior substitutes.

In chapter 2, "Prophecy de Novo: The Nearly Self-Fulfilling Doomsday Forecast," Craig S. Marxsen cites studies that conclude that by 1990 the regulations issued by the Environmental Protection Agency (EPA) had resulted in an estimated 22 percent reduction in manufacturing output and a 21 percent reduction in U.S. gross domestic product compared with what would have been produced in the absence of the regulations—statistics that have considerable significance for families struggling to eke out a living. Even those who believe that the benefits of these regulations have been substantial may be hard pressed to believe that such benefits have exceeded these enormous costs and hence that the regulations can be defended on efficiency grounds.

Legally and Politically Corrosive Policies

The most important claim advanced in various chapters of *Re-Thinking Green* is that environmental bureaucracy has contributed to the corrosion of America's best legal and political traditions, undermining such core institutions as private-property rights, the rule of law, representative government, and individual liberty. The reasons are readily apparent. Throughout its history, the U.S. government has grown during times of crisis (real, imagined, or manufactured), especially military and economic crises. When a crisis occurs, traditional obstacles to the expansion of government give way to political expediency, allowing new government powers, higher taxes, and new restrictions on individual liberty. After the crisis subsides, government may be scaled back to a degree, but the retrenchment is seldom complete. The establishment of new legal, political, and cultural precedents; the softening of ideological opposition to expanded government; and the creation of a new class of lobbyists (composed of politicians, bureaucrats, special-interest groups, and voters who see themselves as beneficiaries of the new powers) create pressures for the maintenance and expansion of the new powers. In short, national crises tend to create a climate conducive to "ratcheting up" the size and scope of government at the expense of individual liberty (Higgs 1987).

In recent decades, sundry environmental "crises" have played a growing role in altering America's legal and political institutions. These crises have inspired much ill-considered and hastily enacted legislation that has crowded out potentially far more effective approaches to dealing with environmental problems. Statutes and regulations rather than common-law rules of nuisance, trespass, and strict liability now govern environmental law—often to the detriment of environmental quality. For example, the U.S. Supreme Court ruled in *Milwaukee v. Illinois* (451 U.S. 304 [1981]) that the common law may not be used to establish emission standards more stringent than those set by the Clean Water Act and its attendant regulations (Meiners and Yandle 1992; Brown and Meiners 2000). Thus, the act forces every jurisdiction and watershed to rely on a standard that is too stringent in some cases and too lax in others.

Or consider another example of how environmental law has changed the U.S. legal system. The Pollution Prosecutions Act of 1990, as Craig Marxsen explains in chapter 18, made the indictment of "environmental criminals" a growth industry for government, especially in prosecuting small business owners with few legal resources. Many of these cases seem absurdly unfair because the laws are not widely known, and the penalties are not proportionate to the harm. A Honolulu wastewater plant manager was sentenced to thirty-three months in prison (his partner received twenty-two months) for discharging only 6 percent more effluent than his permit allowed, and the court refused to hear the claim that the two managers mistakenly believed their actions were lawful under their permit.

Although it is impossible to say with precision how the common law would have evolved in the absence of the Clean Water Act, it is likely that it would have developed means more appropriate for diverse conditions than the act's one-size-fits-all approach. Still, even though the common-law approach to water pollution may be far superior to the Clean Water Act, it has little chance of returning to the forefront because the act has created constituents who are effective in maintaining the legal status quo. (Current antitrust laws also inhibit promising approaches to environmental protection; see Yandle 1998.)

Environmental bureaucracy is also having a major impact internationally. Reports of global climate change, for example, have prompted political leaders around the world to draft and support the Kyoto Accord, which, if ratified, will mandate restrictions on greenhouse gases in an attempt to reduce emissions to 1990 levels. The international scope and broad regulatory sweep of the Kyoto Accord would make it an unprecedented challenge to the traditional concept of national sovereignty. In the United States, the prospect of the adoption of the Kyoto Accord has enlisted intense lobbying efforts not only by those who oppose it, but also by special interests (including some segments of industry) that expect to benefit from its passage. In this respect, however, the Kyoto Accord does not differ from other government regulations that create "winners" who see these regulations as conferring a special advantage to them at the expense of competitors with less political clout.

OVERVIEW

The Seeds of Environmental Bureaucracy

Part I, "The Seeds of Environmental Bureaucracy," contains critiques of two related paradigms that have guided much popular thinking about environmental issues and have helped to spawn environmental bureaucracy. The first is what Marxsen calls "eco-catastrophism"—the view that unless population growth and industrial production are soon reduced substantially, humankind will condemn itself to a world barren of natural resources, buried in pollution, and hence inevitably suffering massive loss of human and nonhuman life.

Doctrines of global environmental calamity are at least as old as the writings of Thomas Malthus two centuries ago, but with the 1972 publication of *The Limits to Growth: A Report for the Club of Rome's Project on the Predicament of Mankind,* eco-catastrophism found a new audience and gained renewed respectability, owing to the book's use of seemingly scientific computer models. Although *Limits to Growth* was, as Samuel P. Hays has noted, "one of the first, and the classic, statement[s] of the dangers of the increasing human load on the finite environment" (2000, 235), it was as flawed as it was influential, according to Marxsen. Economic-growth theorists, for example, found that *Limits to Growth* assumed, rather than established empirically, a connection between its model and reality; failed to incorporate the moderating effects of price increases on resource consumption; and relied on the highly unempirical assumption that technological progress will come to a grinding halt. The book's 1992 sequel *Beyond the Limits* and other studies it inspired also ignored the behavioral effects of price changes and misunderstood the role of technological progress in offsetting diminishing returns in production.

Paradoxically, although *Limits to Growth* failed to convince neoclassical economic-growth theorists of the validity of the notion of inherent limits to growth, the book may have helped, Marxsen argues, to bring about a long-run economic slowdown akin to the decline that it had prophesied. As Mancur Olson (1982) has explained, collective action by coalitions often promotes coalition members' interests at the expense of the public at large. Marxsen, as noted previously, cites studies published in the 1990s that suggest that by promoting regulations that hamper economic growth, successful lobbying by environmental coalitions illustrates Olson's point nicely. If people had a better appreciation of environmental policy's great economic costs, they might come to regard the Cassandras of eco-catastrophism as antisocial perpetuators of poverty rather than as the visionary benefactors of humanity so often portrayed in the mass media and popular culture.

The name of the remedy proposed to manage the alleged limits-to-growth crisis is *sustainable development.* Although most discussions link sustainable development with the goal of preserving environmental quality for the sake of future generations, there is no consensus on its definition—a problem that contributes to an inability to measure it and thus to determine whether a particular policy does much to promote it (Beckerman 2003). Nonetheless, sustainable development is increasingly invoked to rationalize the expansion of environmental bureaucracy at local, national, and international levels.

In chapter 3, "Doomsday Every Day: Sustainable Economics, Sustainable Tyranny," Jacqueline R. Kasun argues that sustainable development suffers from another significant problem as well. According to widely recognized proponents, sustainable development is entirely compatible with policies that others would consider toxic to the *social* environment: birth licenses, the abolition of private landownership, a forced reduction in trade to create economic self-sufficiency at local levels, the resettlement of human populations to create larger wilderness areas, and other extreme measures offered to ameliorate the environmental crisis that these proponents claim awaits us otherwise.

Such extreme proposals may currently circulate only among the environmental avant-garde, but they may be harbingers of policies to come. Proponents of these proposals have written leading textbooks on sustainable development and received support from organizations as influential as the World Bank. If sustainable development entails a political order more compatible with totalitarianism than with a free society, as Kasun suggests, we might well expect that its more liberal proponents eventually will begin to reexamine their assumptions about it.

Global Issues

Part II, "Global Issues," contains examinations of what are often considered the two greatest threats to the natural environment: population growth and global warming. Population growth is widely seen as necessarily destructive of the most important or unique environmental amenities, such as clean air and biodiversity. In chapter 4, "Population Growth: Disaster or Blessing?" the late Peter T. Bauer critiques the more general assumption that population growth is a major obstacle to economic progress in developing countries, an assumption widely held among international development specialists. Bauer also examines whether parents bear the full costs of having and rearing their children or instead impose costs on others and whether pressuring parents to have fewer children is the best means to reduce alleged spillover costs.

Chapters 5 and 6 contrast the operation of markets with regulatory approaches to the problem of climate change. In chapter 5, "After Kyoto: A

Global Scramble for Advantage," Bruce Yandle argues that some organizations have backed efforts to restrict greenhouse-gas emissions not primarily because they believe that doing so is required by a sense of corporate responsibility or environmental stewardship, but because they expect to have a comparative advantage in operating under the new regulatory regime and therefore will be better situated to outperform competitors and reap profits in the marketplace.

Yandle's analysis implies that *political* entrepreneurship (of which corporate support for the Kyoto Accord is but one instance among the many he has studied over the years) may be as important in determining the outcomes in highly regulated markets as traditional consumer-oriented entrepreneurship is in less-regulated markets. The risk of wasteful and anticompetitive political entrepreneurship is therefore an important reason to be leery of bureaucratic approaches to solving public-policy problems, especially where nonbureaucratic alternatives exist or can be encouraged by policy reforms.

What would a nonbureaucratic approach to global warming look like? Because market outcomes are contingent on ever-changing consumer preferences, technologies, and entrepreneurial know-how, it is impossible to predict with precision the effects of such a decentralized market approach to climate change. But a few general possibilities are suggested in chapter 6, "Global Warming and Its Dangers," by J. R. Clark and Dwight R. Lee.

Endangered Species

Part III, "Endangered Species," relates to efforts to preserve vanishing wildlife. In the United States, the main legislation protecting threatened wildlife is the Endangered Species Act (ESA). Despite wide public support for species protection and ample time for policymakers to perfect the 1973 law, instances of species recoveries attributable to the ESA are difficult to find. In fact, more species have been removed from the endangered species list—when the U.S. Fish and Wildlife Service discovered that these species actually were not threatened with extinction—than have been saved by the ESA.

Furthermore, a case can be made that the ESA has actually *harmed* endangered species because of its absolutist goal of saving all endangered species, its failure to offer guidelines for setting priorities, and its disincentives for landowners to report the discovery of members of endangered species on their own property (which in most cases prevents landowners from altering their property), as Randy T. Simmons argues in chapter 7, "The Endangered Species Act: Who's Saving What?"

The problem of landowner disincentives is especially troublesome, Simmons explains, because three-fourths of the species on the endangered list live at least in part on private land and perhaps half of the listed species live exclusively on private land. The loss in property values to landowners is a significant, albeit hidden cost of the ESA, to which should be added the costs of funding the endangered species–related projects carried out by such federal agencies as the U.S. Fish and Wildlife Service, the National Marine Fisheries Services, the Bureau of Land Management, the Army Corp of Engineers ("the world's premier public engineering organization," according to its Web site), and a plethora of agencies at state and local levels. Seldom has one piece of environmental legislation cost so much yet yielded so little in tangible benefits (unless one counts the creation of a bloated, intrusive environmental bureaucracy itself as a benefit).

Fortunately, potential solutions are available, as Simmons explains in chapter 8, "Fixing the Endangered Species Act." He advances eight guiding principles to overcome the failure of policy regarding endangered species—four that emphasize frequently neglected biological truths and four that emphasize politico-economic truths also often overlooked or evaded by environmental bureaucrats, such as the power of creating positive incentives for landowners.

Significantly, parts of Africa show signs of a budding backlash against the disincentives created by inflexible wildlife conservation efforts. Since at least the early 1990s, African proponents of community-based natural-resource management have sought to reform public policy by offering economic-use benefits to encourage local populations to support conservation. To date, however, their efforts have been opposed by influential Western conservationists, as Robert H. Nelson describes in chapter 9, "Environmental Colonialism: 'Saving' Africa from Africans." For example, in Tanzania, widely considered one of the most environmentally progressive countries in Africa, Western environmental organizations have successfully pressured the government to adopt top-down conservation policies that deprive native populations of their long-standing role as environmental stewards. Consequently, dry brush and the deadly tsetse fly have overtaken large tracts of land, crowding out both human beings and other plant and animal species.

Eco-colonialist policies have also damaged the social environment by displacing and impoverishing native populations. This tragedy is especially ironic given that many backers of eco-colonialist organizations consider themselves political progressives who oppose traditional colonialism. Perhaps the oversight occurs because the attraction of eco-colonialism lies further below the surface of human affairs. Whereas traditional colonialism is motivated by mercantilist economic fallacies and myopic geopolitical visions, eco-colonialism, according to Nelson, is rooted in a quasi-religious yearning for a Garden of Eden unspoiled by human intervention.

If unmoved by the loss of human dignity caused by eco-colonialism, its adherents might perhaps experience a crisis of faith from their discovery that the forced removal of Africans from their historical lands has also harmed non-human species. Should this discovery occur, the growing body of literature on community-based natural-resource management cited by Nelson will be available to inform policymakers—and to give solace to the newly contrite.

An especially destructive instance of eco-colonialism is the prohibition of the international trade of ivory, examined in chapter 10, "The Ivory Bandwagon: International Transmission of Interest-Group Politics," by William H. Kaempfer and Anton D. Lowenberg. Against biologists and economists' advice, trade in ivory was made illegal in 1989 by the Convention on International Trade in Endangered Species of Wild Fauna and Flora (CITES), the institutional framework through which countries cooperate to advance species protection. Although the ban on ivory trade was partially lifted in 1997 to allow limited sales from existing ivory stockpiles, widespread trading remains illegal. The ban is a sad but fitting example of how good intentions can give rise to harmful consequences. The ban has harmed not only local villagers (especially farmers whose crops are devastated by free-roaming elephants) and wildlife parks (whose resources are strained by the heavy impact of hard-to-contain elephants), but also the African elephants themselves, which, as noted earlier, are often killed by poachers eager to profit from high black-market ivory prices. The ban has also undermined successful elephant-conservation programs in Zimbabwe and Namibia, which had allowed rural Africans to benefit from elephants through "consumptive utilization."

Given the ban's predictable, undesirable effects and the opposition to it from respected conservationists, why was it adopted? As Kaempfer and Lowenberg explain, three factors were especially important in the ban's eventual adoption: a massive public-relations campaign by animal-rights activists and other supporters of the ban, increased "reputational utility" from supporting the ban, and increased reputational cost imposed on those who might dare to speak against the ban.

The notion that elephants can be viewed as anything other than adorable creatures runs counter to Western perceptions, but in parts of rural Africa it is an undeniable fact that any conservation policy worthy of the name must take into account this notion. Fortunately, according to Kaempfer and Lowenberg, successful elephant-conservation programs in Namibia and Zimbabwe, which give local parties a direct stake in the survival of elephant populations ("consumptive utilization" or "sustainable utilization"), take the local people's economic needs into consideration and can serve as a model for other countries' elephant-conservation policies.

Entrepreneurship, Property Rights, and Land Use

Part IV, "Entrepreneurship, Property Rights, and Land Use," examines some of the key institutions for providing environmental amenities. In chapter 11, "Free Riders and Collective Action Revisited," Richard L. Stroup explains how the free-rider problem lies at the root of government failure, including failures in environmental policy, and why private entrepreneurs often outperform government bureaucrats in delivering the (environmental) goods.

The free-rider problem (of too many people receiving without payment the benefits of goods produced by others and thereby causing the under-provision of those goods) is typically cited as the justification for government provision of "public goods." If, for example, enough people contribute to a cleaner environment, mosquito abatement, or biodiversity, an individual has no economic incentive to contribute to their efforts. But if too many individuals attempt to free ride on others' efforts, there will be too few contributors to ensure that the environmental amenities actually get supplied. The free-rider problem therefore suggests that government provision of public goods is socially efficient—or so the standard story goes.

Stroup, however, shows that the free-rider problem does not end where the public sector begins. Rather, the free-rider problem gives rise to an under-supply of government policies that actually promote the general public's interests. This outcome occurs because information is costly to gather and interpret accurately, and most voters and policymakers prefer to economize on this cost by relying on other people's judgments about public policy. Yet when it becomes too costly for the public to monitor public policy, special-interest groups find that they have a comparative advantage in influencing public policy, a tendency that skews outcomes toward their own ends rather than toward the general public's ends. (Stroup notes, however, that public and private interests sometimes coincide. State forests, for example, are better managed and healthier than federal forests because management of the former is better monitored and influenced by the logging companies and public-school districts that share in the proceeds of timber sales.)

Moreover, Stroup explains, the provision of goods is a complicated process that requires the setting of priorities, attention to costs, alertness to opportunities, and measured risk taking—in short, entrepreneurship. But true entrepreneurship is foreign to public managers because they lack profit-and-loss incentives and thus do not capture the benefits of success or suffer the harm of policy failures. Public managers are free riders in the sense that they bear practically none of costs of their own actions, so they behave accordingly. Voters might want them to act as good environmental stewards (assuming that voters agree about what *good* means in this context), but this wish does not automatically equip public managers with the capacities and incentives to respond appropriately.

Just as the public sector has been overrated as a supplier of environmental amenities, so has the private sector been underrated, as James R. Rinehart and Jeffrey J. Pompe explain in chapter 12, "Entrepreneurship and Coastal Resource Management." Private developers on South Carolina's barrier islands, for example, have discovered how to profit from preserving the habitat of environmentally sensitive coastal flora and fauna and thus to avoid the "tragedy of the commons" that results when unowned or publicly managed common-pool resources (that is, scarce resources for which it is difficult to exclude potential users) become overused.

Under certain conditions, Rinehart and Pompe explain, private developers provide environmental amenities "beyond the requirements of government regulations," but public policy can either help or hinder the establishment of those conditions. For example, government regulations that permit only small-scale development discourage developers from supplying environmental amenities because the economic benefits (increased land values) would spill over onto adjoining properties, thereby depriving the developers of much of the value that their efforts create. By contrast, policies that permit large-scale development enable the developers to capture much more of their efforts' economic benefits, which encourages the provision of such environmental amenities.

The nonprofit organization is also under-appreciated as a supplier of environmental amenities. When a nonprofit is actually in the business of owning and managing natural resources—as opposed to acting only as a lobbyist for anti-development policies—it must weigh alternative uses of its holdings so as to maximize the likelihood of its most desired overall outcome, as Dwight R. Lee argues in chapter 13, "To Drill or Not to Drill: Let the Environmentalists Decide." Both the Audubon Society and the Nature Conservancy, for example, own environmentally sensitive properties on which they allow the commercial extraction of oil and gas deposits, the funds from which they plow back into the acquisition and protection of other sensitive properties.

If a nonprofit environmental organization owned the Alaskan National Wildlife Refuge, would it decline to develop the oil reserves there? Perhaps not. Strong incentives would exist for it to seek a balance between preservation and development. The prospect of internalizing the opportunity cost of locking up the refuge entirely might well entice the organization into managing it in a manner compatible with the organization's overall conservation goals.

John Brätland analyzes offshore petroleum development in chapter 14, "Externalities, Conflict, and Offshore Lands: Resolution Through the

Institution of Private Property." Oil spills or leaks are the major negative externalities associated with offshore land use, but another type of spillover cost must also be taken into account: political externalities. In this regard, Brätland's chapter provides an application of the principle analyzed by Stroup in chapter 11. A negative political externality occurs when political "stakeholders" shift the opportunity costs of their preferences onto other parties through the political process, as occurs when voters or regulators prohibit offshore oil drilling. Brätland argues, however, that both environmental and political externalities can fortunately be resolved through innovative applications of the oft-neglected institution of private-property rights.

Urban Environments

Part V, "Urban Environments," has to do with two trends in municipal land use and industrial planning. In chapter 15, "Is Urban Planning 'Creeping Socialism'?" Randal O'Toole takes a close look at the "smart-growth" movement. Since the 1920s, city planners have used zoning laws to shape the contours of urban life. Their grip is now tightening under the guise of so-called smart growth, a loose collection of policies that promote high-density living and attempt to reverse people's movement from the central city to the suburbs. Smart growth, however, is a deeply flawed paradigm that echoes the mistaken assumptions of the command-and-control policies that once plagued eastern Europe, O'Toole argues.

To create a more livable city, smart growth favors minimum-density requirements, strict design codes, and limits on automobiles and parking, but these policies will likely make city life more hectic, according to O'Toole. Advocates of smart growth in Portland, Oregon, for example, hope to increase its population by 75 percent and to add 120 miles of rail-transit lines, but population growth of this magnitude would overwhelm any reduction in the rate of automobile usage, causing three times more traffic congestion and far slower travel times; air quality would suffer, too, because stop-and-go driving burns fuel less cleanly. O'Toole cites estimates that the plan will increase smog in Portland by 10 percent.

Nor does rail transit offer much hope because it usually carries fewer people than does one lane of freeway. "Although originally promoted as less expensive than highways, rail transit projects being considered by more than 60 U.S. cities typically cost around $50 million per mile—enough to build 2.5 miles of four-lane freeway." Unit costs of other new services—sewerage, water supply, roads, and schools—also increase when all redevelopment costs, including the tearing up of existing low-density infrastructure, are included.

Despite these flaws, smart growth has empowered new constituencies (downtown interests, environmentalists opposed to cars, and bureaucrats who favor regional governance), which may account for its popularity—at least until its undesirable effects become more visible. "Smart growth is a threat to freedom of choice, private property rights, mobility, and local governance," O'Toole concludes. "[It] is clearly an example of creeping social regulation, if not creeping socialism."

The case against smart growth should not be taken as an argument against ecological urban or industrial projects per se, however. In chapter 16, "Eco-Industrial Parks: The Case for Private Planning," Pierre Desrochers examines the industrial symbiosis of eco-industrial parks (EIPs)—communities of companies that recycle each other's by-products or energy—and whether government planners can improve on the successes of private developments in this area.

Private EIPs such as those found in the Danish coastal city of Kalundborg, the Styria province of Austria, the Ruhr region of Germany, or the Houston Ship Channel are advanced as models appropriate for public managers to emulate. But successful emulation may elude these public managers: the planners of Kalundborg, for example, did not plan the system of mutual recycling that has made their city a showcase of industrial symbiosis; each planner planned only for his own company. The symbiosis of recycling between Kalundborg's industries emerged gradually in a piecemeal fashion, as each company sought to serve only its own production needs.

The outcome of industrial symbiosis at Kalundborg exemplifies the point made by Scottish Enlightenment thinker Adam Ferguson that "nations stumble on establishments which are the result of human action, but not of human design." Indeed, industrial recycling is nothing new and may be as old as the Industrial Revolution. Desrochers cites several industrial periodicals that advertised the sale of industrial by-products in the nineteenth and early twentieth centuries, suggesting that resource recovery is not nearly as novel as the idea that public planners can pursue it more effectively than private actors. The central role played by private-property rights and market prices in the operation of successful private EIPs, Desrochers concludes, suggests that short of removing regulatory barriers to resource recovery, government planners can do little to improve on the industrial symbiosis created by a free market.

The By-products of Environmental Bureaucracy

Contributors to part IV, "The By-products of Environmental Bureaucracy," consider how the centralized, regulatory approach to U.S. environmental policy is altering America's legal and political institutions.

In recent years, federal regulators have found an alternative to traditional regulation by rule making or by negotiation: regulation by litigation. The tobacco settlement is probably the most visible instance of regulation by litigation, but the strategy is being employed increasingly in environmental regulation. For example, as Bruce Yandle and Andrew P. Morriss explain in chapter 17, "Regulation by Litigation: Diesel-Engine Emission Control," the EPA sued heavy-duty diesel-engine manufacturers in an effort to get the industry to scuttle technology that it had developed in response to previous EPA dictates. (The EPA apparently considered that technology to be a violation of the spirit of the previous edicts, even though it conformed with the letter of those edicts.) Litigation by regulation has lowered regulatory costs for the EPA, but it has raised costs significantly for the diesel-engine industry, with the added cost of uncertainty probably far exceeding the amount of penalties imposed by courts or through settlements.

Diesel-engine makers settled the suit at a cost of $1 billion, but that outcome did not necessarily translate into cleaner air, Yandle and Morriss argue. "When agency-sponsored suits are piled on top of agency rules," they write, "regulated firms and their customers get caught in a maze so filled with unexpected costs and outcomes that the motivating public-interest goals, if present at the outset, can get lost in a dizzying hurricane of briefs and penalties. After a while, for example, no one seems to be checking on air quality."

The EPA sometimes employs the language of cost-benefit analysis to demonstrate its seemingly tremendous successes, but it does so in a highly misleading manner. It claimed, for example, that from 1970 to 1990 its Clean Air Act programs produced health benefits worth $22.2 trillion at a cost of only $523 billion. Unfortunately, the EPA's study "actually represents a milestone in bureaucratic propaganda," according to Craig Marxsen in chapter 18, "The Environmental Propaganda Agency."

> Like junk science in the courtroom, the study seemingly attempts to obtain the largest possible benefit figure rather than to come as close as possible to the truth. . . . Without the illusory benefit of all the lives saved, the actual benefits of the Clean Air Act were very modest and probably could have been achieved nearly as well with far less sacrifice. The Clean Air Act and its amendments force the EPA to mandate reduction of air pollution to levels that would have no adverse health effects on even the most sensitive person in the population. The EPA relentlessly presses forward in its absurd quest, like a madman setting fire to his house in an insane determination to eliminate the last of the insects infesting it.

Debating Market-Based Environmentalism

Growing concerns about command-and-control approaches to environmental regulations have prompted policymakers to consider replacing them with various market-based instruments, such as excise taxes and tradable emission permits. Part VII, "Debating Market-Based Environmentalism," examines some of the arguments for so-called market-based environmentalism.

In chapter 19, "Market-Based Environmentalism and the Free Market: They're Not the Same," Roy E. Cordato observes that pollution taxes, tradable pollution permits, and other economic incentives are intended to achieve government-mandated outcomes, not market outcomes. The difference between market-based environmentalism and a truly free-market approach, which would rely on private-property rights and the price system, is significant. For example, the tradable pollution permits of the 1990 Clean Air Act Amendments, designed to encourage industry to adopt environmentally friendlier production methods and to improve economic efficiency, are bad policy, Cordato argues, because "they implicitly grant the polluters the right to disregard the property rights of others."

Like the command-and-control policies they are intended to replace, "market-based" environmental policies neglect to confront the government's failure to define and enforce private-property rights with regard to natural resources. Weak or nonexistent property rights give polluters license to harm others or to misuse politically managed resources. In genuinely free markets with private property, according to Cordato, there would be no prior restraint on producers, but because those who harm others or others' property can be sued for damages, natural resources would be used more appropriately. Cordato proposes that instead of attempting to manipulate markets flawed by weak or nonexistent property rights, government policy should focus on creating full-fledged markets with well-defined, well-enforced private-property rights to wildlife habitat, bodies of water, air sheds, and other natural resources. In addition to the environmental benefits of having all natural resources owned privately, with full legal accountability, such a policy would be fully consistent with political liberty, a benefit that Cordato believes is ignored for the most part by both liberals and conservatives. "The primary choice is not between command-and-control and market-based policies," he concludes. "Instead, it is between free-market policies, based on clearly defining and protecting property rights, and socialist—or, perhaps more precisely, mercantilist—policies, based on furthering the societal and personal goals of politicians and special-interest groups. The latter includes both command-and-control policies *and* those labeled 'market-based.'"

In a rejoinder to Cordato in chapter 20, "Market-Based Environmentalism and the Free Market: Substitutes or Complements?" Peter J. Hill argues that although Cordato's criticisms are insightful, market-based regulations nevertheless are preferable to command-and-control regulations. In a world where the costs of adopting a regime of complete private-property rights are significant, market-based regulations may be the best politically viable alternative today. "A more benign reading than Cordato's of the efforts to introduce market-based solutions," writes Hill, "is that the transaction costs of defining and enforcing rights in some cases are very high and that the coercive power of government can be used productively to give us solutions that are better than not taking any action at all."

Environmental Philosophy

The movements and traditions influencing environmental policy have always concerned themselves with more than merely narrow public-policy issues. Magazines devoted to "eco-friendly" living dot our bookstores, and classes on environmental philosophy are standard fare in our colleges and universities—and the breadth of topics they cover is enormous. One organization devoted to environmental philosophy, the International Association for Environmental Philosophy, notes that its field of inquiry includes "not only environmental ethics, but also environmental aesthetics, ontology, and theology, the philosophy of science, ecofeminism, and the philosophy of technology" (www.environmentalphilosophy.org).

Although a survey of environmental philosophy lies outside the scope of *Re-Thinking Green,* a few of its subcategories that bear on public policy are examined in part VIII, "Environmental Philosophy" (others are touched on throughout the volume). One subcategory pertains to existence values.

An *existence value* is the price people hypothetically would pay just to know that a particular wilderness area or species exists in a "natural" state, untouched by human activity. Economists increasingly have invoked the concept of existence value in the hope of bringing scientific clarity to environmental policy debates—and of staving off the criticism that economics is inherently pro-development. According to Robert H. Nelson, however, this trend is unproductive and misunderstood. In chapter 21, "Does 'Existence Value' Exist? Environmental Economics Encroaches on Religion," Nelson argues that concept of existence value is deeply flawed and is rooted not in value-neutral science, but in the religious traditions that have influenced, often without acknowledgment, modern environmentalism. "Economists introduced the idea of existence value in an attempt to solve a new problem facing their profession," writes Nelson. "The problem was real, but the existence-value cure was worse than the disease."

A second set of topics in environmental philosophy considered here pertains to the automobile and its critics. In recent years, the automobile has faced increasing attack both from intellectuals who have blamed it (tirelessly, one might say) for sundry environmental, social, and political ills and from self-styled "direct action" activists who have firebombed car dealerships that sell sport utility vehicles. Environmental bureaucrats have also taken their toll on automobiles, not only directly through the imposition of numerous car-related taxes, but indirectly through policy guidelines—for example, by proposing weekly car-free days for particular city streets and "transportation-demand management" plans that penalize companies for not encouraging their workers to use car pools.

Against this opposition, Loren E. Lomasky argues in chapter 22, "Autonomy and Automobility," that the automobile's benefits far exceed its costs. Lomasky's defense of the automobile is not economic but philosophical (though not utilitarian, as our wording might have suggested). He defends the automobile forcefully yet eloquently by showing how its intrinsic capacity to move us from place to place—automobility—complements autonomy, the distinctively human capacity to be self-directing. "Automobility has value because it extends the scope and magnitude of self-direction," writes Lomasky.

> Moreover, the value of automobility strongly complements other core values of our culture, such as freedom of association, pursuit of knowledge, economic advancement, privacy, and even the expression of religious commitments and affectional preference. If these contentions have even partial cogency, then opponents of the automobile must take on and surmount a stronger burden of proof than they have heretofore acknowledged. For not only must they show that instrumental costs of marginal automobile usage outweigh the corresponding benefits, but they must also establish that these costs outweigh the inherent good of the exercise of free mobility.

The chapters in this book appeared originally in *The Independent Review: A Journal of Political Economy,* a quarterly publication of the Independent Institute. Since its inception in 1996, *The Independent Review* has sought to overcome disciplinary barriers to the understanding of political economy (broadly construed), an aim that we hope this book will also help to accomplish. The editors gratefully acknowledge the many expert referees who reviewed and contributed valuable advice for improving the original submissions to the journal.

REFERENCES

Beckerman, Wilfred. 2003. *A Poverty of Reason: Sustainable Development and Economic Growth.* Oakland, Calif.: Independent Institute.

Brown, Jo-Christy, and Roger E. Meiners. 2000. Common Law Approaches to Pollution and Toxic Tort Litigation. In *Cutting Green Tape: Toxic Pollutants, Environmental Regulations, and the Law,* 99–127. New Brunswick, N.J.: Transaction Publishers for The Independent Institute.

Chase, Alston. 2001. *In a Dark Wood: The Fight over Forests and the Myths of Nature.* New Brunswick, N.J.: Transaction.

Crews, Clyde Wayne, Jr. 2003. *Ten Thousand Commandments: An Annual Snapshot of the Federal Regulatory State, 2003 Edition.* Washington, D.C.: Cato Institute.

Hays, Samuel P. 1959. *Conservation and the Gospel of Efficiency: The Progressive Conservation Movement 1890–1920.* Cambridge, Mass.: Harvard University Press.

———. 2000. *A History of Environmental Politics.* Pittsburgh: University of Pittsburgh Press.

Higgs, Robert. 1987. *Crisis and Leviathan: Critical Episodes in the Growth of American Government.* New York: Oxford University Press.

Meiners, Roger E., and Bruce Yandle. 1992. *The Common Law Solution to Water Pollution: The Path Not Taken.* PERC Reports. Bozeman, Mont.: PERC (April). Available at: http://www.independent.org/publications/article.asp?ID=289.

Mitchell, William C., and Randy T. Simmons. 1994. *Beyond Politics: Markets, Welfare, and the Failure of Bureaucracy.* Boulder, Colo.: Westview Press for The Independent Institute.

Olsen, Mancur. 1982. *The Rise and Decline of Nations: Economic Growth, Stagflation, and Social Rigidities.* New Haven, Conn.: Yale University Press.

Rubin, Charles T. 2000. *Conservation Reconsidered: Nature, Virtue, and American Liberal Democracy.* Lanham, Md.: Rowman and Littlefield.

Stroup, Richard L., and Roger E. Meiners, eds. 2000. *Cutting Green Tape: Toxic Pollutants, Environmental Regulations, and the Law.* New Brunswick, N.J.: Transaction Publishers for The Independent Institute.

Yandle, Bruce. 1998. Antitrust and the Commons: Cooperation or Collusion? *The Independent Review* 3, no. 1: 37–52. Available at: http://www.independent.org/publications/tir/article.asp?issueID=29&articleID=353.

PART I

The Seeds of Environmental Bureaucracy

2

Prophecy de Novo:
The Nearly Self-Fulfilling Doomsday Forecast

CRAIG S. MARXSEN

The Limits to Growth, by Donella H. Meadows and three coauthors (1972), sold nine million copies in twenty-nine languages (Suter 1999, 2). The book awakened anticipation of a cataclysmic end of the world by allegedly seeing that possibility through a new kind of lens. Its prophecy sprang not from an individual claiming divine inspiration, but from the wizardry of the silicone chip processing precise mathematical expressions that defined the "predicament of mankind." The authors claimed that a computer analysis revealed that the final, short countdown had begun and that the apocalypse would occur during the first half of the twenty-first century. Without drastic action, a huge decline in world population would inevitably result from depletion of natural resources, accumulation of pollution, and various effects of overpopulation. The only possible means of prevention was an abrupt halt to global growth—both economic growth and population growth. A 1992 sequel, *Beyond the Limits*, essentially reiterated the claims of the 1972 book. These books have influenced a large following of enthusiasts who remain convinced that growth as we have known it is inherently unsustainable and must give way to a radical imperative to protect the earth.

Robert Higgs (1987, 247–53) has emphasized the political rise of environmentalism in the 1960s and the "energy crisis" of the 1970s as important episodes in a never-ending succession of crises that ratcheted the size of big government ever upward.

Commenting on the Waco massacre that resulted in part from David Koresh's apocalypticism, Charles Krauthammer (1993, 82) argues that the end of the Cold War left a vacuum subsequently filled by the secular eschatology of the ecocatastrophists. Quoting from Al Gore's best-selling book *Earth in the*

Balance, he characterizes Gore's crisis mongering as follows: "if environmentalism does not become 'our new organizing principle,' then 'the very survival of our civilization will be in doubt'" (82). Notwithstanding such expressions of environmental extremism, Gore managed to win a popular majority in the presidential election of 2000.

The expectation of an impending environmental calamity incites political actions that slow global economic growth in a system that preserves itself only by moving forward. In this chapter, I argue that the process of economic growth remains vulnerable to the political fruits of widespread popular belief in the collapse hypothesis set forth in *The Limits to Growth*. Ecocatastrophism dangerously weakens economic performance in several ways and therefore might bring about a persistently declining global standard of living that would create a debacle not unlike the very one the catastrophists fear.

Since the early 1970s, the U.S. economy has suffered a marked slowdown in the growth of multifactor productivity, at least until the mid-1990s. I discuss this stagnation elsewhere (Marxsen 1999); it remains somewhat mysterious, but ecocatastrophism, I contend, has been one of its significant causes.

THE CALAMITY IS NOT THEORETICALLY INEVITABLE

In 1973, William D. Nordhaus carefully assessed the model underlying the *Limits to Growth* forecasts, a "world dynamics" model developed by Jay Forrester, an MIT engineering professor. This model employed a technique already well known in economics, *systems dynamics* (the use of simultaneous difference or differential equations), but without establishing a data-based connection between reality and the model's equation system (Nordhaus 1973, 1182). To validate his model, Forrester relied on his personal judgment of its plausibility rather than on empirical verification or comparison with established growth theory (Nordhaus 1973, 1182–83). Emphasizing this shortcoming, Nordhaus titled his 1973 critique "World Dynamics: Measurement Without Data."

Forrester enhanced popular acceptance of his findings by presenting them as computer output, the then-still-mysterious computer serving as a sort of room-filling talisman. When Robert Solow reviewed *The Limits to Growth*, however, he scornfully charged that its conclusions were logically so close to its assumptions that one hardly needed a computer to obtain them (1973, 43). The 1972 model (and its 1992 successor, too) postulates the depletion of vital exhaustible stocks, such as natural resources and the environment's waste-disposal capacity, at increasing rates, with no moderating mechanism as exhaustion approaches. Hence, Solow explained, anyone can see that the system will bounce off its ceiling and collapse. Later, Peter Huber was kinder to Forrester, judging him to be an excellent "systems dynamics" modeler (1999, 8) but nonetheless emphasizing Forrester's mistake of grossly underestimating the available stocks of energy and other resources, which he constrained by virtually freezing human ingenuity (9–10).

Both *The Limits to Growth* and its 1992 sequel focus on computer-generated simulations of the time path of world population, industrial output, food production, natural-resource exhaustion, and pollution. In both renderings, an inevitable decline in industrial capacity and population dominates the twenty-first century. The models take the form of a system of simultaneous nonlinear difference equations, most of which are of first order (Nordhaus 1992, 5). In other words, the value of a variable such as population at time $t + 1$ is given by an equation containing the values of several variables at time t. As Nordhaus reiterated in 1992, the equations and the definitions of variables in both the 1972 and 1992 versions of the book were invented de novo rather than taken from established relationships in relevant fields of study; worse, they were not justified by any attempt to verify statistically the behavioral equations involved (14). The *Limits* models thus ignored the relevant economic theory refined since Charles Cobb and Paul Douglas published their famous mathematical rendition of the "theory of production" in the *American Economic Review* in 1928. In fact, Cobb and Douglas themselves explicitly proposed "including the third factor of natural resources" in their equations (165). Growth theory as a field of economics culminated in the Nobel Prize–winning efforts of Robert Solow, who carefully incorporated the traditional Cobb-Douglas production function into his analysis. Modern growth accounting, as represented by the work of Edward Denison, John Kendrick, and others, has expanded and elaborated the Cobb-Douglas production function at the heart of its methodology. Forrester completely ignored all this scholarship.

In response to the shortcomings of *The Limits to Growth*, Nordhaus considers the issues in the context of established economic theory and growth accounting (1992, 15), using a Cobb-Douglas growth model similar to the following illustration:

$$\text{Output} = (\text{technology})\,(\text{labor})^{0.6}\,(\text{resources})^{0.1}\,(\text{land})^{0.1}\,(\text{capital})^{0.2}$$

As Nordhaus explains, both labor and capital can keep up with, for example, annual population growth of 1 percent. However, output will grow by only 0.8 percent as a result. The following expressions give an idea of the calculus involved: labor grows 1 percent if we multiply it by 1.01; output grows by 0.6 percent when labor grows by 1 percent because $1.01^{0.6}$ is approximately equal to 1.006. Likewise, 1 percent capital growth has an effect illustrated by $1.01^{0.2}$(equal to approximately 1.002). Finally, 1.006 times 1.002 is approximately equal to 1.008, which illustrates the combined effect that makes output grow by 0.8 percent. Furthermore, resources (natural resources and environmental waste-disposal capacity) are decreasing by approximately 0.5 percent per year, tending to reduce output at a 0.0005 annual rate. The technology term therefore must increase at a rate of 0.0025 per year in order to boost output growth up to 1 percent per year and make it keep up with population growth (Nordhaus 1992, 15–16). Historically, in fact, technology has increased at a rate between 0.01 and 0.02 per year, which is why mankind has enjoyed material progress rather than the onset, many years ago, of Malthus's version of what *The Limits to Growth* portrays (Nordhaus 1992, 16).

Nordhaus's modern growth accounting illustration shows output rising as natural-resource inputs are falling. Growth comes from the increase of "multifactor productivity," the economist's way of describing the effect of technological progress. With a little more than a quarter of a percent per year of technological progress, income per capita rises unendingly so that tribulations bring no economic collapse. If multifactor productivity grows slower, income per capita sinks persistently. A growing world population with a declining income per capita indeed might become increasingly vulnerable to massive loss of life, much as people in poor countries prove more vulnerable to natural disasters. The violent outcomes of the *Limits* models, however, require that well-established growth-accounting models abruptly cease to fit reality or technological progress suddenly turns negative and falls precipitously. A great political debacle seems implicit in the *Limits* model, such as might erupt from some future crisis of intense ecological anxiety more than from any real environmental causes.

CONTINUED GROWTH REMAINS POSSIBLE

Mark Sagoff (1995) writes that some of the foremost mainstream economic thinkers dismiss the idea of limits to growth because knowledge (technology) effectively substitutes for natural resources and services of the natural environment. According to Sagoff, Solow found that past growth depended "simply on the rate of (labor-augmenting) technological change" and that "most of the growth of the economy over the last century had been due to technological progress" (1995, 611). Sagoff notes that ecological specialists such as Herman Daly, on the other hand, argue that further growth will soon prove impossible because of limits imposed by sources of raw materials and sinks for wastes. Paul and Anne Ehrlich, Daly, Robert Costanza, and Donella Meadows (1992) have mounted arguments against the growth model of neoclassical economics, but Sagoff shows that these arguments fail. He cites impartial assessments of the sufficiency of exploitable energy resource reserves, and he explains how price signals have shifted forestry and fishing from extractive to farming approaches. Potential food supplies seem more than adequate and are amenable to technological augmentation. Even the seemingly most serious chemical threats to the environment, such as chlorofluorocarbons (CFCs), demand little more than shifting out of activities that create risks out of proportion to their benefits.

Sagoff concludes that ecological economics fails at every turn to show that growth is unsustainable.

Carlos Davidson (2000), a biologist with an economics background, takes issue with what he perceives as Sagoff's agnosticism concerning the existence of significant environmental destruction relevant to humankind's well-being. He perceives that Sagoff risks overstatement of the environment's robustness, and he argues that human activities clearly damage the environment, but not in a way that is likely to lead to catastrophe. According to Davidson, environmental damage is not so much like pulling rivets out of an airplane as it is like pulling threads out of a tapestry. The tapestry becomes more and more threadbare and damaged looking, but it never reaches some critical threshold of cataclysmic failure. The ecosystem is brimming with redundancy, and problems such as reductions in biodiversity do not threaten the viability of the simpler system that results. Like an old carpet, an increasingly damaged and dirty environment would show no tendency to resolve the deterioration trend catastrophically.

Herman Daly takes issue with a theory he imputes to economists, the theory that man-made capital is a substitute for natural capital (land and natural resources) (Sagoff 1995). Daly regards the idea of substitutability between natural and man-made capital as akin to substituting sawmills for diminishing forests. Daly, however, does not appear to understand economic growth theory fully. In the context of "limits to growth," neoclassical growth theory focuses on the inability of capital accumulation to advance economy-wide real income per capita perpetually in the absence of technological progress. The law of diminishing returns dominates capital accumulation. Holding constant the percentage of total output or income saved, no matter how high the level, causes an increase in the burden of depreciation until it absorbs all the savings that otherwise would have increased the capital stock per worker still further. This phenomenon brings us to steady-state growth in the kind of two-factor production function models with which Solow pioneered neoclassical growth theory (see Solow 1970). The law of diminishing returns means that the boost to output, and therefore to savings, that comes from accumulating more capital diminishes. Capital becomes a decreasingly suitable substitute for land, natural resources, or labor, much as Daly insists. The boost to total depreciation cost (which increases linearly), however, does not diminish as we accumulate more and more capital per worker. In the long run, it is simply impossible to sustain a rate of capital accumulation that exceeds the rate of growth of the labor force, if the technology variable remains constant. Without technological progress, the eventual and inexorable decline of income per capita must occur if land and natural resources enter the production function as Nordhaus suggests, in spite of even the highest imaginable saving rates. The widening of capital becomes as futile a strategy as a hunter-gatherer tribe's trying to grow forever by continually enlarging their stock of bows and arrows. Nor is human capital exempt from this effect because it, too, is subject to depreciation, just as tangible, physical capital is. Daly exalts the law of diminishing returns as the ultimate and insurmountable constraint that, because of technological progress, it never actually became in the past.

Richard Brinkman and June Brinkman (2001) distinguish growth and development, where development is the technological progress component and has enjoyed the attention of institutional economists. They focus on the "new" endogenous growth models—the so-called "aK" growth models associated with Paul Romer and Robert Lucas—that reject Solow's treatment of technology as an exogenous variable (507). If, in an "aK" model, capital accumulation itself breeds enough technological progress to offset diminishing returns to capital, then convergence to steady-state growth never happens in a sufficiently frugal society. One basic message of "aK" growth models is that copious saving can help perpetuate growth. The Brinkmans insist, however, that technological progress is actually a cultural variable. They argue that because technological progress is culturally determined, the United States now may be suffering from a long-term malaise or economic decline that endogenous growth models cannot illuminate (2001, 520–22). The Brinkmans surely must gain encouragement, however, from considering Japan's extremely high saving rate that continued through the 1990s despite the country's economic stagnation.

We need not enter the controversy over exogenous versus endogenous growth models to benefit from the Brinkmans' reflections. Endogenous growth models were not intended to rule out stagnation originating from cultural constraints such as overzealous environmentalism. Endogenous growth theory focuses on the power of capital accumulation, human capital investment, and other outcomes of private and public choice to stimulate technological progress (Romer 1994, 3–22). One can accept most tenets of endogenous growth theory without rejecting the idea that technological progress also is determined culturally. Endogenous growth theory hardly seems in conflict with the thesis of this chapter.

Neoclassical growth theorists seem inclined to preserve a greater aura of mystery around the technological progress variable. Substituting a simple function of capital for the technology variable obscures the profound role that the law of diminishing returns plays, and exogenous growth theorists resist that as well as what they perceive as too easy explanations of technological progress, even including theories that emphasize facets of freedom. Solow himself, presumably insisting on rigor, expresses skepticism toward correlations that explain growth by using dependent variables such as openness to trade and "political and social and legal things," arguing that these correlations might be getting matters backward ("Three Nobel Laureates" 2000). He reasons that because countries are growing satisfactorily, they might be more tolerant of free trade, for example, so that free trade is more the effect than the cause of growth. For Solow, theoretical efforts to "endogenize" technological progress remain unproved, and policies to educate workers, encourage innovation, and get out of the way are the best founded we have identified thus far ("Three Nobel Laureates" 2000). A meticulous continuation and expansion of work done by Edward Denison and John Kendrick might better satisfy Solow's apparent desire for rigor in identifying growth's causes.

In this chapter, however, I focus on a narrower issue: the corrosive effect of ecocatastrophism that appears capable of overwhelming the processes working in growth's favor, whether those processes are well understood or not.

MANCUR OLSON RECOGNIZED GROWTH'S TRUE ENEMY

Without explaining what causes growth, Mancur Olson had much to say about what gets in growth's way, and this was the theme of his 1982 book, *The Rise and Decline of Nations: Economic Growth, Stagflation, and Social Rigidities*. Collective action by numerous coalitions results in a variety of restrictions—laws, regulations, and so forth—that come about as each group tries to further the interests of its particular members without concern for the common good of society. Examples range from protectionist trade barriers to the Jim Crow laws of the southern United States before passage of the Civil Rights Act. In Olson's view, in any dynamic society it is just a matter of time before stagnation sets in. Great upheavals such as wars or natural disasters sometimes break the grip of special interests, and a new "golden age" of growth results. A few years or decades later, however, stagnation returns.

Ironically, the very people who fear that environmental limits will bring about a catastrophic collapse are among the most successful in forming coalitions and instigating restrictions that retard growth. In particular, environmental organizations have played a very prominent role in creating and propelling a vast body of environmental regulations that perhaps have been the most costly and productivity sapping of all the regulations in effect in the United States in the last thirty years. James Robinson (1995) found convincing evidence that environmental regulations account for the bulk of the slowdown in productivity growth that plagued the U.S. manufacturing sector between 1974 and 1986. He concluded that because of the burden of environmental regulations, productivity was 11.4 percent lower than it could have been by 1986 (411, 414). His figures imply that in the absence of environmental regulations, manufacturing output would have grown between 1974 and 1986 by an annual factor of 1.01 percent faster than its actual growth (Marxsen 2000, 76). Had the manufacturing sector enjoyed such a boost to annual growth from 1970 to 1990, manufacturing output would have been 22.35 percent higher than it actually was by 1990 (Marxsen 2000, 76).

Michael Hazilla and Raymond Kopp found that by 1990, environmental regulation had reduced real gross national product (GNP) by 92.41 percent as much as it had reduced manufacturing output alone, largely through ripple effects (Marxsen 2000, 76). They thus established a ratio of proportionality between effects on the economy-wide GNP (or on gross domestic product [GDP]) and effects on the manufacturing sector alone, and this ratio points toward a shocking inference from Robinson's findings. If manufacturing output would have been 22.35 percent higher in 1990 without Environmental Protection Agency (EPA) regulation, then real GDP probably could have been 20.65 percent higher than it actually was in 1990, ceteris paribus (Marxsen 2000, 76). Continued growth of GDP per capita caused by the entry of females into the labor force and the influx of huge amounts of foreign capital obscured an economy-wide stagnation of U.S. multifactor productivity from 1973 to 1995. Thus, the environmental movement unwittingly stifled technological progress in an effort to save us from its computer-spawned prophecy.

A jump in productivity growth is prominent in aggregate data for the period from 1995 to 2000. However, just six sectors accounted for almost all of this jump—computer manufacturing, semiconductors, telecommunications, retail, wholesale, and securities. The other fifty-three economic sectors, taken as a group, had almost no productivity growth from 1995 to 2000 (McKinsey Global Institute 2001, 1). A recent study conducted by a team that included an advisory board chaired by Robert Solow reported a decline of U.S. total factor productivity for the period from 1995 to 1999 in the 70 percent of the economy that lies outside the six "jumping" sectors (McKinsey Global Institute 2001, 5).

Multifactor productivity contracted in that segment at a compound annual rate of –0.3 percent, compared with a growth rate of +0.4 percent in the previous period, between 1987 and 1995. Stagnation apparently continued for those trying to produce almost anything other than computers, their parts, or the networks of wires that link them together.

RENT SEEKING IN THE "NEW ECONOMY"

The process Mancur Olson described remains largely invisible to most people. The restrictions instituted by special-interest groups are roundabout. For example, petroleum refiners cannot expect to prevent new competitors from entering the petroleum refining industry by advocating legislation based on the rationale that they would enjoy monopoly profits. However, incumbent petroleum producers might just cease resisting and, instead, support passage of laws such as the Clean Air Act of 1970. That legislation treats existing producers differently from new entrants, who must conform to "new source" emission standards. An apparent expression of corporate environmental responsibility, such advocacy even has public relations value. The environmental law actually constitutes a stout barrier to entry for new competitors. The benefits incumbent producers get in the form of secure monopoly power are more than worth whatever hardships the law imposes on existing firms. Moreover, the petroleum producers have no need for moral remorse. In their hearts, they know that all they really did was deflect and capture a political thrust originally designed without regard for the harm it was going to do them.

Bruce Yandle (1999b) explains such behavior by means of his 1983 "bootleggers and Baptists" model of regulation. Bootleggers, always under attack from Baptists, found it expedient to promote secretly the Baptists' efforts to outlaw the sale of alcoholic beverages on Sunday. The law would affect only the vendors who legally sold such beverages and would increase the demand for their illegal substitutes. The *Limits to Growth* vision of environmental collapse fosters public demand for environmental carnivals. Yandle (1999a, 36), reflecting on the 1997 Kyoto Protocol, argues that Kyoto was a rent seeker's festival of cartelization more than an effort to reduce carbon emissions. The agreement would produce differential effects across countries, industries, and firms, and the opportunities for gaining markets or sheltering existing ones were sufficiently valuable to make participants' strategic efforts worthwhile (20). At Kyoto, the environmentalists played the role of the Baptists, adding a moral dimension for the rent seekers, the equivalent of the bootleggers, feasibly to exploit in a political direction (Yandle 1999a, 29). The old adage that "politics makes strange bedfellows" is thus illustrated, but the result is hardly good for society as a whole. Yandle concludes that if global warming is a genuine threat, the Kyoto Protocol is not a useful mechanism for allaying it (1999a, 36). Compared with "business as usual," full Kyoto compliance would reduce atmospheric carbon dioxide levels by an almost undetectable 0.39 percent by 2010 (Yandle 1999a, 23). Unfortunately, likely future reductions of GDP would be more readily detectable and substantially larger in percentage terms, according to several studies Yandle cites (1999a, 25–29). The apparent influence of environmental idealists in efforts such as Kyoto becomes enormous because computerized telecommunications equipment brings to bear, in real time, the tremendous power of a multitude of multinational rent seekers.

The information technology revolution provides a mechanism somewhat analogous to the Krell machines furtively underlying the mysterious monster of the *Forbidden Planet.* In that 1956 Warner Studios sci-fi classic, Dr. Morbius (Walter Pidgeon) is unaware that his subconscious whims are directing an invisible creature powered by hundreds of networked nuclear reactors and that the creature inexplicably has destroyed all of his former space colony companions (except his wife and daughter). Likewise, the misguided environmentalist today directs the power of countless rent seekers, collected by computer/telecommunications technology, to form a monster that destroys productivity growth.

The computer revolution has given rise to electronic networks with the potential to serve as the nervous system of a great global leviathan we only can begin to imagine today. Productivity growth, however, is not the only threatened victim of such a monster. It easily can single out individual humans as, for example, ecological offenders.

CATASTROPHIST SEVERITY

David Koresh and his followers, guided by their own fatidic perspective, seemingly distinguished themselves from ordinary Bible-study groups by developing a propensity for violence. Likewise, the ecocatastrophists' apocalyptic vision promotes an emphasis on the use of force and state-sponsored brutality for persecuting people, most of whom are guilty only of trying to use resources efficiently. Pietro S. Nivola and Jon A. Shields (2001) emphasize the economic hurtfulness of the manifold excesses of America's "zero tolerance" and "adversarial legalism" about environmental matters, describing an ongoing environmental-protection campaign that too often has defied common sense. Let us consider, then, the intolerance emanating from America's apocalyptic green Machiavellianism.

Rancorous enforcement has come to distinguish the U.S. criminal justice system, suggesting some fundamental change in American attitudes. In May 2001, *The Economist* reported that the United States had more people in prison than any other country in the world, recently passing Russia to become the world's leading jailer ("Coming to a Neighborhood near You" 2001). Patrick McCormick, a Christian ethics professor, notes that the United States has a national incarceration rate five to eight times as great as other industrial democracies (2000, 509). He marvels that in 2000 the United States had half a million more prisoners than China and held one-quarter of the entire world's prison population (509). The numbers suggest that Americans feel more vengeful now. McCormick's purpose is to portray the U.S. "war on drugs" as an unjust war. Statistically at least, Americans used to seem more tolerant of one another and less tolerant of such vindictiveness on the part of government.

Since the publication of the original *Limits to Growth* in 1972, the U.S. government has begun to use prison sentences to force those in business to obey environmental laws. The nation appears desperate to stop what polluters are doing to the earth. The EPA initiated 75 percent of its 1993 cases against individuals, not corporations (Litvan 1994, 30). Environmental crime cases from 1991 through 1993 led to the conviction of 353 people, who were sentenced to an average of 8.5 months in prison (30). From 1983 to June 1995, the Department of Justice indicted 406 corporate defendants and 1,052 individual defendants, obtained 732 convictions of individuals and 331 of organizations, collected more than $298 million in criminal penalties, imposed 558 man-years of sentenced imprisonment, and actually obtained 351 man-years of confinement—all for environmental crimes (Millner 1995, 37). With a 400 percent increase in the EPA's budget for its environmental crimes section over the first five years of the 1990s, the scheduled number of gun-carrying EPA environmental police reached 200 in 1996 (Litvan 1994, 29). In 1994, Congress debated bills to increase prison sentences from three years to five years for knowing violations of environmental laws.

On the surface, at least, some of the cases lamented in the print media seem absurdly unfair. For example, Dennis Marchuk, a lawyer involved in real-estate development, received a two-year prison sentence for storing a large number of bags of asbestos at his Marcus Hook industrial center, after removing them from old buildings on the site, stashing them in basement foundations, steam tunnels, and cavities under roads (Roberts 1993, 3–4). Marchuk reportedly never suspected his actions exposed him to such criminal prosecution (3). As he discovered, however, small businesses make easy targets because they lack the money to fight or the upscale lawyers on retainer. The Pollution Prosecutions Act of 1990 drove most of the EPA's rising effort to put businessmen into prison (4). One cannot but feel sorry for two Dallas-area businessmen who thought discharging untreated wastewater into the Irving, Texas, water system would subject their companies merely to a fine. Judge Jerry Buchmeyer sentenced the owner of one company to twenty months in prison plus a year of probation. The vice president of a second company received two years in prison plus three years on probation (Hemphill 1993, 30).

Mariani and Weitzenhoff, plant managers of an eastern Honolulu wastewater treatment plant (Cohn-Lee 1994, 1351–52), violated the Clean Water Act and were found guilty of exceeding the plant's EPA discharge permit by "allowing discharge of waste-activated sludge directly into the ocean" (Kole and Lefeber 1994, 38). Over a fourteen-month period, they exceeded the permit limit by just 6 percent (Cooney' 1996b). On August 8, 1994, the Ninth Circuit of Appeals declared that the government needed only to prove that Mariani and Weitzenhoff knew that they were discharging; the government did not have to prove the men knew they had exceeded their permit's limit (Cohn-Lee 1994, 1356). Moreover, the court refused even to allow the men to defend themselves on the basis that they did not know they were violating their permit (Cooney 1996a). The court would not let the men present in their defense the claim that they mistakenly believed "their actions were authorized under their EPA permit, in order to prevent a catastrophic failure of their sewage treatment plant" (Cooney 1996a). One of the men received twenty-one months in prison, the other thirty-three months; then both were given "upward adjustments of six months on their sentences for 'obstruction of justice'" (Kole and Lefeber 1994, 38). In January 1995, the U.S. Supreme Court, without comment, refused to overturn their convictions (513 U.S. 1128). The courts completely ignored the fact that farmers now spread between one-third and two-thirds of North American and European sewer sludge on farmland as fertilizer (MacKenzie 1998, 26).

Keith A. Onsdorff, an attorney specializing in the defense of individuals and corporations charged with criminal environmental offenses, expresses serious misgivings about the direction U.S. environmental regulation has taken in recent years: (1996, 14). Whereas legal debates have focused on the issue of defendant knowledge that a crime was committed, Onsdorff argues that federal environmental law has "lost its moral compass" (14). Radical departure from our nation's commitment to just and proportionate criminal law is obvious, especially in light of the absence of proof of significant harm to the environment in our present statutory scheme (14).

The exact definition of "obstruction of justice" seems confusing to someone not trained as a lawyer. Folk wisdom prohibits defending a spill on Aunt Bertha's carpet by making light of the harm done. But the Mariani and Weitzenhoff case calls to mind another case, *United States v. Goldfaden*, in which the defendant pleaded guilty to unlawful industrial waste discharge and was sentenced to thirty-three months in prison. Hoping to receive a reduced sentence for cooperation, he tried to show that he had not committed a severe offense. Rather than reducing the defendant's sentence, the court added months to it, finding him guilty of "obstruction of justice" by testifying untruthfully (Kole and Lefeber 1994, 38). Perhaps the judge had read *The Limits to Growth* and regarded Goldfaden as a man contributing to the destruction of the earth.

A two-year sentence tormented a man who had contaminated his garbage with ordinary dry-cleaning fluid ("Dry Cleaner Gets Fine" 1996, 2C). Eric A. Bradley, thirty-eight, of Decatur, Georgia, improperly disposed of filters and chemicals in an exterior trash bin in November and December 1994 at the dry-cleaning business he owned in Columbus, Ohio, at the time. Judge David W. Fais of the Franklin County Common Pleas Court said he had no alternatives, although everyone thought Bradley would be sentenced only to probation in addition to his $10,000 fine. Alexander Volokh and Roger Marzulla (1996) have explained how the Comprehensive Crime Control Act of 1984 included strict mandatory sentencing guidelines for environmental offenders. In an effort to narrow disparities in sentences, the U.S. Sentencing Commission created a "Sentencing Table" that adjusted for knowledge of the vulnerability of victims, abuse of a position of trust, use of special skill to facilitate the crime, acceptance of responsibility, criminal history, and so forth. Volokh and Marzulla emphasize that the environmental sentencing guidelines are so tough that they often produce sentences longer than the maximums prescribed by the actual laws enforced.

Guidelines evidently force judges commonly to impose prison time in cases such as Bradley's (Volokh and Marzulla 1996). Jonathan Adler (1993) has discussed the absurdity of the EPA's classification of perchloroethylene (common dry-cleaning fluid) as "hazardous waste" in light of the relative safety and low toxicity that originally led to its acceptance as the dry-cleaning agent of choice.

Startled by the severity of environmental enforcement, the business community scrambled to comply with pollution laws. Public Employees for Environmental Responsibility (PEER) reported in 1998 that the Clinton administration allegedly had undertaken 52 percent fewer environmental-infraction prosecutions and obtained 60 percent fewer convictions during the 1994–96 period, compared with the 1989–91 days of wrath ("Soft on Enviro Crime" 1998, 4). President Clinton, however, escalated government enforcement efforts among those who, in the eyes of ecocatastrophists, were destroying the earth. PEER subsequently found 1996–98 pollution prosecutions only 27 percent lower than in 1989–91, convictions 38 percent lower, and the conviction rate 10 percent lower, with a mere 25 percent shortfall in the number of defendants sentenced to prison (Mokhiber 1999, 7)—a 52 percent *increase* in Clinton's pollution prosecutions in the 1996–98 period compared with the 1994–96 interval. Russell Mokhiber characterizes this increase as "some improvement" (1999, 7) in Clinton's toughness on environmental crime. Groups fearful of our imagined ecocatastrophe, impelling Clinton, helped boost America to its premier position among gulag nations.

AMERICA'S TASTE FOR INSIDIOUS MOVIES

Richard Stroup and Roger Meiners (2000, 1–22) have emphasized that the U.S. liability system has gone wrong. Not only do Stroup and Meiners observe that liabilities currently existing under hazardous-waste laws exceed $1 trillion (2000, 6–7), but they contend that such "calculated costs greatly understate the liability problem" (12). They find that the underlying perverse changes in the U.S. legal system have resulted from "the electorate's susceptibility to thinly supported claims of environmental crisis" (17). Bruce Benson has discussed the destabilization of property rights that bizarre changes in America's tort law have produced—changes such as the 1980 Superfund legislation (2000, 129–50). Tort frenzy perforates modern corporations as if they were public lands when oil rights went to anyone who could stake a claim (137). Passionate lynch mob juries persuaded by the ecocatastrophist argument or a judge fully convinced that pollution is destroying the earth seem poorly qualified to referee the optimal internalization of environmental externalities that Ronald Coase is acclaimed for proposing. Instead, they become promoters of "junk science in the courtroom," as Peter Huber (1991) describes it.

Americans' misguided zeal for punishing environmental offenders appears in their acclaim for several recent motion pictures. Two are worthy of attention because they are likely to be more familiar than the details of less-dramatized real-world trials. The movies *Erin Brockovich* (2000) and *A Civil Action* (1998) stand out especially. For Erin Brockovich (played by Julia Roberts), the villain is Pacific Gas and Electric, which leaked water containing chromium 6 into the groundwater. In *A Civil Action*, John Travolta plays a lawyer who sues Beatrice Foods and W. R. Grace, bringing "a civil action" against them for pouring solvents on the ground and allegedly polluting the groundwater. Both Pacific Gas and Electric and W. R. Grace paid millions in out-of-court settlements. Pacific Gas and Electric paid out the largest individual-action settlement in history

($333 million). Yet, according to a lengthy EPA report available on the Internet (Grevatt 1998), chromium 6, or hexavalent chromium, a common rust inhibitor, turns out to be harmless when people ingest it orally in small quantities. The solvents disposed of by W. R. Grace were not likely even to have been present in the water consumed by its alleged victims. Michael Fumento (1998, 2000) has analyzed these two groundwater contamination cases, and he argues convincingly that both lack merit. Chromium 6, though suspected of being carcinogenic when inhaled, does not prove to be carcinogenic at all when ingested orally (Grevatt 1998, 48). The human body converts it to a chemical commonly found in vitamin supplements. Referring to an EPA report, Fumento (2000) summarizes a profile that makes it almost certain that hexavalent chromium did not cause a single one of the symptoms prompting the lawsuit in Erin Brockovich's case. Likewise, he (1998) explains that W. R. Grace's pollutants could not have reached the plaintiffs' wells in time to cause their diseases, and research fails to show that the pollutants are even capable of causing such diseases.

Shortly before his death, Mancur Olson (1996) delivered a famous speech about why some nations are rich and others poor. Poor nations lack institutions that make property rights secure over the long run, so the gains from capital-intensive production elude them. Olson adds, "Production and trade in these societies is [*sic*] further handicapped by misguided economic policies and by private and public predation." (22). According to Tillinghast-Towers Perrin (a management consultant for insurance issues), the annual cost of the U.S. tort system, including payments to injured people, legal fees, and administrative expenses, was no less than $165 billion in 1999, a figure equal to approximately 2 percent of GDP or double the percentage in most other industrial countries (France 2001, 115).

The information technology revolution is advancing the art of selecting targets and strategies that effectively will manipulate a jury to make an apparently frivolous and emotionally driven decision that often defies science. Like doctors perfecting some new medical treatment, lawyers collaborate over the Internet. A successful courtroom presentation creates a marketable script that other lawyers play repeatedly to many audiences, just as a door-to-door cookware vendor repeatedly makes a set sales pitch. The public's resistance to tort reform is perhaps promoted by the widespread belief that damage to the environment is pushing mankind toward the calamity foretold in *The Limits to Growth.* The public had little sympathy for Hooker Chemical (later a subsidiary of Occidental) when it suffered its Love Canal legal crisis. Exxon fared little better when it spilled oil in Prince William Sound. Michelle Malkin (1996, 34) later reported, however, that the chemicals in Love Canal actually did not appear to have caused any health effects at all. And when Gregg Easterbrook (1995) visited Prince William

Sound several years after the great oil spill, he found that the only lasting damage seemed to be that caused by the mandated cleanup actions.

Jan Schlichtmann, the real-world lawyer who inspired the movie *A Civil Action*, having recovered only $8 million in expenses in his 1986 first attempt, returned to sue W. R. Grace a second time. Affiliated with a large San Francisco law firm, he pressed a class-action product-liability suit alleging that Zonolite Attic Insulation contains hazardous amounts of asbestos (Breslau and Welch 2001, 48). Defending against more than 58,000 asbestos-related lawsuits had weakened W. R. Grace before Schlichtmann's second attack (48). On December 31, 2000, there were 124,907 bodily injury claims, 7 property damage lawsuits, and 9 class-action attic insulation lawsuits pending against Grace (W.R. Grace & Co. Takes Asbestos Charge and Warns of Bankruptcy 2001, 16). Grace declared bankruptcy on April 2, 2001. It followed into Chapter 11 more than two dozen other companies including Babcock & Wilcox, Armstrong World Industries, Owens Corning, and Burns & Roe Enterprises—companies that produced or used similar products (16). Technological progress toward energy efficiency seems also to be a wounded victim.

Spectators probably felt remote from the W. R. Grace bankruptcy, as a general bear market preceded it. The investor in an index fund most likely never noticed that the S&P 500 dropped W. R. Grace out of the index itself in December 2000. Likewise, shortly after Julia Roberts received her Oscar and while Erin Brockovich was enjoying what remained of her $2 million in loot, Pacific Gas and Electric declared bankruptcy under Chapter 11 on April 6, 2001. Maybe the average investor thought that in part the stock market was foreseeing the cataclysm foretold in *The Limits to Growth* and thus never suspected that our tort system was playing a significant causal role. We almost certainly can attribute to widespread acceptance of ecocatastrophism some of the public acclaim for such destructive and unjust predation.

Destroying long-respected corporations and inhibiting innovation are not the only effects of this ill-conceived and pervasive attitude. Veneration of the excesses of environmental tort suits and overblown concern for insignificant traces of contaminants in the groundwater greatly hinder rational waste-disposal efforts in the United States. Landfill standards and regulations seem excessively rigorous and unnecessarily costly inasmuch as the maintenance of waterproof covers is probably adequate for aquifer protection. Joint-and-several liability deters greater use of landfilling as a waste-disposal method. Much manure and sewer sludge is disposed of presently in ways that contribute to surface-water pollution, whereas regulations make its burial prohibitively expensive. Paradoxically, landfilling of garbage, sewer sludge, manure, and other kinds of waste might offer a realistic solution to the problem of increased anthropogenic carbon dioxide in the atmosphere. In the United States, yard waste is commonly banned from landfills even though landfilling of the totality of our carbonaceous waste potentially might provide an artificial carbon sink that might halt the rise in atmospheric carbon dioxide altogether (Marxsen 2001). Government, hysterically pressed to harass by overregulation every reasonable waste-disposal activity, actually is an unintentional promoter of some of our most serious pollution problems.

CONCLUSION

The American propensity to demonize polluters and punish them severely derives to an extent from an ill-founded conviction that all manner of environmental offenses are bringing us closer to the apocalypse foretold in *The Limits to Growth*. The ecocatastrophists' predictions lack sound theoretical foundations. Demagogic prophecy has encouraged excesses and extremism on the part of environmental catastrophists. *The Limits to Growth* seems to epitomize the belief that drives misguided Americans to pervert justice and suspend mercy in order to subdue polluters.

Environmental catastrophism has driven a massive expansion of the regulatory state, and environmental regulations may have proved sufficient to wipe out almost fully the U.S. multifactor productivity growth that comes from technological progress. Unfortunately, our now stunted productivity growth was theoretically all that stood between modern civilization and the inexorable decline and deterioration that population growth and natural-resource depletion themselves might theoretically bring about. Paradoxically, the otherwise dubious prophetic vision of *The Limits to Growth* remains potentially akin to a self-fulfilling prophecy, threatening to help usher in a slow-motion version of the very scenarios of collapsing modern society that its models portray.

REFERENCES

Adler, Jonathan H. 1993. *Taken to the Cleaners: A Case Study of the Overregulation of American Small Business.* Cato Institute Policy Analysis no. 200 (December 22). Washington, D.C.: Cato Institute. Retrieved May 28, 2001, from http://www.cato.org/pubs/pas/pa-200.html.

Benson, Bruce L. 2000. Rent Seeking on the Legal Frontier. In *Cutting Green Tape: Toxic Pollutants, Environmental Regulation, and the Law,* edited by Richard L. Stroup and Roger E. Meiners, 129–50. New Brunswick, N.J.: Transaction Publishers for The Independent Institute.

Breslau, Karen, and Craig Welch. 2001. Another Civil Action. *Newsweek* (February 5): 48.

Brinkman, Richard L., and June E. Brinkman. 2001. The New Growth Theories: A Cultural and Social Addendum. *International Journal of Social Economics* 28, nos. 5–7: 506–25.

A Civil Action. 1998. Touchstone Pictures, directed by Steven Zaillian, starring John Travolta, running time 115 minutes. Buena Vista Home Entertainment, D6592, Stock #16790. Based on Jonathan Harr, *A Civil Action.* New York: Random House, 1995.

Cobb, Charles W., and Paul H. Douglas. 1928. A Theory of Production. *American Economic Review* 18, no. 1 (March): 139–65.

Cohn-Lee, Richard G. 1994. Mens Rea and Permit Interpretation Under the Clean Water Act. *Environmental Law* 24 (July): 1351–70.

Coming to a Neighbourhood Near You. 2001. *The Economist* 360, no. 8220 (May 5–11): 23–24.

Cooney, John F. 1996a. *Recent Developments in Criminal Prosecution of Public Utility Executives and Employers under the Environmental Protection Laws.* Retrieved March 9, 1996, from http://venable.com/govern/cooney.htm.

———. 1996b. *What You Don't Know Can Imprison You.* Retrieved November 3, 2001, from http://venable.com/articles/environ/whatyou.htm.

Davidson, Carlos. 2000. Economic Growth and the Environment: Alternatives to the Limits Paradigm. *BioScience* 50, no. 5 (May): 433–40.

Dry Cleaner Gets Fine, Sentence in Waste Case. 1996. *Columbus Dispatch,* March 5, 2C.

Easterbrook, Gregg. 1995. *A Moment on the Earth: The Coming Age of Environmental Optimism.* New York: Viking.

Erin Brockovich. 2000. Universal Pictures and Columbia Pictures, a Jersey Films Production, written by Susannah Grant, directed by Steven Soderbergh, starring Julia Roberts, Albert Finney, and Aaron Eckhart.

France, Mike. 2001. The Litigation Machine. *Business Week* (January 29): 114–23.

Fumento, Michael. 1998. Disney Pollutes. *Forbes* (December 28): 52.

———. 2000. "*Erin Brockovich,* Exposed." *Wall Street Journal,* March 28, A30.

Grevatt, Peter C. 1998. *Toxicological Review of Hexavalent Chromium.* Case no. 18540-29-9 (August). Washington, D.C.: U.S. Environmental Protection Agency. Retrieved March 14, 2001, from http://www.epa.gov/iris/toxreviews/0144-tr.pdf.

Higgs, Robert. 1987. *Crisis and Leviathan: Critical Episodes in the Growth of American Government.* New York: Oxford University Press.

Hemphill, Thomas A. 1993. Penalties for Polluters: Finding a Fair Formula. *Business and Society Review* 87 (fall): 29–32.

Huber, Peter W. 1991. *Galileo's Revenge: Junk Science in the Courtroom.* New York: Basic.

———. 1999. *Hard Green: Saving the Earth from the Environmentalists: A Conservative Manifesto.* New York: Basic.

Kole, Janet C., and Hope C. Lefeber. 1994. The New Environmental Hazard: Prison. *Risk Management* 41, no. 6: 37–40.

Krauthammer, Charles. 1993. Apocalypse, with and without God. *Time* 141, no. 12 (March 22): 82.

Litvan, Laura M. 1994. The Growing Ranks of Enviro-Cops. *Nation's Business* 82, no. 6: 29–32.

MacKenzie, Debora. 1998. Waste Not. *New Scientist* (August 29): 26.

Malkin, Michelle. 1996. Love Canal Fiasco Gives Regulatory Success a Bad Name. *Washington Times*, January 1, 34.

Marxsen, Craig S. 1999. Why Stagnation? *B>Quest* 4:1. At http://www.westga.edu/~bquest/1999/stag.html.

———. 2000. The Environmental Propaganda Agency. *The Independent Review* 5, no. 1 (summer): 65–80.

———. 2001. Potential World Garbage and Waste Carbon Sequestration. *Environmental Science and Policy* 4, no. 6 (December): 293–300.

McCormick, Patrick T. 2000. Just Punishment and America's Prison Experiment. *Theological Studies* 61, no. 3 (September): 508–32.

McKinsey Global Institute. 2001. *U.S. Productivity Growth 1995–2000.* October. Washington, D.C.: McKinsey Global Institute. Retrieved January 26, 2005, from http://www.mckinsey.com/knowledge/mgi/productivity.

Meadows, Donella H., Dennis L. Meadows, and Jorgen Randers. 1992. *Beyond the Limits.* Post Mills, Vt.: Chelsea Green.

Meadows, Donella H., Dennis L. Meadows, Jorgen Randers, and WilliamW. Behrens III. 1972. *The Limits to Growth.* Washington, D.C.: Potomac Associates, New American Library.

Millner, Glenn C. 1995. An Ounce of Prevention: Reduce Your Risk of Environmental Criminal Prosecution. *Industry Week* 244, no. 14: 37.

Mokhiber, Russell. 1999. Clinton's Lack of Convictions. *Multinational Monitor* 20, no. 12 (December): 7–9.

Nivola, Pietro S., and Jon A. Shields. 2001. *Managing Green Mandates: Local Rigors of U.S. Environmental Regulation.* Washington, D.C.: AEI-Brookings Joint Center for Regulatory Studies. Retrieved February 15, 2002, from http://www.aei.brookings.org/publications/books/green.pdf.

Nordhaus, William D. 1973. World Dynamics: Measurement Without Data. *The Economic Journal* 83, no. 332 (December): 1156–83.

———. 1992. Lethal Model 2: *The Limits to Growth Revisited. Brookings Papers on Economic Activity* 2: 1–59.

Olson, Mancur. 1982. *The Rise and Decline of Nations: Economic Growth, Stagflation, and Social Rigidities.* New Haven: Yale University Press.

———. 1996. Big Bills Left on the Sidewalk: Why Some Nations are Rich, and Others Poor. *Journal of Economic Perspectives* 10, no. 2 (spring): 3–24.

Onsdorff, Keith A. 1996. What the Weitzenhoff Court Got Wrong. *Journal of Environmental Law and Practice* 4, no. 1 (July–Aug): 14–18.

Roberts, William L. 1993. Law Enforcement Putting Teeth into Environmental Punishment. *Philadelphia Business Journal* 12, no. 29: 3–4.

Robinson, James C. 1995. The Impact of Environmental and Occupational Health Regulation on Productivity Growth in U.S. Manufacturing. *Yale Journal on Regulation* 12 (summer): 387–434.

Romer, Paul M. 1994. The Origins of Endogenous Growth. *Journal of Economic Perspectives* 8, no. 1 (winter): 3–22.

Sagoff, Mark. 1995. Carrying Capacity and Ecological Economics. *BioScience* 45, no. 9 (October): 610–21.

Soft on Enviro Crime. 1998. *Multinational Monitor* 19, no. 12 (December): 4.

Solow, Robert M. 1970. *Growth Theory: An Exposition.* New York: Oxford University Press.

———. 1973. Is the End of the World at Hand? *Challenge* (March–April): 39–50.

Stroup, Richard L., and Roger E. Meiners. 2000. Introduction: The Toxic Liability Problem: Why Is It Too Large? In *Cutting Green Tape: Toxic Pollutants, Environmental Regulation, and the Law,* edited by Richard L. Stroup and Roger E. Meiners, 1–26. New Brunswick, N.J.: Transaction Publishers for The Independent Institute.

Suter, Keith. 1999. The Club of Rome: The Global Conscience. *Contemporary Review* 275, no.

1602 (July): 1–5.
Three Nobel Laureates on the State of Economics. 2000. *Challenge* 43, no. 1 (January– February): 6–32.
Volokh, Alexander, and Roger Marzulla. 1996. *Environmental Enforcement: In Search of Both Effectiveness and Fairness.* Reason Public Policy Institute Policy Study no. 210 (August). Los Angeles: Reason Public Policy Institute.
W.R. Grace Takes Asbestos Charge and Warns of Bankruptcy. 2001. *Engineering News Record* 246, no. 6 (February 12): 16. News site: http://www.enr.com/.
Yandle, Bruce. 1999a. After Kyoto: A Global Scramble for Advantage. *The Independent Review* 4, no. 1 (summer): 19–40.
———. 1999b. Bootleggers and Baptists in Retrospect. *Regulation* 22, no. 3: 5–7.

3

Doomsday Every Day

Sustainable Economics, Sustainable Tyranny

JACQUELINE R. KASUN

What will be the Clinton "legacy" is an intriguing question that comes up in various contexts. Certainly there have been unprecedented political and sexual scandals. The last American administration of the twentieth century will also be remembered for its devotion to feminism, environmentalism, multiculturalism, and one-worldism. Less well known, but arguably more world-changing in its effects, is the administration's dedication to the concept of sustainable development.

"Sustainable development" was the galvanizing theme of the 1992 Earth Summit in Rio de Janeiro. Based on the work of the Brundtland Commission in 1987, the goal of sustainable development has been enthusiastically promoted by the World Bank, the U.N. Development Fund, the U.N. Environment Programme, and the United Nations agencies promoting "world governance." It inspired President Clinton's Council on Sustainable Development. It has precipitated an avalanche of World Bank publications, such as the fourteen volumes of the *Environmentally Sustainable Development Proceedings* series of the 1990s, transforming untold acreages of forest into official paper. The phrase occurs frequently in the Chinese Communist press, usually in conjunction with news about the progress being made in the family planning program (Hong 1998). The two topics—sustainable development and "family planning"—are linked throughout the literature.

Economists have struggled, without much success, to reconcile the various definitions that have been offered for "sustainable development." Herman Daly, an economist who has been involved since the beginning, says not to worry—lots of good ideas can't be defined (1996, 2). Daly, long associated with the World Bank, has written the seminal works in the field and is now joined by a host of authors producing textbooks for the college generation. Instruction in "sustainable economics" suffuses or replaces introductory economics courses at a number of institutions.

Whatever it is, sustainable development promises to transform life on this planet. The Rio conference produced agreements on everything from land-use planning (including "sustainable mountain development") and greenhouse gases to, of course, birth control. There were agreements on "human settlements," "sustainable agriculture," "biodiversity," and on and on in its "Agenda 21" and its Climate Convention and its Convention on Biological Diversity (*Agenda 21* 1992). Though Congress did not adopt the program, the Clinton administration proceeded as if it had, adopting new federal regulations and appointing a President's Council on Sustainable Development, made up of federal officials and prominent environmentalists, to pursue the agenda with vigor.

The Clinton Council on Sustainable Development issued its own version of Agenda 21, declaring that we must "change consumption patterns," "restructure" education, "conduct a high-visibility public awareness campaign . . . to adopt sustainable practices," "create a network of conservation areas for each bioregion . . . based on public/private partnerships" (so much for private property), "realign social, economic and market forces . . . to embrace conservation," "use building codes [to secure] . . . environmental benefits," have "local . . . community planning . . . to develop a common vision," create "a council of . . . key stakeholders to . . . achieve sustainable management of forests," and "promote development of compact . . . neighborhoods" (good-bye, suburbs) (President's Council 1995).

Moreover, it decreed that "population must be stabilized at a level consistent with the capacity of the earth to support its inhabitants," whatever that capacity might be (President's Council 1995). The definitions may be elusive, but the program is uniform throughout the literature. It is to create massive, new bioregional conservation areas; control land use, consumption, and markets; re-educate the masses; and control population.

The Sierra Club announced at the U.N. Population Conference in Cairo in 1994 that "local activists" of the club in the United States were working "in a consensus-based . . . process to establish . . . thresholds for . . . population and consumption impact on the local ecoregion. . . . Addressing local carrying capacities will improve the quality of life for all and help develop sustainable communities" (Sierra Club 1994). The club didn't specify what action those local activists would take if it turns out that local populations exceed carrying capacity, but, as will be shown, other devotees of sustainability have done so.

Since the Rio conference, more than 130 countries have created new bureaucracies to implement Agenda 21 and its requirements for sustainable development, according to the Earth Council, whose founder is Maurice Strong, director of the Rio conference and now assistant secretary general of the United Nations (Earth Council 1997). Many local and regional compacts for sustainable development exist in the United States, stretching from Florida through Missouri to Santa Cruz and Humboldt County, California. Henry Lamb of the Environmental Conservation Organization has described some of them, including the statewide plans for Florida and Missouri (1998).

Sustained by foundation money and federal grants, rarely mentioning Agenda 21, salaried environmental activists are convening unsuspecting local citizens to engage in the "visioning" process to plan for the sustainable community in their future. Vice President Gore's Clean Water Initiative and the Clinton administration's American Heritage Rivers Initiative are nurturing the process by encouraging local "watershed councils" to make comprehensive plans for their regions.

HERMAN DALY'S APOCALYPTIC VISION

Probably not many of these souls have read the works of Herman Daly or Maurice Strong, the Rio documents, or the modern college textbooks in sustainable economics. If they had, they might be less eager to help. Daly, an economist, first came to national attention during the 1970s when the Joint Economic Committee of Congress published his plan for reducing births by government licensing. As in China, the government would issue the licenses in the restricted numbers requisite for achieving its population targets, and persons attempting to give birth without licenses would be punished. Unlike the Chinese system, the licenses could be bought or sold, as in the modern schemes for emissions control (Daly 1976).

People of common sense hearing such schemes tend to find them fantastic and amusing. But the World Bank was so enchanted by Daly's notions that it gave him a job as a senior economist in the Environment Department. In 1990 he and a theologian co-author, John B. Cobb, Jr., published their comprehensive plan for the salvation of the world, *For the Common Good: Redirecting the Economy towards Community, the Environment and a Sustainable Future.* Disputing major teachings of economics, the authors called for university "reform" to reduce the influence of economics and increase attention to the "social and global crisis" (357–60). That reform, of course, is now going forward. Like other leaders of mass movements, they argued that logical reasoning is greatly overdone and called for "a conscious shift toward . . . relativisation"

(359). Such a shift also is rapidly occurring. Daly's hostility toward economics is not unique; many aspiring world-changers have seen economics, with its emphasis on logical reasoning based on fact, as the enemy of their plans.

Daly and Cobb called for the conversion of "half or more" of the land area of the United States to unsettled wilderness inhabited by wild animals (255), the abolition of private land ownership (256–59), a giant forced reduction in trade and a change to self-sufficiency at not only the national level but at local levels also (229–35, 269–72), government controls to reduce output to "sustainable biophysical limits" (whatever those might be) (143), and the resettlement of a large portion of the population to rural areas (264, 311)—remember Cambodia and Pol Pot, who has been called "the ultimate deep ecologist."

Moreover, they wanted a prohibition of the movement of private wealth (221, 233)—so much for any escape from the sustainable paradise—the abolition of direct elections, except for local officials who would in turn elect higher officers of the government (177), and, of course, complete population control by means of birth licenses. The intent was to promote the "biospheric vision" in the spirit of "deep ecology," which sees the need for a "substantial decrease in the human population" to promote "the flourishing of nonhuman life" (377). They added that this necessary reduction in the "human niche," a phrase echoed in subsequent United Nations documents, might be achieved either by a fall in population or by a decline in resource consumption (378).

Daly and Cobb understood that these vast changes would require some readjustments in attitudes, to say the least, and saw hope in the "influence of ecological and feminist sensitivities" (377). Not only have those attitude adjustments materialized, but academic economics, identified by Daly as the enemy, has also been remarkably helpful, producing quantities of new books and courses on sustainable development and related topics. Generous grants from government, foundations, and international agencies have encouraged this outpouring.

The justification for these massive changes in human life on the planet lay in what Daly and Cobb called "the wild facts"—that is, the alleged extinction of species, the ozone hole, the greenhouse effect, acid rain, and the imminent exhaustion of oil supplies. The last, of course, has disappeared from the current list of portending calamities; but never mind, we now have deforestation and the methane crisis. In any event, the bottom line was that we suffer from an excessively human-centered point of view, and people should be taught to adopt the "biospheric vision" (376) in recognition of our "community with other living things" in the spirit of "deep ecology."

Daly and Cobb provided no evidence of any of the catastrophes they listed and even acknowledged some uncertainty about the "precise physical effects" (416). Nevertheless, they insisted that the impending crises were "facts" that could not be denied. Scientific disputes over these matters have expanded since then, prompting the True Believers to develop new arguments.

Some of us may wonder whether the work we do makes any difference in the scheme of things. Daly and Cobb need have no such concerns. Their words, phrases, and arguments now appear throughout the United Nations documents on the sustainable society and the literature of sustainable economics. And Daly, now at the University of Maryland, has reiterated his vision in a 1996 book, *Beyond Growth: The Economics of Sustainable Development*. Together with Robert Costanza, Daly now directs the International Society for Ecological Economics, based in Solomons, Maryland.

STEVEN HACKETT'S CONTRIBUTION

The nature of current college instruction in the field can be seen in the textbook *Environmental and Natural Resources Economics: Theory, Policy, and the Sustainable Society* (1998), by Steven C. Hackett, who teaches economics at Humboldt State University. As in Daly's case, Hackett's justifications for proposing fundamental social change are the imperiled biosphere and "the continued growth of human population," which causes "loss of biodiversity" and "deteriorating . . . wilderness areas" (12, 13), and many other ills.

On these points, there is serious debate, as the author admits. He insists nevertheless on "the potential for catastrophic change in the global climate . . . rising sea levels . . . inundation of . . . low-lying areas . . . desertification of . . . grain-producing areas . . . mass hunger . . . and . . . rapid loss of biodiversity" (12). These dire forecasts, of course, have been featured on television for a generation and will probably not unduly alarm modern students. Nor will these hardened young consumers of doomsday prophecies be surprised to learn that population growth threatens the "habitats of many of the world's species of animals and plants . . . the integrity of the world's remaining temperate zone wilderness areas, coral reefs and other marine ecosystems, and tropical rainforests" (12, 13).

Descriptions of these expected calamities recur throughout the book, repeating what college students have heard from Peter Jennings, Ted Turner, Al Gore, and Zero Population Growth throughout their young lives (Singer 1999). Global warming portends "hundreds of millions or more people leaving Bangladesh, the Nile Delta, and coastal China . . . summer droughts . . . heat waves . . . reduce[d] soil fertility" (190–91). According to Hackett, the distinguished scientists (including a former president of the National Academy of Sciences) who dispute this scenario (Seitz 1998) have ulterior motives; he says many of them are in the pay of the coal and oil industries (192). Never mind the flood of grants going from the Department of Energy, the World Bank, and other sources to Hackett's side.

"Deforestation" is a dire threat, according to Hackett, although Food and Agriculture Organization data show that forests occupy 30 percent of the world's land area, a fraction that has not declined since 1950 (U.N. Food and Agriculture Organization 1950–1994). In the United States the forest cover of one-third of the land has not declined since 1920, but the annual growth has more than tripled, according to the U.S. Forest Service (U.S. Forest Service 1992). The National Wilderness Preservation System grew from 9 million acres in 1964 to 104 million in 1994 (National Wilderness Preservation System 1994).

The "rate of extinctions" is a matter of great concern to Hackett, but here again many questions arise. For one thing, there are no data. As David Jablonski, who also believes in the decline, has noted, "we have no idea how many species there are or how many are endangered" (Stevens 1991). Species such as the blue whale (Baskin 1993) and the black-footed ferret (Lamberson 1994), once reported as nearing extinction, turn out to be more numerous than previously thought. The vast extent of unexplored wilderness throughout the world means that human beings are very far from being able to take a census or even make a decent guess about the numbers of other species. Also, if the earth really is warming, that change should be very good news for the species, because many of them thrive especially well in warm climates.

This is not to argue that nothing should be done about the obvious cases of excessive hunting and abuse of the non-human creatures. The reports—one hopes they are false—of the massive kills during the big-game hunts of the Duke of Edinburgh, who heads the World Wildlife Fund, are sickening.

Hackett describes the causes of the impending environmental collapse. First, there is social and economic injustice. Certainly no one can deny that the world has more than enough injustice. That it is a main cause of environmental problems, however, is not clear. When he reports that "the wealthiest 20 percent of the world's people receive 82.7 percent of the world's income," while "two-thirds of the world's people live on the equivalent of $2 or less per day" (13), he seriously distorts economic reality. An economist, of all people, should understand that income bears some relation to productivity. The people of Bangladesh are not desperately poor because the people in the United States enjoy a high standard of living in their relatively free and peaceful society. Bangladesh suffers from a huge, corrupt, foreign-aid-dependent bureaucracy that milks and strangles its people's productivity.

At another point in the book, Hackett points out the major problems in measuring gross product and thus in comparing it for different countries, but these difficulties do not deter him from making this comparison between the rich and the poor.

Also maddeningly unworthy of an economist is Hackett's statement that "the South African government must also provide for the basic needs of the very poor, mostly black, people . . . including medical care, water, . . . housing, and schools" (301). Does Hackett not realize that the people always and everywhere provide for themselves as well as for their government? The people raise food; they build houses, hospitals, and schools; they nurse the sick and teach and pay taxes. What the government should do, but often doesn't, is to allow the people to work and produce in peace and safety. Hackett's patronizing attitude, so common among various world-changers, toward people, this view of them as the helpless wards of government, is profoundly disturbing in a textbook on economics.

IS THE EARTH OVERPOPULATED?

Overpopulation, according to Hackett, is a major cause of our doleful condition. Having softened the obviously elitist implications of the diagnosis by professing his concern for injustice, he can get on with the real message. The prolific people of the less developed countries are wreaking havoc on their "fragile environments," engaging in "deforestation . . . migration to . . . polluted urban areas . . . massive environmental degradation" (13), and so forth. Unmentioned are the government policies that create these disasters, such as the destructive taxation of farmers' productivity, the government monopolies that underpay and overcharge the people, the confiscation of traders' stocks and pack animals, the endless wars financed by foreign aid.

Hackett doesn't mention the large current declines in fertility and population growth rates throughout the world. United Nations figures show that seventy-nine countries with 40 percent of the world's population now have fertility rates too low to prevent ultimate population decline in those countries (U.N. Population Division 1996). But this evidence gives little comfort to Hackett, who quotes estimates showing that "2 to 5 hectares of productive land are needed to support . . . the average person . . . in an industrialized country [whereas] . . . the world has only 1.5 hectares per capita of ecologically productive land . . . and . . . only 0.3 hectare per capita are suitable for agricultural production" (263). In other words, not only does the less developed world have far too many rapidly multiplying people, but *population in the industrialized countries is several times too large.*

As he does throughout the book, Hackett hedges by saying that we don't really know our "carrying capacity," but the undergraduate reader is going to learn that, whatever that capacity may be, there are already far, far too many people on the earth. In a like vein, Paul Ehrlich, famous for his unblemished record of wrong forecasts, has said the world has "perhaps" five times as many people as it can tolerate (Ehrlich 1989).

Let us not imagine, therefore, that the advocates of the sustainable society are merely talking about cleaning up pollution and giving birth control pills to people in Africa, Asia, and Latin America. Although present State Department and U.N. efforts to restrain the increase of dark-skinned people are very strenuous indeed, they are seen as not nearly enough. Hackett quotes Devall on the desirability of "a substantial decrease of the human population" (20). And he describes the "coercive fertility-control" in China (234) and the proposals of Daly and Cobb and Kenneth Boulding for birth quotas. Spokesmen for the Clinton administration, such as Timothy Wirth, have specified that world population control must include the United States (Wirth 1996). Notice, too, that all of the sustainable society documents call for "population stabilization," without saying whether that is to occur at a population size larger or smaller than the present population.

We hope no guilt-ridden students rush to jump out of our overladen lifeboat before, first, asking why Hackett, Daly, Cobb, and Ehrlich have not done so already and, second, hearing some other information. Again according to Ehrlich and other more reliable sources, human beings actually occupy between 1 and 3 percent of the world's land area (Vitousek et al. 1986). The entire world population could be put into the state of Texas, leaving the rest of the world devoid of people. The population density of that giant city of Texas would be about 20,000 persons per square mile, which is somewhat higher than in San Francisco but lower than in Brooklyn (5.9 billion world population divided by 262,000 square miles of land in Texas implies 22,500 persons per square mile, or 1,200 square feet per person).

Farmers use less than half of the world's arable land (Revelle 1984). The world food supply has increased a great deal faster than population since 1950, according to the Food and Agriculture Organization (U.N. Food and Agriculture Organization 1996). This increase, however, has left millions in Bangladesh and elsewhere still hungry, for the reasons already mentioned. Recent studies at the Council for Agricultural Science and Technology show that farmers could feed a future population of 10 billion by using less cropland and producing less silt and pesticide runoff than at present, thus leaving more land for nature (Waggoner 1994). And the prospects for the world population's ever reaching 10 billion grow dimmer by the hour (U.N. Population Division 1998).

Although many population scholars note that fertility declines when output and income grow, Hackett presents Ehrlich's claim that "Mexico and Brazil . . . have undergone periods of income growth with little or no reduction in birthrates" (232). In fact, the crude birthrate (births per 1,000 population) in Mexico fell by more than 40 percent between 1950 and 1995, according to U.N. data, and in Brazil it fell by more than 50 percent (U.N. Population Division 1996). In 1950–55 the typical Mexican woman was having almost seven children during her lifetime; in 1990–95 the number was three. In Brazil in 1950–55, the typical woman was having six children during her lifetime; by 1990–95, the number was 2.44 (U.N. Population Division 1996). The reason these changes are probably permanent is that the world is rapidly urbanizing, and rearing children is much more difficult and entails a higher opportunity cost in urban settings, where women can and often do work outside their homes.

Hackett misses another important fact in his chart of the "demographic transition" (233). Not only does the birthrate fall as development proceeds, but the death rate rises, after an initial decline. The reason is that eventually the population grows older on the average as fewer babies are born, and death rates are higher for elderly groups. Thus, the crude death rate in elderly Sweden is 11.4 per 1,000 population, whereas in youthful Mexico it is only 5.2 (U.N. Population Division 1996). These two events—the decline in the birthrate and the rise in the death rate—work together to reduce population growth.

MARKET FAILURE?

Hackett has little hope that existing institutions can steer the earth away from the looming catastrophes. As for markets, they "*reinforce* self-interested behavior" (29). One searches Hackett's book in vain for any sign of understanding Adam Smith's "invisible hand" that leads men to serve one another and to economize in their use of resources as they pursue their own self-interest. There is no sign that Hackett has ever read the great economist John Maurice Clark, who called the market "our main safeguard against exploitation" because it performs "the simple miracle whereby each one increases his gains by increasing his services rather than by reducing them" (1948). He seems unaware of Walter Eucken's perception that markets break up the great concentrations of economic power (1950) or F. A. Hayek's (1948) and Ludwig von Mises's (1949) realization that markets provide otherwise unavailable information about the scarcity of the resources that are the focus of his concerns.

This is not to argue that markets will solve all economic problems. Well-known and much-discussed problems of externalities, public goods, and common pool resources, sometimes arise, as Hackett notes. But the nonmarket economies of this century have provided vivid object lessons in the pitfalls of "communitarian" planning, and the work of James Buchanan, Gordon Tullock, and others has pointed up the perverse incentives that infest the public sector as it goes about trying to correct "market failure."

At times Hackett acknowledges that public ownership and management do not always produce ideal results, but for the most part he sees the market as the villain and concludes that our best hope lies in "cooperative rather than noncooperative decision making" (91). It is a conclusion he draws from game theory, and it leads to his hopes for "sustainable development" through small-group negotiations. On this issue, more later.

A glaring defect of the market, according to Hackett, is that "issues of fairness, ethics, and spirituality may not be commensurable with monetized costs or benefits. Can we compare the value of a unique sacred place to the revenues and jobs created by logging, mining, or razing the site?" (97). The young activists trashing a congressman's office in Hackett's own area, Humboldt County (*Times Standard* 1997, 1998), which is three-fourths covered with trees (Lammers 1998), have decided that a stand of privately owned trees is uniquely sacred and that logging amounts to "razing." They have made that judgment even though tens of thousands of acres of old-growth redwoods exist in parks and reserves where they will never be cut (Lammers 1998) and this particular stand will certainly be replanted—the company plants thousands of trees a year, and a redwood will grow six feet a year out of its own stump (as I've seen in my own ten acres of redwoods). Do the rhetoric and violence of these inflamed youngsters constitute a preferable basis for decision making in this case?

Not surprisingly, Hackett finds private property highly suspect: "It is clear that systems centered around private property . . . can conflict with the common good" (26). After a brief discussion of John Locke and proposals for protecting natural resources by assigning private property rights to them, Hackett points students to a patron saint of the French Revolution: "From Rousseau's perspective . . . private property rights . . . alienate people from nature . . . [and] lead to inequality . . . and wars." He quotes the great man: "Competition and rivalry . . . opposition of interests . . . and always the hidden desire to profit at the expense of others. All these evils were the first effect of property" (25–26).

Such an indictment demands a response. Private owners did not hunt the buffalo almost to extinction. And it was not a private property system that sent millions to the gulag. When the Ethiopian government socialized the privately owned donkeys, most of them perished (Deressa 1985). I keep the off-road vehicles out of my private forest. And the biblical good shepherd was not the government or the assembly of "stakeholders" in the "sustainable community"; he was the *owner* of the sheep. Where does the common good lie in these decisions? And, most important, Who decides what the common good is? In fairness, also, Hackett might have mentioned the bloodbath that Rousseau's ideas encouraged. Like Devall, Daly, and other environmental utopians of our own time, Rousseau distrusted reason and argued for going "back to Nature." Ever the romantic, he sent his five children to a foundling home (Gauss 1972).

Economists have long noted that voluntary trade must make its participants better off or they wouldn't engage in it, whether they are children trading the contents of their trick-or-treat bags or Mexicans buying used bottles from California to turn them into gravel. Adam Smith and David Ricardo, and even Sir Dudley North before them, saw it as the solution to the uneven distribution of resources. Hackett, however, like Daly and Cobb, whom he quotes at length, lists many objections to trade. It "may . . . allow rich countries to import pollution-intensive, resource-intensive, and endangered-species products they do not wish to produce themselves and to export their toxics and trash" (225). It "tends to erode livable wages, the bargaining power of unions, and environmental and other standards of communities" (226). It "undermines sustainability" (227) and "has put great pressure on . . . endangered wildlife" (229).

Nevertheless, Hackett concludes that although there are "important questions" about how much and what kind of trade to allow, "it is neither practical nor desirable to eliminate trade completely" (230). What a relief. Clearly, however, what is left will be a far cry from free trade, just as all other human activity will be far from free in the "sustainable society."

Throughout the world, controllers and would-be controllers have seen, to use Smith's phrase, the human "propensity to truck, barter, and exchange" as a resource to be exploited or suppressed for the benefit of those in power. From mercantilist England, France, and Spain to the recent Soviet Union and modern Ethiopia, governments have sought to channel this propensity, always with the result of impoverishing their subjects. To illuminate the ill effects of trade controls was the main task of Smith's *Wealth of Nations*. That modern proponents of the "sustainable society" should be so eager to revive such controls should give us pause—doubly so because these people intend to reduce human consumption, and they understand very well that trade restrictions do impoverish people.

Like Daly, Hackett takes a dark view of what he calls "mainstream economics." Students who have studied economics, according to Hackett, are less altruistic than other students (28). Economics itself, he maintains, tends to reduce everything to a monetary cost-benefit comparison without recognizing "intrinsic" values. In his view, however, not all intrinsic values are equally worthy of recognition. Individual rights are especially suspect. By contrast, the "sustainability ethic holds the interdependent health and well-being of human communities and earth's ecology over time as the basis of value" (209), and is therefore clearly superior to the viewpoint of mainstream economics.

ECONOMICS AND ETHICS

Private property, the market, and economics itself, it would seem, are the bad fruit of a bad tree, the disordered ethical system of contemporary society. Hackett blames the shortcomings of economics on its "teleological ethics"—that is, the end justifies the means—attributing the idea to "religious philosophers" (21). This reference enables him to take a swipe at both religion and economics. Evidently, Hackett either never had catechism or was inattentive when Sister told him the end does not justify the means. His example is "utilitarianism," which he describes as the "normative base" for "much of the traditional economic perspective" (21). His straw man is Jeremy Bentham, a nineteenth-century eccentric who had his body stuffed and put in a glass case after he died so it could be on view for University College, London, undergraduates for all time (Mack 1972).

Bentham's mechanical pleasure-pain calculus has amused students for generations, but other men—Smith, Jean Baptiste Say, Ricardo, Carl Menger, Alfred Marshall, and others—did the serious work of showing how the market reveals and reconciles the varied and conflicting desires of multitudes of individuals, channeling their self-interest to the service of others in their pursuit of individual gain.

These monumental themes receive barely a glance from Hackett, who remains intent on showing the failures of market calculations and the need for more sublime direction by persons imbued with the spirit of the sustainable community and tutored in sustainable economics. To illustrate, Hackett poses the "question of whether an action (for example, policy protecting old-growth forest) is to be judged on its intrinsic rightness or based on the measurable benefits and costs that might result" and "the proper balance between individual self-interest and the common good," again undefined (17–18).

There ensues a discussion of the "fundamentals of ethical systems," beginning with "deontological ethics," which judges an action by "its intrinsic rightness" (19). As an example, Hackett quotes at length from the "ecosophy," or "earth wisdom," of Bill Devall, George Sessions, and Arne Naess:

> The well-being and flourishing of human and non-human life on Earth have value in themselves. . . .
>
> The flourishing of human life and cultures is compatible with a substantial decrease of the human population. The flourishing of non-human life requires such a decrease.
>
> Those who subscribe to the foregoing points have an obligation . . . to . . . implement the necessary changes. (20, quoting Devall 1988)

Clearly, this call is not for minor adjustments in lifestyle. A "substantial decrease of the human population" is no small thing. Our "obligation . . . to . . . implement the necessary changes" is a profoundly serious matter. This proposal is not a nickel-and-dime deal. True, Hackett is only quoting Devall at this point, but his discussion makes it clear that Devall's insistence on "intrinsic rightness" is a far more beautiful thing than the crass monetary valuations of "utilitarian" economics.

To make the issue perfectly clear, Hackett offers an example. Suppose an endangered species is threatened by development. Guess what will happen in a "society that views the existence of a species as being of intrinsic value" (à la Bill Devall). Then guess what will happen if a monetary cost-benefit comparison determines the outcome. Obviously, all economists, except an enlightened few, should be taken out and shot.

Nowhere in Hackett's discussion of ethics does he refer to the Judeo-Christian tradition of stewardship—the admonition to "keep" the earth (Gen. 2:15), the prescribed days of rest for men and beasts (Deut. 5:14), the prescribed years of rest for the land (Lev. 25:4), the love of nature with its "Leviathan" taking its sport in the sea and its "coneys" among the rocks (Ps. 104), its cedars of Lebanon (Ps. 92), its hills that "rejoice on every side" and its valleys that "laugh and sing" (Ps. 65), and the strict injunctions against the worship of nature and the human sacrifice that often accompanied it (Deut. 17:3, 20:2–6; 2 Kings 17; Job 31:26).

Modern economic reasoning does not destroy these values any more than modern atmospheric science destroys the beauty of a sunset. Certainly, the sin of greed has always beset the race, as has idolatry. Just as certainly, modern economics has its idolaters as well as its Midases, but such corruption is nothing new on earth. Economic reasoning enables us to compare alternatives. It enables us to see that a society following the romanticism of Devall or Daly would probably be no more attractive or healthful than the one we have. One of the greatest tragedies of our time is not that undergraduates study economics but that they study so little of the great civilizing themes of our heritage—our great literature, art, and music, our legacies from the ancient Greeks, our tradition of human rights and our history of the struggle for liberty —and that they know so little about Christianity or Judaism. Thus deprived, they are left vulnerable, not so much to "utilitarianism" as to environmental lunacy.

Worse yet, as John Grobey, professor of economics and a senior colleague of Hackett at Humboldt State University, has noted, the result must be to deprive young people of the traditional birthright of youth—hope for the future. Taught from their earliest years that their own burgeoning humanity is destroying the earth and all of nature, the youth of today face a more depressing prospect than perhaps any previous generation. No wonder the doubling of the suicide rate among children aged ten to fourteen since 1980 (U.S. Bureau of the Census 1997). No wonder the epidemic of school shootings. No wonder the recent case in Humboldt County in which a young man on trial for attempted murder gave as his defense "overpopulation, dwindling resources and the certain doom of the planet" (Parker 1998).

The changes in "basic economic, technological, and ideological structures" called for by Devall obviously threaten traditional views of individual rights to life, liberty, and property. The question that occurs to a mainstream economist at this point is, Just which individuals will be given the awesome responsibility of determining the "common good" and the best interests of the community and the ecology? And what will happen to human beings, stripped of individual rights, who get in the way of the grand march to the sustainable community? Hackett gives hints but no answers. He acknowledges the seminal work of Daly, but without mentioning Daly's call for massive resettlement of populations. The question remains: Is the centuries-long pilgrimage from Magna Carta through *Areopagitica* and the Bill of Rights to Selma to be renounced now in the name of the environment? Will this denouement be the Clinton legacy?

NO PRICE IS TOO GREAT

Having demolished economics, private property, the market, and individual rights, Hackett poses the question, "So what is the nature of our economy, and what should we do to change it?" (27) Traditionally, textbook writers have not set out to change society or to incite their students to change it, but Hackett's is not a traditional book. His answer is, "The discipline of ecological economics has recently organized itself around the integration of ecology (nature's household) with economics (humankind's household), an integration that is central to the concept of a sustainable society" (209).

Accordingly, Hackett offers a long list of ways to use "ecological economics" to bring about the sustainable society, including solar cookers, wind machines, hydrogen-powered vehicles, "eco-labeling," encouragement of local small businesses, and eco-tourism (presumably reserved for a select few in the new regime of heavy taxes on gasoline and restrictions on access to conservation areas). Undergirding everything would be new taxes, government regulations, and subsidies. In a word, applying "ecological economics" entails a comprehensive government network of social and economic controls to reduce us—those of us who remain after the population has been forcefully "stabilized"—to a preindustrial standard of living (270–84). (No doubt our leaders, who jet from one international conference to another to plan our future, would be exempt from the constraints applied to the rest of us.)

Hackett is honest enough to acknowledge that government requirements for alternative energy in California have given rise to energy prices 50 percent above the national average (279). And he is economist enough to admit that, if we are indeed "running down the natural environment" (256), steeply rising resource prices will force the shift to less depleting and polluting technologies. The problem is, as other devotees of planning have claimed, the market may be "too late" to prevent "irreversible" destruction (277). The evidence, however, does not suggest that government is quicker than the market to recognize problems (Stroup and Meiners 1999). Witness the numerous examples of government projects that have become ecological disasters, including Aswan and Chernobyl and the Great Leap Forward.

"Most important to the success of more sustainable production and consumption," Hackett concludes, "is for people to become convinced that existing systems are destructive" (272). His book certainly does its bit along those lines. To shore up his case for wrenching changes, Hackett advises that conventional cost-benefit analysis is inadequate, that "there is no good ethical argument for using a pure rate of time preference other than zero" (241). He omits the standard procedure of risk analysis, which takes into account the probability (or improbability) of uncertain events. In plain English, we are to treat nightmarish visions of the far-distant future as if they were present reality. Stop arguing. It's an emergency. Do as we say, now.

Sealing his case, he argues, "Preservation has option value—it gives us time to learn about the possible services that are provided to people by the rain forest" (110). Never mind that trees grow and that reforestation is also occurring. Preserve it, except, of course, when it's going into masses of U.N. publications on the sustainable society.

Hackett makes it sound as if the sustainable society will be brought about by local meetings of "stakeholders" negotiating over local issues. But undergirding these cozy negotiations will be "regulations, taxes, subsidies, and direct funding of clean technology" (277). Of course, the Sierra Club will be there to help.

Here is the rub. To avert a highly problematic future disaster, much disputed by competent scientists, Hackett and his soul-mates in the United Nations and the Clinton Council on Sustainable Development would require human beings to submit to a gigantic present sacrifice of freedom, human dignity, and material welfare in a regime controlled by unelected officials of a global eco-bureaucracy. Have we learned nothing from the utopian horrors devised for us during the past century?

People do love nature. The tremendous expansion of national parks and conservation areas during this century testifies to that love. The environmental movement itself is an expression of our determination not to let the industrial age destroy the oceanic Leviathan and the cedars of Lebanon. The real danger now, however, is not that we stand on the verge of destroying nature but that, stampeded by environmental terrors on every hand, we are plunging over the cliff into totalitarianism.

REFERENCES

Agenda 21 and Other UNCED Agreements. 1992. Retrieved August 30, 1997, from the World Wide Web at http://www.igc.apc.org/habitat/agenda21.

Baskin, Yvonne. 1993. Blue Whale Population May Be Increasing off California. *Science* 260 (April 16): 287.

Clark, John Maurice. 1948. *Alternative to Serfdom*. New York: Knopf.

Daly, Herman E. 1976. Transferable Birth Licenses. *In U.S. Economic Growth from 1976 to 1986: Prospects, Problems, and Patterns*, Vol. 5, *The Steady State Economy*, by the Joint Economic Committee, U.S. Congress, Washington, D.C., 1976, 26–28.

———. 1996. *Beyond Growth: The Economics of Sustainable Development*. Boston: Beacon Press.

Daly, Herman E., and John B. Cobb, Jr. 1990. *For the Common Good: Redirecting the Economy towards Community, the Environment and a Sustainable Future*. London: Green Print (Merlin Press).

Deressa, Yonas. 1985. The Politics of Famine. Biblical Economics Today 8 (April–May).

Devall, Bill. 1988. *Simple in Means, Rich in Ends*. Salt Lake City, Utah: Peregrine Smith.

Earth Council. 1997. *Report Highlights*. Retrieved January 7, 1999, from the World Wide Web at etwork.org/global/reports/reg_meet_sum/report.htm.

Ehrlich, Paul R. 1989. Our Earth Is Past the Point of No Return. Newsday, February 6.

Eucken, Walter. 1950. *The Foundations of Economics*. Edinburgh: Hodge.

Gauss, Christian. 1972. Rousseau. *Encyclopedia Americana*. Vol. 23, pp. 723–25.

Hackett, S. C. 1998. *Environmental and Natural Resources Economics: Theory, Policy, and the Sustainable Society*. Armonk, N.Y.: M. E. Sharpe.

Hayek, Friedrich A. 1948. *Individualism and Economic Order*. Chicago: University of Chicago Press, 1948.

Hong, W. 1998. Tianjin Hold National Family Planning Meet. *Tianjin Tianjin Ribao*, daily newspaper of the Tianjin Municipal CPC Committee, January 6.

Lamb, H. 1998. The Cost of Sustainable Development. *eco-logic*, March–April, pp. 16–21.

Lamberson, Roland. 1994. Scholar of the Year Lecture (on mathematical modeling). Humboldt State University, May.

Lammers, Phyllis A. 1998. *Humboldt County's Forest Resource Dollars*. Eureka, Calif., May.

Mack, Mary Peter. 1972. Bentham. *Encyclopedia Americana* Vol. 3, pp. 559–60.

National Wilderness Preservation System. 1994. Fact Sheet. Retrieved March 13, 1998, from the World Wide Web at http://www.nps.gov/partner/nwpsacre.html.

Parker, Rhonda. 1998. Jury Reflects on Plaza Stabbing Case. *Eureka (Calif.) Times Standard*, May 23, p. A1.

President's Council on Sustainable Development. 1995. Draft Report, reprinted excerpts. *eco-logic*, November–December.

Revelle, Roger. 1984. The World Supply of Agricultural Land. In *The Resourceful Earth: A Response to Global 2000*, edited by Julian L. Simon and Herman Kahn. Oxford: Blackwell.

Seitz, Frederick. 1998. Letter. *Petition Project*. Retrieved August 20, 1998, from the World Wide Web at http://www.oism.org/pproject/.

Sierra Club. 1994. *Population and Consumption: The Sierra Club Has a Vision for Both*. Report distributed at the International Conference on Population and Development, Cairo, September.

Singer, S. Fred. 1999. *Hot Talk, Cold Science: Global Warming's Unfinished Debate* (revised second edition). Oakland, California: The Independent Institute.

Stevens, William K. 1991. Species Loss; Crisis or False Alarm? *New York Times* (Science Times), August 20, p. B-5.

Stroup, Richard L. and Roger E. Meiners. 1999. *Cutting Green Tape: Toxic Pollutants, Environmental Regulation and the Law*. New Brunswick, N.J.: Transaction Publishers for The Independent Institute.

Times Standard. 1997–98. Eureka, Calif. Various issues.

U.N. Food and Agriculture Organization. 1950–94. *Production Yearbook*. Rome.

———. 1996. Food Requirements and Population Growth. *World Food Summit Technical Background Documents* 1-05. Rome. Vol. 1, p. 6.

U.N. Population Division. 1996. *World Population Prospects: The 1996 Revision*. New York.

———. 1998. *Revision of the World Population Estimates and Projections*. New York.

U.S. Bureau of the Census. 1997. *Statistical Abstract of the United States*. Washington, D.C.: U.S. Government Printing Office, table 139.

U.S. Forest Service. 1992. *Forest Resources of the United States*. Washington, D.C.: U.S. Department of Agriculture.

Vitousek, Peter M., Paul R. Ehrlich, Anne H. Ehrlich, and Pamela A. Matson. 1986. Human Appropriation of the Products of Photosynthesis. *BioScience* 36 (6): 369.

Von Mises, Ludwig. 1949. *Human Action: A Treatise on Economics*. New Haven: Yale University Press.

Waggoner, Paul E. 1994. *How Much Land Can Ten Billion People Spare for Nature?* Ames, Iowa: Council for Agricultural Science and Technology, February.

Wirth, Timothy E. 1996. Text of speech to "Soap Summit II." New York, September 7.

PART II
Global Issues

4

Population Growth

Disaster or Blessing?

PETER T. BAUER

The twenty-third General Population Conference of the International Union for the Scientific Study of Population, which met in Beijing in October 1997, focused on overpopulation as a serious threat to human survival and a major cause of poverty. Warren Buffet, Bill Gates, corporations, governments, and international organizations are dedicating and promising to dedicate enormous resources to reverse the threat of overpopulation. But population density and poverty are not actually correlated.

Poverty in the Third World is not caused by population growth or pressure. Economic achievement and progress depend on people's conduct, not on their numbers. Population growth in the Third World is not a major threat to prosperity. The crisis is invented. The central policy issue is whether the number of children should be determined by the parents or by agents of the state.

Since the Second World War it has been widely argued that population growth is a major, perhaps decisive obstacle to the economic progress and social betterment of the underdeveloped world, where the majority of mankind lives. Thus Robert S. McNamara, former president of the World Bank, wrote: "To put it simply: the greatest single obstacle to the economic and social advancement of the majority of peoples in the underdeveloped world is rampant population growth. . . . The threat of unmanageable population pressures is very much like the threat of nuclear war." And many others have made similar statements.

THE APPREHENSIONS REST ON FALSE ASSUMPTIONS

These apprehensions rest primarily on three assumptions. First, national income per head measures economic well-being. Second, economic performance and progress depend critically on land and capital per head. Third, people in the Third World are ignorant of birth control or careless about family size; they procreate regardless of consequences. A subsidiary assumption is that population trends in the Third World can be forecast with accuracy for decades ahead.

Behind these assumptions and, indeed, behind the debates on population are conflicting views of mankind. One view envisages people as deliberate decision makers in matters of family size. The other view treats people as being under the sway of uncontrollable sexual urges, their numbers limited only by forces outside themselves, either Malthusian checks of nature or the power of superior authority. Proponents of both views agree that the governments of less developed countries (LDCs), urged by the West, should encourage or, if necessary, force people to have smaller families.

National income per head is usually regarded as an index of economic welfare, even of welfare as such. However, the use of this index raises major problems, such as demarcation between inputs and outputs in both production and consumption. Even if an increase in population reduced income per head, a matter to which I shall return later, such a reduction would not necessarily mean that the well-being either of families or of the wider community had been reduced.

In the economics of population, national income per head founders completely as a measure of welfare. It ignores the satisfaction people derive from having children or from living longer. The birth of a child immediately reduces income per head for the family and for the country as a whole. The death of the same child has the opposite effect. Yet for most people, the first event is a blessing, the second a tragedy. Ironically, the birth of a child is registered as a reduction in national income per head, while the birth of a calf shows up as an improvement.

The wish of the great majority of mankind to have children has extended across centuries, cultures, and classes. The survival of the human race evinces that most people have been willing to bear the cost of rearing two or more children to the age of puberty. Widely held ideas and common attitudes reflect and recognize the benefits parents expect from having children. The biblical injunction is to be fruitful and multiply. Less well known in the West is the traditional greeting addressed to brides in India, "May you be the mother of eight sons." The uniformly unfavorable connotation of the term *barren* reflects the same sentiment. The practice of adoption in some countries also indicates the desire for children. All this refutes the notion that children are simply a cost or burden.

Some have argued that high birth rates in the LDCs, especially among the poorest people, result in lives so wretched that they are not worth living, that over a person's lifetime, suffering or disutility may exceed utility; hence, fewer such lives would increase the sum total of happiness. The implication is that external observers are qualified to assess the joys and sorrows of others; that life and survival have no value to the people involved. This outlook raises far-reaching ethical issues and is unlikely to be morally acceptable to most people, least of all as a basis for forcible action to restrict people's reproductive behavior, especially when one recalls how widely it was applied to the poor in the West only a few generations ago.

Nor is this opinion consistent with simple observation, which suggests that even very poor people prefer to continue living, as shown, for instance, by their seeking medical treatment of injuries and illnesses. Clearly, the much-deplored population explosion of recent decades should be seen as a blessing rather than a disaster, because it stems from a fall in mortality, a prima facie improvement in people's welfare, not a deterioration.

Much of the advocacy of state-sponsored birth control is predicated on the implicit assumption that people in high-fertility Third World countries do not know about contraceptives and that, in any case, they do not take into account the long-term consequences of their actions. But most people in the Third World do know about birth control, and practice it. In the Third World, fertility is well below fecundity; that is, the number of actual births is well below the biologically possible number. Traditional methods of birth control have been widely practiced in societies much more backward than contemporary Third World countries. Throughout most of the Third World, cheap Western-style consumer goods have been conspicuous for decades, whereas condoms, intrauterine devices, and the Pill have so far spread only very slowly. This disparity suggests that the demand for modern contraceptives has been small, either because people do not want to restrict their family size or because they prefer other ways of doing so.

It follows that the children are generally wanted by their parents. Of course, a woman who does not want many children may have to bow to the wishes of her husband, especially in Catholic or Muslim societies. Attempting to enforce changes in mores in such societies raises issues that I cannot pursue here. In any event, this matter does not affect my argument. Children are certainly avoidable.

Nor are people in LDCs generally ignorant of the long-term consequences of their actions. Indeed, young women often say that they want more children and grandchildren to provide for them in their old age. The readiness to take the long view is evident also in other decisions, such as planting slow-maturing trees or embarking on long-distance migration.

EXTERNALITIES

Under this heading, the first question is whether parents bear the full costs of having and rearing their children. If they do not bear those costs fully, they will have more children than they would otherwise. Then, according to the usual assumptions of welfare economics, the satisfaction the parents gain from the additional children would be less than the additional burden, some of which others must bear. It is often assumed that parents in the Third World do not bear the full costs of having children, in particular the costs of health care and education, and that in fact taxpayers bear a substantial part of those costs. These particular costs, however, are unlikely to be heavy in LDCs. They are likely to be lower relative to the national income than in the West. For instance, schools are often simple, inexpensive structures. For social and institutional reasons, basic health services are extensively performed by medical auxiliaries and nurses rather than fully qualified doctors. In any event, if the adverse externalities warrant remedial action, such action should take the form of changes in the volume, direction, and financing of the relevant public expenditures rather than imposed reductions in family size.

The extended family provides a further example of the same negative externality. Parents may have more children if they know that other members of their extended family will bear part of the cost. However, as just noted, the burden falling on others is likely to be small. Moreover, the extended family is embodied in the mores of much of the less developed world. Any effect of the operation of the extended family in this context will diminish or disappear if the extended family system gives way with modernization, a prospect to which I shall return.

Congestion in cities is sometimes instanced as an adverse externality resulting from population growth. But the rapid growth of the cities, especially the capitals, derives from their pull. This in turn reflects the limitations of rural life to many people and the higher incomes and other benefits available or expected in the cities. The income differences increase when policies benefiting the urban population, such as have been widely adopted, depress rural earnings. That the growth of large cities results from these influences is evidenced by the large conurbations in sparsely populated LDCs such as Brazil and Zaire and by the rapid urbanization of LDCs. In any case, undesirable crowding in large cities is not a function of their size or growth, much less of the growth of the national population: it is the inevitable consequence of fixing the prices of housing and transport without regard to their true scarcity.

Similar considerations apply to the supposed adverse external effects of population growth on the environment, including deforestation, soil erosion, and depletion of fish stocks. The assignment of property rights and free-market pricing can optimize the rate of use of forests, soils, fisheries, other presently open-access resources.

Altogether, it is highly unlikely that population growth would cause major adverse externalities, let alone externalities warranting the placing of pressure on people to have fewer children.

Despite the practically exclusive preoccupation with purported adverse externalities of population growth, population growth often has favorable external effects. It can facilitate the more effective division of labor and thereby increase real incomes. In fact, in much of Southeast Asia, Africa, and Latin America, sparseness of population inhibits economic advance. It retards the development of transport facilities and communications, and thus inhibits the movement of people and goods and the spread of new ideas and methods. These obstacles to enterprise and economic advance are particularly difficult to overcome. At the more advanced stages of development, significant positive externalities arise from greater scope for the division of labor in economic activity in science, technology, and research.

I shall argue later that even if it were shown that adverse externalities are significant and outweigh the positive externalities, that condition would call for policies quite different from placing pressure on parents to have fewer children.

DOES POPULATION GROWTH REDUCE INCOME PER HEAD?

Even if population growth is unlikely to reduce welfare, is it likely to reduce conventionally measured income per head? It seems commonsensical that prosperity depends on natural resources, namely, land and mineral resources, and on capital and that population growth reduces the per capita supply of these determinants of income. Indeed, if nothing else changes, an increase in population must reduce income per head in the very short run.

This truism, however, reveals nothing about developments over a longer period. Then, productivity depends on other influences, which can be elicited or reinforced by an increase in population. Such influences include the spread of knowledge, division of labor, changes in attitudes and habits, redeployment of resources, and technical change. In short, economic analysis cannot demonstrate that an increase in population must entail a reduction of income per head over a longer period.

There is ample evidence that rapid population growth has certainly not inhibited economic progress either in the West or in the contemporary Third World. The population of the Western world has more than quadrupled since the middle of the eighteenth century, yet real income per head is estimated to have increased at least fivefold. Much of this increase in incomes took place when population was increasing as fast as or even faster than it is currently in most of the less developed world.

Similarly, population growth in the Third World has often gone hand in hand with rapid material advance. In the 1890s, Malaya was a sparsely populated area of hamlets and fishing villages. By the 1930s it had become a country with large cities, active commerce, and extensive plantation and mining operations. The total population rose through natural increase and immigration from about 1.5 million to about 6 million, and the number of Malays from about 1.0 million to about 2.5 million. The much larger population enjoyed much higher material standards and lived longer than the small numbers of the 1890s. Since the Second World War a number of LDCs have combined rapid population increase with rapid, even spectacular economic growth for decades on end, including Taiwan, Hong Kong, Malaysia, Kenya, the Ivory Coast, Mexico, Colombia, and Brazil.

Conventional views on population growth assume that endowments of land and other natural resources are critical for economic performance. This assumption is refuted by experience in both the distant and the more recent past. Amid abundant land, the American Indians before Columbus were extremely backward while most of Europe, with far less land, was already advanced. Europe in the sixteenth and seventeenth centuries included prosperous Holland, much of it reclaimed from the sea, and Venice, a wealthy world power built on a few mud flats. At present, many millions of poor people in the Third World live amid ample cultivable land. Indeed, in much of Southeast Asia, Central Africa, and interior of Latin America, land is a free good. Conversely, land is now very expensive in both Hong Kong and Singapore, probably the most densely populated countries in the world, originally with very poor land. For example, Hong Kong in the 1840s consisted largely of eroded hillsides, and much of Singapore in the nineteenth century was empty marsh. Both places are now highly industrialized and prosperous. The experience of other countries, both in the East and in the West, teaches the same lesson. Poor countries differ in density. For example, India's population density is some 750 people per square mile whereas Zaire's density is approximately 40 people per square mile. And prosperous countries differ in density. Japan's density is some 850 people per square mile whereas U.S. density is approximately 70 people per square mile. All these instances suggest the obvious: the importance of people's economic qualities and the policies of governments.

It is pertinent also that in both prosperous and poor countries the productivity of the soil owes very little to the "original and indestructible powers of the soil," that is, to land as a factor in totally inelastic supply. The productivity of land results largely from human activity: labor, investment, science, and technology.

The wide differences in economic performance and prosperity between individuals and groups in the same country, with access to the same natural resources, also make clear that the availability of natural resources cannot be critical to economic achievement. Such differences have been, and still are, conspicuous the world over. Salient examples of group differences in the same country include those among Chinese, Indians, and Malays in Malaysia; Chinese and others elsewhere in southeast Asia; Parsees, Jains, Marwaris, and others in India; Greeks and Turks in Cyprus; Asians and Africans in East and Central Africa; Ibo and others in Nigeria; and Chinese, Lebanese, and West Indians in the Caribbean. The experience of Huguenots, Jews, and Nonconformists in the West also makes clear that natural resources are not critical for economic achievement. For long periods, these prosperous groups were not allowed to own land or had their access to it severely restricted.

Mineral resources have often yielded substantial windfalls to those who discovered or developed them or expropriated their owners. Latin American gold and silver in the sixteenth century and the riches of contemporary oil-producing states illustrate the prosperity conferred by natural resources. But the precious metals of the Americas did not promote economic progress in pre-Columbian America, nor did their capture ensure substantial development in Spain. The oil reserves of the Middle East and elsewhere were worthless until discovered and developed by Westerners, and it remains a matter of conjecture whether they will lead to sustained economic advance in the producing countries.

Population growth as such can induce changes in economic behavior favorable to capital formation. The parents of enlarged families may work harder and save more in order to provide for the future of their families. In LDCs as in the West, poor people save and invest. They can sacrifice leisure for work or transfer their labor and land to more productive use, perhaps by switching from subsistence production to cash crops. Poor and illiterate traders have often accumulated capital by working harder and opening up local markets.

FAMINE AND UNEMPLOYMENT

Despite the repeated warnings of doomsayers, there is no danger that population growth will cause a shortage of land and hence malnutrition or starvation.

Contemporary famines and food shortages occur mostly in sparsely populated subsistence economies such as Ethiopia, the Sahel, Tanzania, Uganda, and Zaire. In these countries, land is abundant and, in places, even a free good. Recurrent food shortages or famines in these and other LDCs reflect features of subsistence and near-subsistence economies such as nomadic style of life, shifting cultivation, and inadequate communications and storage facilities. Those conditions are exacerbated by lack of public security, official restrictions on the activities of traders, restrictions on the movement of food, and restrictions on imports of both consumer goods and farm supplies. Unproductive forms of land tenure such as tribal systems of land rights can also bring about shortages. No famines are reported in such densely populated regions of the less developed world as Taiwan, Hong Kong, Singapore, western Malaysia, and the cash-crop-producing areas of West Africa. Indeed, where a greater density of population in sparsely populated countries brings about improved transport facilities and greater public security, it promotes emergence from subsistence production.

Nor should population growth lead to unemployment. A large population means more consumers as well as more producers. The large increase of population in the West over the last two centuries has not brought about persistent unemployment. Substantial unemployment emerged when population growth had become much slower in the twentieth century. And when, in the 1930s, an early decline of population was widely envisaged, that development was generally thought to portend more unemployment because it would reduce the mobility and adaptability of the labor force and diminish the incentive to invest.

The experience of the contemporary less developed world confirms that rapid increase of population does not result in unemployment and also that the issue cannot be discussed simply on the basis of numbers and physical resources. Until recently, population grew very rapidly in densely populated Hong Kong and Singapore without resulting in unemployment. Singapore has far less land per head than neighboring Malaysia, yet many people move from Malaysia to Singapore in search of employment and higher wages, both as short-term and long-term migrants and as permanent settlers.

The idea that population growth results in unemployment implies that labor cannot be substituted for land or capital in particular activities and also that resources cannot be moved from less labor-intensive to more labor-intensive activities. The idea implies that the elasticity of substitution between labor and other resources is zero in both production and consumption. But the development of more intensive forms of agriculture in many LDCs, including the development of double and treble cropping, refutes such notions, as do frequent changes in patterns of consumption.

WHAT DOES THE FUTURE HOLD?

Dramatic long-term population forecasts are often put forward with much confidence. Such confidence is unwarranted. It is useful to recall the population forecasts of the 1930s, predicting a substantial decline of population, primarily in the West but to some extent worldwide. Articles by prominent academics appeared under such headings as "The End of the Human Experiment" and "The Suicide of the Human Race." Yet within less than one human generation, the population problem had taken on exactly the opposite meaning. The scare remained, but its algebraic sign was changed from minus to plus.

Today, only the roughest forecasts of population trends in the Third World are warranted. The basis for confident predictions for the Third World, or even for individual LDCs, is far more tenuous than it was for the spectacularly unsuccessful forecasts of long-term population trends in the West in the 1930s. In much of the Third World there is either no registration of births and deaths or a very incomplete one. Estimates of the population of African countries differ by as much as a third or more; for populous countries such as Nigeria, this discrepancy means tens of millions of people. Estimates of the population of the People's Republic of China, the most populous country in the world, also differ substantially.

In the coming decades, major political, cultural, and economic changes will occur in much of the Third World. These changes are unpredictable, and so are people's responses. For instance, contrary to expectations, the economic improvement in recent decades in some Third World countries has resulted in higher fertility. Similarly, a decline of mortality in many LDCs has not been accompanied by the decline of fertility that had been widely expected in the belief that people had many children to replace those who died young. Moreover, in some of these countries urban and rural fertility rates are about the same, whereas in others the rates differ widely. The relationship of fertility to social class and occupation also varies much more in the Third World than in the West. The foregoing considerations should put into perspective such widely canvassed and officially endorsed practices as forecasting to the nearest million the population of the world for the year 2000 or beyond.

One demographic relationship of considerable generality does bear upon population trends in LDCs. Professor Caldwell, a leading Australian demographer, has found that systematic restriction of family size in the Third World is practiced primarily by women who have adopted Western attitudes toward childbearing and child rearing, as a result of exposure to Western education, media, and contacts. Their attitude toward fertility control does not depend on income, status, or urbanization but on Westernization. In this context,

Westernization means the readiness of parents to forgo additions to family income from the work of young children and also to make increased expenditure on education, reflecting greater concern with the material welfare of their children.

Caldwell's conclusion is more plausible and more solidly based than the widely held view that higher incomes lead to reduced fertility. It is true that in the West and the Westernized parts of the Third World, higher incomes and lower fertility are often, though by no means always, associated. But it is not the case that higher incomes and smaller families reflect greater ambition for material welfare for oneself and one's family. Rather, both the higher incomes and the reduced fertility reflect a change of preferences. By contrast, when parental incomes are increased as a result of subsidies or windfalls, without a change in attitudes, the parents are likely to have more children, not fewer. This last point pertains to the proposals of many Western observers who, without recognizing the contradiction, urge both population control and also more aid to poor people with large families.

Some broad, unambitious predictions of Third World population prospects may be in order. Although the speed and extent of Westernization are uncertain, the process is likely to make some headway. Some decline in fertility will result. But the large proportion of young people and the prevailing reproductive rates will ensure significant increases in population in the principal regions of the Third World over the next few decades. The population growth rate for the Third World as a whole is unlikely to fall much below 2 percent per year and may for some years continue around 2.5 percent, the rough estimate for the 1980s. It is therefore likely to remain considerably higher than the rate of growth in the West, Japan, and Australasia. Therefore, over the years, the population of the West, Japan, and Australasia will shrink considerably relative to that of Asia, Africa, and Latin America.

It is unlikely that Third World population growth will jeopardize the well-being of families and societies. But if their well-being were for any reason to be seriously impaired by population growth, reproductive behavior would change without official pressure. There is, therefore, no reason to force people to have fewer children than they would like. When such pressure emanates from outside the local culture, it is especially objectionable. It is also likely to provoke resistance to modernization generally.

CONCLUSION

The central issue of population policy is whether individuals and families or politicians and national and international civil servants should decide how many children people may have.

Advocates of officially sponsored population policies often argue that they do not propose compulsion but intend only to extend the options of people by assisting the spread of knowledge about contraceptive methods. But people in LDCs usually know about both traditional and more modern methods of birth control. Moreover, in many Third World countries, especially in Asia and Africa, official information, advice, and persuasion in practice often shade into coercion. In most of these societies, people are more subject to authority than in the West. And especially in recent years, the incomes and prospects of many people have come to depend heavily on official favors. In India, for example, promotion in the civil service, allocation of driving and vehicle licenses, and access to subsidized credit, official housing, and other facilities have all been linked at times to restriction of family size. Forcible mass sterilization, which took place in India in the 1970s, and the extensive coercion in the People's Republic of China are only extreme cases in a spectrum of measures extending from publicity to compulsion.

Policies and measures pressing people to have fewer children can provoke acute anxiety and conflict, and they raise serious moral and political problems. Implementation of such policies may leave people dejected and inert, uninterested in social and economic advance or incapable of achieving it. Such outcomes have often been observed when people have been forced to change their mores and conduct. It is widely agreed that the West should not impose its standards, mores, and attitudes on Third World governments and peoples. Yet, ironically, the most influential voices call for the exact opposite with regard to population control.

There is one type of official policy that would tend to reduce population growth, extend the range of personal choice, and promote attitudes and mores that foster economic advance and improvement of the well-being of the population. That policy is the promotion of external commercial contacts, especially contacts with the West, by the people of LDCs. Such contacts have been powerful agents of voluntary change in attitudes and habits, particularly by eroding those harmful to economic improvement. Throughout the less developed world, the most prosperous groups and areas are those with the most external commercial contacts. And such contacts also encourage voluntary reduction of family size. Thus, extension of such contacts and the widening of people's range of choice promote both economic advance and reduction in fertility. In these circumstances, a reduction of family size is achieved without the damaging effects of placing official pressure on people with regard to their most private and vital concerns. Yet policies of this kind are not on the agenda of those who advocate reducing population growth in LDCs.

5

After Kyoto

A Global Scramble for Advantage

BRUCE YANDLE

The 1997 Kyoto Protocol is generally viewed as having set carbon emission reduction goals for the developed world for the purpose of avoiding global warming, a tragedy of the commons.[1] But instead of avoiding the plaguing free-rider problem that works in favor of tragedy, the protocol allows uncontrolled growth in carbon emissions from the developing world, which will more than offset reductions elsewhere. Strictly speaking, the Kyoto Protocol is not about avoiding a tragedy of the commons. It involves much more than that.

Apart from expectations that meeting the terms of the protocol will not reduce future emissions, the scientific basis for Kyoto's massive and costly undertaking is far from settled (Singer 1999). Competing models that seek to link human behavior to climate change yield mixed results, which is to say that temperature changes may be caused by solar activity, clouds, or ocean temperature changes. There is also contradictory evidence on temperature change itself. It is not a foregone conclusion that temperatures, though increasing in the last few years, are rising systematically. Nor is it clear that human activity contributes to the current temperature increases.

Putting the scientific questions to one side and viewing the protocol on its own terms, there is serious question whether it is primarily about carbon emission reductions and global warming or about something else. An analysis of the agreement and of the post-Kyoto strategizing suggests that control of global warming is largely symbolic, which does not gainsay its vital importance to environmental groups. The real effects of the protocol relate to cartelization and efforts by interest groups and countries to gain competitive advantage in a globally competitive world. Global warming may be just the right wrapping for a major rent-seeking package (Yandle 1998).

To support this assertion, or to refute the null hypothesis, it is necessary to (1) provide evidence from the agreement itself that cartelization is a real possibility and that differential effects across countries, industries, and firms offer significant incentives for political agents to raise rivals' costs; (2) show that the resource costs of meeting the terms of the protocol are large enough and the opportunities for gaining markets or protecting existing ones fruitful enough to justify the cost of cartelization; and (3) provide clear evidence that countries, firms, and industries are indeed behaving strategically in efforts to use Kyoto as a rent-seeking endeavor. To accomplish these three tasks, I first discuss the global commons problem and provide background information on the Kyoto agreement that shows differential effects across countries. I then survey research on the costs of Kyoto, showing that the resource costs and transfers involved are massive and identifying potential winners and losers in particular product and geographic markets. I provide anecdotal and statistical evidence regarding the behavior of energy firms, trade associations, and countries to demonstrate the crucial importance of Kyoto's environmental symbolism in achieving cartelization goals. Finally, I offer some thoughts on Kyoto and alternate policies.

THE GLOBAL COMMONS AND THE KYOTO "SOLUTION"

The Commons Problem

Most environmental problems begin with a commons, an unrationed resource that tends to be overexploited. Steps taken to ration activity on the commons are justified as being necessary to avoid the tragedy of the commons (Hardin 1968; Anderson and Leal 1991). Garrett Hardin and others remind us that property rights and other rationing institutions can emerge, converting a tragedy to triumph. But the Buchanan-Tullock analysis of political instruments chosen for accomplishing this feat warns us that rent-seeking actions taken to define regulatory strategies can earn a high private return (Buchanan and Tullock 1975). In their story, special interest groups and the politicians who "solve" the commons problem can share monopoly rents if the appropriate regulatory remedy is applied.

Appealing to the commons problem when considering global warming is both logical and politically useful. Although its scarcity is still not well established, the upper atmosphere is clearly an unrationed resource. Any steps taken to alter this nonproperty arrangement will require global political action. Yet the commons imperative is by its nature a centralizing force at any level of human activity. Decentralized spheres of private action and exclusive private rights—what might be termed individual or local sovereignty—are reduced as collective action expands. In addition, as the scope of action expands, special-interest groups seek to have wealth redistributed in favorable ways. Of course, successful steps taken to avoid real tragedies can generate social gains by precluding the destruction of valuable assets. But as institutions form, there are trade-offs to consider. Rent-seeking costs may be larger or smaller than the gains from avoiding a tragedy of the commons.

As the dimensions of the commons to be managed expand beyond the community to include the state, then the region, the nation, and finally the world, the diverse rules that govern heterogeneous communities give way to ever more homogeneous regulations that can restrict competition in the name of environmental protection. Customs, traditions, and institutions such as the common law tend to be pushed to one side as statutes and treaties form a more extensive social order. Local sovereignty is compromised as collective decision-making is delegated to state, national, and then international bodies. And the potential rents to be earned by competing countries, far-flung global firms, and accommodating politicians achieve significant value, but not without cost.

Another trade-off accompanies the transfer of sovereignty from smaller to larger communities (Yandle 1997, 29–30; Ostrom and Schlager 1996, 46). When remotely determined homogeneous rules are imposed on diverse communities, some efficiencies in resource use, where more readily measured costs are weighed against locally perceived benefits, are exchanged for the avoidance of more remote costs, not easily observed at local levels. As control becomes more remotely determined, the efficiency losses tend to increase. It is obviously important that the marginal benefits from avoiding global costs exceed the ever-increasing marginal costs of such local efficiency losses.

The Kyoto Protocol is a case in point. The December 1997 agreement to reduce greenhouse gases, endorsed by representatives from 174 nations, joins an estimated 180 other environmental treaties on deposit with the UN Secretary General (Committee to Preserve American Security and Sovereignty 1998). Kyoto and the other treaties differ fundamentally from national decisions to legislate in the interest of cleaner air or water. First, the protocol establishes a relatively homogeneous rule—greenhouse gas reductions based on 1990 emissions—for scores of heterogeneous communities. Next, constitutional constraints and domestic rules of law that normally protect property rights and sharpen the spur of competitive behavior in domestic economies can be relaxed in the name of avoiding global warming. Coordinated output restrictions, ostensibly for environmental protection, are viewed benevolently.

How the Protocol Evolved

The Kyoto Protocol is an evolved agreement rooted in the notion that developed countries, which are necessarily large energy users and greenhouse gas producers, should bear the brunt of reducing emissions in the name of avoiding costly climatic changes. This idea, discussed formally at Toronto in June 1988 and considered by the U.S. Congress in 1989 in a proposed bill, the Global Warming Prevention Act, was fundamental to commitments reached in 1992, when representatives of 160 nations attended the Rio de Janeiro Conference on Environment and Development (Manne and Richels 1991, 88).

Efforts to contain greenhouse emissions were strengthened at the 1995 Conference of Parties to the Rio de Janeiro Agreement, yielding the Berlin mandate, which stressed the importance of gaining national commitments to greenhouse gas reductions. Then, an ad hoc group that met in Geneva in 1995 and again in 1996 called for binding mandates for thirty-eight developed countries known as Annex I, including primarily the members of the Organization for Economic Cooperation and Development (OECD) and eastern European states. Cooperation and emission reporting were expected of developing countries, and a stronger commitment was expected from eastern European states in transition, but no quantifiable emission reduction commitments were called for. A follow-up meeting for resolving issues left on the Kyoto table took place in Buenos Aires in November 1998.

The Kyoto Protocol, endorsed in Kyoto on December 11, 1997, still awaits ratification by the U.S. Senate, which has indicated that it will not ratify the treaty until the developing world makes reduction commitments.[2] The protocol sets 1990-based emission reductions for greenhouse gases (primarily carbon dioxide) for the Annex I countries, to be achieved by the "commitment period," 2008–2012. At that time, emissions will be averaged across the designated years to determine compliance.

For the protocol to become binding, fifty-five countries must ratify it—which implies that seventeen non–Annex I countries must give their approval—and these ratifying countries must account for at least 55 percent of the desired emission reductions. In April 1998 the European Union members officially ratified and signed the Kyoto treaty, accepting an 8 percent reduction of carbon emissions over the next thirteen years (Leopold 1998). The United States, which had accepted a 7 percent reduction, still had not ratified the agreement, and the likelihood of its doing so was slight. In April, Japan, Australia, Brazil, Canada, Norway, and Monaco signed. Argentina and Pacific island nations had signed earlier.

As of June 1998, forty countries, representing 38.9 percent of the total pledged emission reductions, had officially committed themselves to the

December 1997 protocol (List of Signatories 1998). At the November 1998 Buenos Aires meeting, Argentina and Kazakhstan agreed to join the Annex I countries in cutting emissions (Fialka 1998), and the United Nations exerted pressure on African countries to join the list of volunteers (BBC News 1998).

The United States was left holding the high trump card. By the terms of the protocol, the parties to the convention have until March 15, 1999, to make their commitments binding. Because it accounts for 36 percent of the industrialized nations' emissions and 23 percent of the world's, the United States alone can prevent the agreeing nations from reaching the 55 percent reduction in emissions required for the agreement to become binding. But whether the protocol is finally ratified or not, Kyoto-inspired restrictions are already in the works.

The Magnitude of the Task

How large is the Kyoto challenge? Consider the tons of emissions to be reduced. In 1990 the Annex I countries, with the United States leading, produced roughly 64 percent of all greenhouse gases, which then totaled 6 billion tons annually.[3] The developing countries, led by China, produced the remaining 36 percent. Forecasts of emissions for the year 2015 place total emissions at 8.45 billion tons, with the developing countries producing 52 percent of the total; the developed countries by then would be minority players. By the year 2100, 19.8 billion tons of greenhouse emissions are expected, with the developing world producing 66 percent of the total.

Data on projected atmospheric concentrations of carbon dioxide illustrate the Kyoto challenge in yet another way. In the absence of any intervention—what some call business as usual—concentrations will rise from 1990 levels of 353 parts per million (ppm) to 383.5 ppm by 2010 (Business Roundtable 1998). With full Kyoto compliance, which means that the developed countries achieve their reduction targets while the rest of the world—largely, China and India—is unconstrained, year 2010 concentrations are projected to reach 382.0 ppm, which is roughly 8 percent higher than 1990 concentrations. Comparison of projected year 2010 business-as-usual concentrations with full-Kyoto-compliance levels shows a 0.39 percent reduction in concentrations, an amount that would be undetectable.

Massive Costs and Differential Effects

The Kyoto agreement contains emission reduction targets for the thirty-eight Annex I countries that average about 5 percent. Table 1 shows the relevant data for country groups and individual countries. As indicated, commitments

Table 1: World Total Carbon Emissions, Annex I Countries (millions of metric tons per year)

Region/ Country	1990 Emissions	2010 Emissions	Kyoto vs. 1990	Kyoto Target	Kyoto Gap	Gap to 1990
Japan	308	466	94%	290	176	–57.3%
United States	1,337	1,803	93%	1,243	560	–41.9%
Canada	137	182	94%	129	53	–33.8%
Western Europe	1,016	1,208	92%	935	273	–26.9%
Australia	100	127	108%	108	19	+19.0%
Former USSR	1,029	872	100%	1,029	(157)	NA
Eastern Europe	309	306	94%	290	16	–5.0%

Source: Business Roundtable 1998, 12.

range from an 8 percent reduction for western European countries to an 8 percent increase for Australia. The table also shows the implied Kyoto emission target and the gap that results when the target level of emissions is subtracted from predicted uncontrolled year 2010 emissions. The gap amounts to a total reduction of 940 million metric tons of carbon.[4] Applying a cost per ton of $23 to $300 for removing these emissions—the wide range of estimates found in various studies to be discussed later—one obtains a total cost that ranges from $230 billion to $3 trillion. These potentially massive amounts represent costs that some people will pay and revenues that other people will receive. For the terms of Kyoto to be met, the transactions must be completed by the years 2008–2012. The last column of the table shows the implied percentage reduction for each area. This indicates the relative burden. Two countries, Japan and the United States, stand out with respect to the burden to be borne. The differential effects are significant.

Provisions of the Kyoto agreement allow the trading of emission credits across the world. Though controversial and still embryonic, such trading figures into the cost estimates just mentioned. If an industrial firm in the United States, for example, could purchase emission reductions from sources here or in another country at a price less than its control cost, the purchase of those carbon credits could reduce total control costs. But for purchases to occur, there must be suppliers as well as an effective institutional infrastructure to accommodate the trading.

The data on tons of emissions to be reduced by country and region illustrate the challenge faced by the United States, which is responsible for 560 million metric tons of reductions. Assuming emission reduction trading were operational, the states of the former USSR and eastern Europe would offer a fertile market for the purchase of credits. But the total tons of emissions to be reduced by the United States is large when compared with the excess available to the former USSR and with total emission magnitudes elsewhere. Domestic trading, which would take place at higher cost, could supplement a world market for emission credits if one should arise before the 2008–2012 deadline. But what if a group of countries could cartelize and exclude the United States?

Throughout the Kyoto negotiations, leaders from the European Union (EU) opposed credit trading as a substitute for taking action in each country. At the same time, the EU leadership favored an exclusive EU carbon-credit trading market, which the United States opposed. Finally, emission trading entered the agreement, and the EU achieved its goal: a trading cartel that allows the EU flexibility in setting reduction goals for member states but excludes the United States and Japan, the two countries with the most to gain by trading and the most to lose if the trading option is foreclosed.

ESTIMATES OF THE COST OF KYOTO

Academic Studies

Academic economists have done substantial research on the economics of controlling greenhouse gases (Manne and Richels 1991; Nordhaus 1991; Pearce and Barbier 1991; Whaley and Wigle 1991; Jorgenson and Wilcoxen 1993; Kosobud et al. 1994; Larsen and Shah 1994; Sinclair 1994; Holtz-Eakin and Selden 1995; Carrato, Galeotti, and Gallo 1996; Chen 1997). Among the studies are estimates based on large national econometric models, reports focused on Europe, and several studies that examine the relative costs associated with different regulatory instruments including emission taxes and permit trading.

When reviewed together, the academic studies indicate that carbon emission reductions of the magnitude called for by the Kyoto Protocol cannot be obtained at low cost. For example, the Jorgenson-Wilcoxen (1993) study assumes that U.S. carbon emissions would be held to 1990 levels and—incorrectly, as it turns out—that a 14.4 percent reduction in year 2020 emissions would achieve 1990 levels. (More current data indicate that a reduction of at least 36 percent would be required.) The authors apply emission taxes as the control instrument. Even with an assumed reduction much lower than that required by Kyoto, the study indicates that coal would be taxed at $11.01 per ton, oil at $2.31 per barrel, and natural gas at $0.28 per thousand cubic feet. The resulting revenues would yield $26 billion annually to the federal government. The authors find that coal would sustain a 40 percent price increase and an associated 26 percent decline in production. They predict that the rate of GDP growth would decline by an amount that ranges from a small fraction of a percentage point to one percentage point, relative to baseline growth.

Making somewhat similar assumptions and again applying emission taxes as the control instrument, David Pearce and Edward Barbier (1991) examine the U.K. economy. They estimate that a 67 percent increase in taxes for coal, along with 40 percent for gas and 54 percent for oil, would be required to achieve 20 percent reduction of U.K. carbon emissions by 2005. Pearce and Barbier emphasize that unilateral action can accomplish very little. If there is an uncontrolled component of the world economy, domestic energy use and carbon emissions may fall, but importation of substitutes could lead to carbon emission replacement by exporting countries.

Allan Manne and Richard Richels (1991) come close to meeting some of the terms of Kyoto in their global analysis of the cost of achieving 20 percent reductions in emissions by the year 2030. They accurately assume that the industrialized world accepts the reduction goal and the developing world does not. Their estimates of GDP effects indicate that the United States would sustain rising GDP losses across the control period, hitting a net loss of 3 percent by the year 2030. Losses for the OECD countries are shown to be much lower, reaching 1 to 2 percent in 2030. Mexico and other oil-producing countries sustain even larger losses, and China has the largest losses of all, losing 10 percent of GDP in the last half of the twenty-first century. They show that the price of coal increases fourfold and, remarkably, that substitution effects cause the demand for oil to increase, not decrease. This study illustrates Europe's relative gain from the protocol and helps to explain China's adamant opposition to it.

Other academic studies offer similar findings. Most relevant for present purposes, the research tells us (1) that the potential cost of the Kyoto Protocol to the United States would be far greater, both in total and in relative terms, for the United States than for other industrialized countries; (2) that a number of major European countries, including the United Kingdom and Germany, have already made adjustments that lighten their load; (3) that producers of substitutes for coal have much to gain; and (4) that the costs and the effects of compliance would differ significantly across countries, depending on the nature of the control instruments and the permit markets.

Economic Studies by Consulting Firms, Government, and Trade Associations

As the details of the emerging Kyoto Protocol became more predictable, groups with the most to lose or to gain, government agencies, and consulting firms came forward with more focused studies of Kyoto's impact. In some cases, the findings help us to understand more about the stakes involved for special-interest groups. The government studies arrive at cost estimates that vary far more widely than those in either the academic research or the work of private consulting groups. The estimate by the President's Council of Economic Advisers, for example, indicates that achieving the carbon reductions required by the protocol would impose very low costs on the U.S. economy.

The consulting firm DRI/McGraw-Hill provided an economic impact study for the United Mine Workers/Bituminous Coal Association, itself an interesting alliance (Impact of Carbon Mitigation Strategies 1997). The study contains scenarios in which greenhouse gas emissions would be stabilized at 1990 levels and at 10 percent lower levels by 2010. Instead of taxes, the study assumes government-issued marketable permits to be the control instrument. The report estimates that permit prices would range from $180 to $280 per ton of carbon across the control period if the goal were to maintain 1990-level emissions. Permit prices would be higher in the 10 percent reduction case. The permit prices reflect the incremental cost of reducing emissions by a ton of carbon. The study shows coal prices rising sevenfold, electricity prices about 100 percent, and retail gasoline prices 40 to 50 percent. Estimated employment losses reach 1.4 million jobs for the years 2000–2020, and GDP growth would be reduced by one percentage point at most as the economy adjusted to the constraints.

In a study commissioned by the American Petroleum Institute, WEFA assumes U.S. carbon emissions stabilize at 1990 levels by the year 2010 (National Impacts 1998). The study rejects the possibility of an intercountry permit market, owing to the lack of legal infrastructure, and assumes a U.S. permit market instead. The analysts then assume that uncontrolled emissions would be 27 percent above 1990 levels by 2010, and 46 percent above the target by 2020. To achieve the necessary reductions, carbon permit prices would have to rise across the control period from $100 per ton per year to $300.

According to WEFA, the emission reductions would lead to a 30–55 percent increase in consumer prices, with energy-intensive sectors sustaining shocks comparable to those associated with the Arab oil embargoes of the 1970s. Real GDP would fall 2.4 percent below baseline 2010 estimates, that loss alone amounting to $227 billion 1992 dollars. Cumulative GDP losses across the years from 2001 to 2020 would total $3.3 trillion, and employment decline would exceed 22.8 million workers by the year 2010.

With regard to international competitiveness and trade, the study notes:

> One key reason for the lower level of real GDP is reduced global competitiveness. Because the imposition of the carbon target and permit system is not borne equally by all countries . . . U.S. exports are relatively more expensive on the world market, while the prices of many imported products will fall. As a consequence, exports are lowered dramatically, while imports are increased substantially. (National Impacts 1998, 4–5)

The report notes that chemicals, paper, textiles and apparel, and computer and electronic parts production would be severely affected.

The U.S. Department of Energy has prepared a number of studies of Kyoto Protocol effects, with widely varying estimates of costs. The agency's first 1997 study (Office of Policy and International Affairs 1997) assumes a control target of achieving 1990 emission levels by 2010; it considers both a U.S. and an international market for tradable permits. In a strictly domestic market, permit prices would rise to $150 per ton of carbon emissions. With a world market, the price would be $40. Assuming a domestic control scenario, U.S. coal consumption falls by 50 percent by 2010, and the price triples. Total GDP losses across the period have a present value of $418 billion, and revenues from permits reach $400 billion by 2010.

In sharp contrast to this DOE study, a 1997 report prepared by the same agency's Office of Energy Efficiency and Renewable Energy, known as the "three labs" study (Interlaboratory Working Group 1997), indicates that Kyoto's costs would be inconsequential. This happy outcome is associated with the successful outcomes of yet-to-be-specified federal programs to create energy-efficient technologies and encourage switching from high- to low-carbon fuels. The report's authors do not estimate the costs of the federal programs or explain how those programs would accomplish the predicted feat in just thirteen years.

In a 1998 report, the DOE reversed itself again, indicating that gasoline prices would rise approximately 66 cents per gallon and electricity prices 86.4 percent above business-as-usual baseline prices (U.S. Department of Energy 1998). Unlike the "three labs" study but more in line with DOE's first 1997 study, this more recent analysis predicts that GDP in the year 2010 would be $397 billion lower than its baseline forecast.

Faced with conflicting evidence from the DOE on Kyoto's economic effects and with other major studies showing substantial costs, Congress was apparently confused. Along with other experts, Janet Yellen, chairwoman of the President's Council of Economic Advisers, gave testimony to Congress in March 1998, before the October 1998 DOE report was published (Yellen 1998). Yellen's analysis indicates that Kyoto will be practically costless to the U.S. economy.

To arrive at this optimistic conclusion, she makes a number of assumptions that implicitly expand the time span of her analysis far beyond the 2010–2012 compliance period. Specifically, she assumes that improved forestry practices and reforestation, which obviously take decades to complete, would cheaply offset carbon emissions and that U.S. firms and government agencies would participate in assisting developing countries in "clean economic development," thereby gaining emission credits. In addition, she assumes that U.S. carbon emitters would participate in a world market for trading emission credits, a process for which no institutional framework had been created at the time of her testimony. Beginning with a baseline estimate of $240 per ton for reducing carbon from emissions, Yellen applies the effects of her assumed cost savings and arrives at a final cost of $23 per ton for removing carbon.

Finally, one last study deserves comment. Ronald J. Sutherland, senior economist for the American Petroleum Institute, prepared an interesting analysis of Kyoto policy options that starts with a relatively simple econometric model for explaining carbon emissions in developed countries (Sutherland 1998). Accepting the agreement's binding time constraint, Sutherland shows that it would be practically impossible for the United States to achieve Kyoto's goals, especially if nuclear energy production were ultimately replaced by gas-fired turbines, as now predicted. He forecasts three important effects. First, clean nuclear fuel would be replaced with dirtier natural gas. Second, continued income growth would lead to a predictable increase in the demand for energy, and therefore to more carbon emissions. And third, the price increases necessary for achieving Kyoto goals, like those estimated by WEFA and the DOE, would simply not be accepted.

Using estimates of the price elasticity of demand from ten large-scale studies, Sutherland shows that gasoline prices would have to rise from $1.25 per gallon now to $4.23 per gallon in 2010 to meet Kyoto's U.S. goals. This increase, he believes, would be politically impossible to achieve. Creating an even greater challenge, post-2010 prices would have to rise continuously to offset increases in demand generated by rising incomes. Unlike the more optimistic findings of those who implicitly abandon Kyoto's time constraints, Sutherland demonstrates the hopelessness of relying on technical change, prices, and controls to reach Kyoto's U.S. goal for 2010.

Focusing on the studies that come closest to meeting the actual constraints imposed by Kyoto, we can deduce that the incremental cost of reducing emissions by a ton would range from $100 and $300. Given the number of tons to be cut, total reduction costs would amount to $1 trillion to $3 trillion. Coal obviously would be the hardest-hit industry. All fossil fuels would rise in price, but some, such as natural gas, would become relatively cheaper. In short, the resources to be transferred in credit markets would be massive, as would be the substitution effects across energy commodities and energy-intensive industries.

THE STRUGGLE FOR ADVANTAGE

My theory of Bootleggers and Baptists, a subset of the economic theory of regulation, calls attention to coalitions that seem to prevail when environmental and other social regulation is being formulated (Yandle 1983).[5] Although powerful interest groups still matter, this theory tells us that at least two interest groups must work in the same the direction. One group, the Bootleggers, look to regulation to enhance their market position or to limit competition. The other group, Baptists, adds a moral dimension to the regulatory cause that happens to solve the Bootleggers' problem. Traditionally, for example, both Baptists and Bootleggers have supported laws that limit the legal sale of alcoholic beverages.

In the post-Kyoto period, we should expect to find environmentalists playing the role of the Baptists. The theory suggests that we should find allies among the Bootlegger population—countries, industries, and firms that foresee a greener bottom line by supporting the "green" position. Within industries, some firms have specialized assets or outputs favored by rules that raise the cost of competing assets and products. With the Kyoto agreement, countries such as the United Kingdom are positioned to exploit carbon reductions already made and to raise the costs of competing economies. Other countries can become low-cost suppliers of carbon reduction offsets. But unlike conventional regulatory arrangements that involve one national government regulating domestic industries, Kyoto presents us with the unusual situation of countries behaving like firms, strategically positioning themselves to benefit while gaining protection and credibility from international environmental groups that embrace Kyoto as a necessary part of their crusade.

Post-Kyoto Episodes

Consider the following anecdotal evidence. In January 1997, Enron Corporation, which was a major provider of low-carbon natural gas, announced that it was forming the Enron Renewable Energy Corporation in an effort to "take advantage of the growing interest in environmentally sound alternatives of power in the $250 billion U.S. electricity market" (Salisbury 1998b). The new division was created to develop nontraditional energy products, which suffer a competitive disadvantage in current energy markets. Recognizing this disadvantage, Tom White, Enron Renewable Energy CEO, supported President Clinton's $6.3 billion plan to fight global warming, which included $3.6 billion in tax credits to spur the production and purchase of renewable energy and related technologies (Salisbury 1998b).

Of course, taxpayer subsidies can become habit-forming. On April 9, 1998, the National Corn Growers Association announced a major lobbying effort to prevent congressional efforts to eliminate the 5.4-cent-per-gallon subsidy to producers of corn-based ethanol (National Corn Growers Association 1998). The association's newsletter claimed that "ethanol is good for the economy, good for the environment, good for America," a theme endorsed in a speech by Mary Nichols, U.S. EPA Assistant Administrator for Air and Radiation, when she addressed the National Ethanol Conference in Des Moines, Iowa: "One area where I think we can do more together is the area of climate change and global warming" (Stark 1998). The celebration of ethanol did not mention that ethanol production may use more energy than it provides or that the federal government's $600 million annual ethanol subsidy assisted the production of beverage as well as industrial alcohol (Bandow 1997). On May 6, 1998, Republican leaders salvaged the subsidy, partly in the name of global warming prevention (Pianin 1998). Global warming appears to have saved the day for the corn producers.

Following the corn producers' cue, U.S. soybean producers heralded the environmental benefits associated with blends of diesel fuel and soybean oil (National Biodiesel Board 1998). In efforts to gain regulatory approval of biodiesel as an "alternative fuel" to substitute for ordinary diesel, which would insure the industry's participation in the Department of Energy's alternative fuel program, the lobbying organization indicated that "biodiesel helps reduce the effects of global warming by directly displacing fossil hydrocarbons" (National Biodiesel Board 1998).[6]

As regulation theory predicts, not every member of an industry expects to benefit by Kyoto restrictions. For example, Dean Kleckner, president of the 4.8-million-member American Farm Bureau Federation, opposed the protocol "because of its potential harm to U.S. farmers." Kleckner reflected the concerns of farmers who expect to see Kyoto-induced higher prices for food, fertilizer, and fuel (Farm-State Senators Skeptical 1998). The differential effects generated by regulation explain the formation and destruction of political coalitions.

The Breakup of Anti-Kyoto Coalitions

The Kyoto Protocol provides a setting in which some Bootleggers become converted to Baptists. One of the larger anti-Kyoto groups, the Global Climate Coalition (GCC), which was formed by major oil producers and hundreds of other firms, attempted to debunk Kyoto's weak scientific underpinnings and emphasized the expected economic costs of the protocol. However, some of the members began to see a silver lining around the Kyoto cloud. In June 1998, Shell Oil announced its departure from GCC. Friends of the Earth representative Anna Stanford claimed credit for Shell's green conversion, declaring: "We're delighted that our hard work has paid off, that Shell has bowed to public pressure and seen that the future lies in fighting climate change and investing in green energy" (Shell Withdraws from Global Climate Coalition 1998). Shell responded that "there are enough indications that CO_2 emissions are having an effect on climate change" (Magada 1998) and that the firm was "promoting the development of the gas industry particularly in countries with large coal reserves such as India and China" (Magada 1998). With Kyoto's help, firms with strategically located supplies of clean natural gas could improve their bottom lines while improving their green image.

British Petroleum's (BP's) earlier decision to part company with the GCC made Shell's departure a bit easier. Following serious discussions with leaders of the Environmental Defense Fund and the World Resources Institute, the Baptists in this case, John Browne, CEO of BP, indicated that firms such as BP should play a "positive and responsible part in identifying solutions" to the global warming problem (British Petroleum to Take Action 1997). Anticipating increased demand for oil as a cleaner coal substitute, BP also announced significantly increased investment in the development of solar and alternative energy technology.

In November 1998, the American Automobile Manufacturers Association, speaking for General Motors, Ford, and Chrysler, refused to help pay for Global Climate Information Project television ads opposing Kyoto (Look Who's Trying to Turn Green 1998). (The project is another major coalition of firms, labor unions, and farmers.) Shortly thereafter, the World Resources Institute gathered executives from GM, BP, and Monsanto to pledge support for Kyoto. Bill Ford, the newly named president of Ford Motor Company, stated: "There is a rising tide of environmental awareness. Smart companies will get ahead of the wave. Those that don't are headed for a wipeout" (Look Who's Trying to Turn Green 1998). Of course, pure environmentalism might explain the behavior of firms leaving the anti-Kyoto club, but other incentives may be involved, too.

In October 1998, Senate Bill 2617 was introduced, amending the Clean Air

Act to "provide regulatory relief for voluntary early action to mitigate greenhouse gas emissions."[7] The proposed legislation, which received strong bipartisan support, would have provided advanced carbon-reduction credits for actions to reduce emissions. If enacted, the rule change could have provided immediate and massive bottom-line benefits to carbon-emitting firms. For example, from 1996 to 1998, Mobil Oil has cut CO_2 emissions by 1 million tons (Salisbury 1998a). At $300 per ton, that reduction becomes a potential asset worth $300 million. In 1999, U.S. electric utility industries were undertaking actions to cut emissions by 47 million tons of CO_2 in two years. The potential side payments associated with this much emission reduction would have been enough to make a firm revise its anti-Kyoto stance and then push for restrictions and an emission-credit market.

Trimming the Budding Permit Market

Strategic actions taken to limit the scope of the budding permit market would redound to the benefit of U.S. firms that gain "advanced credit" for actions already taken. With the market constrained to the United States, credit prices would be higher than otherwise. British Deputy Prime Minister John Prescott may not have realized that his efforts to make the United States feel the pain of Kyoto could play into the hands of American firms with pre-endowed credits. Early on, Prescott showed disgust at the idea that U.S. agents would "buy tradable greenhouse emission permits from Russia" (Raven 1998). He stated that "Europe has always been clear that while we accept the trading possibilities in this matter, they should not be used as a reason for avoiding taking action in your own country" (Raven 1998). However, the European Commission's June 3, 1998, communication setting out principles for the November Buenos Aires meeting of Kyoto parties sent a somewhat confusing signal: "It is recognized that the flexible mechanisms can play an important role in meeting commitments at least cost, thereby safeguarding the competitiveness of EU industry. The existence of the EU bubble does not prevent the Community from fully participating in international emission trading" (Climate Change 1998, 2). The EU expected to gain internally by allocating burdens differentially across countries while meeting an overall goal and to gain externally through managed emission trading. Still, the Communication endorses Prescott's position regarding U.S. trading strategies:

At Buenos Aires, discussions on emission trading should focus on ensuring the establishment of strict rules and for setting minimum requirements that any Party or private entity needs to fulfill in order to participate in international trading. . . . It is also necessary to define the Protocol's use of the word "supple-

Table 2: EU Member State Emission Reduction Goals: 1990–2010

Country	Percent Change from 1990 Levels
Luxembourg	–28.0
Denmark	–21.0
Germany	–21.0
Austria	–13.0
United Kingdom	–12.5
Belgium	–7.5
Italy	–6.5
Netherlands	–6.0
France	0
Finland	0
Sweden	+4.0
Ireland	+13.0
Spain	+15.0
Greece	+25.0
Portugal	+27.0

Source: Friends of the Earth 1998.

mental" in respect to the contribution of the flexible mechanisms. In principle a ceiling should be set for trading to ensure that the main reduction in emissions are by domestic efforts. (Climate Change 1998, 2)

Analyzing the EU Allocation Scheme

In June 1998, EU leaders met to negotiate each member's reduction allocation (Friends of the Earth 1998). The draft proposal called for the individual reductions shown in table 2, which lists proposed emission reductions in descending order.

Modeling the Proposal

To explain these proposed targets, I fitted a regression equation, using as a dependent variable the number of tons of carbon emissions to be added or reduced in the 1990–2010 time period for each European Union country. This was calculated by multiplying the percentages in table 2 by each country's 1987 level of carbon emissions.[8] Independent variables included in the linear model are the 1995 GDP per capita in U.S. dollars, the 1995 ratio of carbon emissions to GDP (kg/US$), the 1998 population of the country, and the year when each country joined the European Union.

The politics of redistribution suggest that higher-income countries will absorb more of the environmental cleanup load than lower-income countries. The estimated regression coefficient of the GDP-per-capita variable should be negative. On the other hand, the cost of emission reductions will be higher for countries that have "cleaner" GDP. Generally speaking, the lower the amount of CO per dollar of output, the more costly it will be to reduce emissions by another ton. The estimated regression coefficient of the carbon-to-GDP variable should be negative. Thus, for example, France, although its large economy produces a large amount of emissions and the country has a relatively high income per capita, also has a relatively clean GDP. Therefore, France is being asked only to maintain its 1990 emission level.

In the regression equation, the population-size variable permits a test of the redistribution argument that smaller groups value a one-ton allowance more than larger groups, all else equal. The argument is based on the public-choice logic that a given concession to a less populous country is worth more per capita than the same concession to a more populous country (Chen 1997). This implies that the estimated regression coefficient of the population variable should be negative. The year of EU entry is included in the regression equation to test for the effect of side payments from older to newer members. The threat of exit is always greater for newer members than for older ones. Side payments in the form of emission allowances help to keep newer members in the fold. By this argument, the estimated regression coefficient of the year-of-entry variable should be positive. A competing argument maintains that newer members are not as well seasoned in the bargaining process as older members. In that case the estimated regression coefficient of the year-of-entry variable should be negative.

The Estimate

The following ordinary-least-squares equation, with intercept suppressed, was estimated:

$$tons = -1.477\ percap - 52046\ CO/GDP - 449.128\ pop + 31.173\ year$$

$$(-2.934) \qquad (-2.255) \qquad (-4.707) \qquad (2.980)$$

$$R^2 = 0.63 \quad F(3,11) = 9.04$$

Student's t-ratios are shown below the coefficients. As hypothesized, the proposed reductions follow a rough pattern that yields fewer tons of emission growth for higher-income countries, fewer tons for countries with higher carbon-to-GDP ratios, lower allowances for higher-population countries, and higher allowances for newer members of the European Union. For example, Spain, a recent EU member, has a lower income per capita and a "dirtier" GDP than France, but contributes a smaller total carbon load to the environment. Spain is allowed to increase emissions by 15 percent over the 1990 baseline amount, whereas France is to achieve and then maintain the 1990 emission level.

Recognizing the limits of an analysis based on such a small sample, it may still be instructive to interpret the equation's coefficients. At the mean of the estimate, a $1,000 increase of per capita GDP is associated with a 1.47-ton reduction in allowed carbon emissions. An increase of one gram per GDP dollar of carbon emissions (the sample mean is 420 grams) is associated with 52 additional tons of allowed emission growth. An increase of 1 million in population is associated with a loss of 449 tons of emission allowances, and an increase in the year of membership from, say, 1974 to 1975, is associated with an increase of 31 tons in allowances.

Because it is always cheaper to reduce emissions when concentrations are higher, the EU proposal establishes future sellers and buyers of tradable emission-reduction permits. Nation-states with little room for emission growth or lower emission concentrations will buy permits from states with dirtier carbon streams and larger allowances for emission growth. The allocation scheme enables us to predict that wealth will flow generally from northern to southern European countries, at least for trades within the European bubble. The EU plan of encouraging bubble trading within the EU and managing external trades places the central government in the traditional protectionist position of controlling exports and imports. But, in this case, the goods traded are permits (emission reductions), not commodities. An examination of the residuals in the regression equation for emission allowances, to identify allowances more than one standard deviation from the mean value, shows France gaining—and therefore being a potential seller of permits—and Germany and Portugal losing disproportionately in the political process that determined emission allowances.

FINAL THOUGHTS

The Kyoto Protocol is perhaps one of the most far-reaching international accords to be reached in modern times. And for all we know, global climate change may be one of the most serious threats faced in modern times. But whereas the climatic theory and evidence of global warming are shaky or at least controversial, the economic theory and evidence that help explain Kyoto policies are well established.

At the outset of this chapter, I asserted that Kyoto was more about cartelization and rent-seeking than about actions to reduce carbon emissions. To support such a proposition, one must show that the protocol itself contains a framework that allows, if it does not encourage, cartelization; that the scheme for reducing emissions gives rise to differential effects; that the magnitude of costs involved and opportunities for cartelization are large enough to offset the cost of organizing cartels; and that evidence of actions already taken buttresses the cartelization hypothesis.

Having considered the foregoing detailed discussion of the protocol, its costs, its differential effects, and a variety of related activities, the reader can judge whether the evidence is compelling. Agreeing that Kyoto is more about rent-seeking cartel efforts than about heat-reducing climate control does not negate the possibility that global warming is a real phenomenon that deserves attention. But if global warming is a genuine threat, the analysis presented here suggests that the Kyoto Protocol is not a useful mechanism for allaying that threat. What then might be a more suitable response? Several policy actions come to mind.

First, an independent research organization funded by tax revenue should be commissioned to measure, monitor, and report on temperature changes and related conditions. The monitoring organization should be insulated from the political process. Specifically, it should not be a part of the United Nations or the U.S. Environmental Protection Agency. In addition to monitoring and reporting, the organization should analyze data and provide evidence of the extent to which warming is related to human activities as opposed to natural atmospheric changes.

Second, if there is persuasive evidence that human activity is causally related to global warming, nations must determine whether on balance the warming entails more costs than benefits. If global warming is beneficial for some countries but harmful to others, actions can be considered for accommodating migration and other remedies for negatively affected people as well as for the affected natural environment. If human-induced global warming is found to be harmful on balance, then policies should be adopted to alter the actions that contribute significantly to the problem. Kyoto-type restrictions of carbon emissions may become relevant, but the mechanisms for achieving such reductions must include all people and must allow for side payments and other measures that enable developing countries to have access to energy.

Third, if global disaster is pending, access to accurate information becomes a top priority. As argued by Friedrich Hayek (1945), the price system is the most effective and lowest-cost information system available to mankind. Political actions that disguise or distort accurate prices should be eliminated. In particular, subsidies and taxes that affect important relative prices should be eliminated.

Fourth, incentives to encourage changes in investment, the discovery of new technologies, and the migration of capital and people should be enhanced. Capital-gains taxes should be repealed. Depreciation schedules should be shortened. Immigration policies should be liberalized. Markets and borders should be opened.

Of course, in making such policy recommendations, one presupposes a public-interest theory of government, in which politicians desire to minimize social costs and resist the pressures that emerge when special interests struggle for advantage. Unfortunately, the Kyoto story I have just told provides little if any reason to believe that public-interest policy recommendations, even if logically sound and empirically well founded, will receive serious consideration.

NOTES

1. For a lay summary of the protocol, see Sparber and O'Rourke 1998. For the complete text, see United Nations Framework Convention on Climate Change 1998.
2. A nonbinding resolution (S.R. 98) was passed in the Senate by a vote of 95 to 0 on July 25, 1997, requesting that the executive branch sign an agreement only if a commitment is made by developing countries to reduce emissions (Freedman 1997).
3. These data are from Antonelli and Schaefer (1997, 18) and are drawn from reports of the Intergovernmental Panel on Climate Change.
4. These data are taken from "The Kyoto Protocol: A Gap Analysis" (1998, 32), and were developed by Dr. Bert Bolin, a member of the Intergovernmental Panel on Climate Change.
5. The economic theory of regulation is associated with Stigler (1971) and Peltzman (1976). Posner (1974) added elements to it.
6. Although differential effects may explain the different positions taken by members of the oil industry, the situation faced by coal producers was more clear-cut. Coal producers and related unions are among the most vocal in their opposition. They succeeded in obtaining West Virginia legislation prohibiting the state's division of environmental protection from "proposing or implementing rules regulating greenhouse gas emissions from industrial sites." But when Governor Cecil Underwood signed the bill, he indicated that the state "should continue to encourage the development and implementation of technologies that allow the clean burning of coal" (Governor Signs Bill 1998).
7. Communication with Cato Institute, November 30, 1998.
8. Data on emissions from Larsen and Shah (1994, 843). For 1995, data for Luxembourg were not available in The World Bank Tables and were constructed by interpolation from Larsen and Shah.
9. Data for the independent variables are for 1995, except for population, which is for 1998, and are taken from World Bank 1998.

REFERENCES

Anderson, Terry, and Donald R. Leal. 1991. *Free Market Environmentalism.* San Francisco: Pacific Research Institute for Public Policy.

Antonelli, Angela, and Brett D. Schaefer. 1997. The Road to Kyoto. *Backgrounder,* October 6. Washington, D.C.: Heritage Foundation.

Bandow, Doug. 1997. Ethanol Keeps ADM Drunk on Tax Dollars. October 2. Washington, D.C.: Cato Institute. http://www.cato.org (May 5, 1998).

BBC News. 1998. World: Africa Conference to Prepare African Countries for Cut in CO2 Emissions. October 22. http://news.bbc.co.uk.

British Petroleum to Take Action on Climate Change. 1997. EDF Letter, September. http: //www.edf.org. (July 11, 1998).

Buchanan, James M., and Gordon Tullock. 1975. Polluters' "Profit" and Political Response. *American Economic Review* 65:139–47

Business Roundtable. 1998. *The Kyoto Protocol: A Gap Analysis.* Washington, D.C.: Business Roundtable.

Carrato, Carlo, Marzio Galeotti, and Massimo Gallo. 1996. Environmental Taxation and Unemployment: Some Evidence on the "Double Dividend Hypothesis in Europe." *Journal of Public Economics* 62 (1–2): 141–81.

Chen, Zihoi. 1997. Negotiating an Agreement on Global Warming. *Journal of Environmental Economics and Management* 32: 170–78.

Climate Change. 1998. European Union, DN: IP/98/498. http://europa.eu.int. (July 28).

Committee to Preserve American Security and Sovereignty. 1998. *Treaties, National Sovereignty, and Executive Power: A Report on the Kyoto Protocol.* Alexandria, Va.

Farm-State Senators Skeptical of Climate Plan. 1998. Reuters. http://www.yahoo.com (March 9).

Fialka, John J. 1998. Two Developing Nations Agree to Reduce "Greenhouse" Emissions in Breakthrough. *Wall Street Journal,* November 12, p. A32.

Freedman, Allan. 1997. Senate Sends Signal to Clinton on Global Warning Treaty. *Congressional Quarterly* 28 (July 26): 4.

Friends of the Earth. 1998. EU Greenhouse Gas Deal Too Much Hot Air. Press release. http: //www.foe.co.uk (July 11).

Governor Signs Bill Critical of Kyoto Protocol. 1998. http://www.state.wv.us (July 11).

Hardin, Garrett. 1968. The Tragedy of the Commons. *Science* 162: 1243–48.

Hayek, Friedrich A. 1945. The Use of Knowledge in Society. *American Economic Review* 35 (4): 519–30.

Holtz-Eakin, Douglas, and Thomas M. Selden. 1995. Stoking the Fires? CO2 Emissions and Economic Growth. *Journal of Public Economics* 57: 85–101.

Impact of Carbon Mitigation Strategies on Energy Markets, the National Economy, Industry, and Regional Economies. 1997. Report prepared for UMWA/BCOA by DRI/McGraw-Hill.

Interlaboratory Working Group on Energy-Efficient and Low-Carbon Technologies. 1997. *Scenarios of U.S. Carbon Reductions.* Washington, D.C.: Office of Energy Efficiency and Renewable Energy, U.S. Department of Energy.

Jorgenson, Dale W., and Peter J. Wilcoxen. 1993. Reducing U.S. Carbon Dioxide Emissions: An Assessment of Different Instruments. *Journal of Policy Modeling* 115 (5–6): 491–520.

Kosobud, Richard F., Thomas A. Daly, David W. South, and Kevin G. Quinn. 1994. Tradable Cumulative CO2 Permits and Global Warming Control. *Energy Journal* 15 (2): 213–32.

Kyoto Protocol to the United Nations Framework on Climate Change. 1997. http://www.cnn.com (July 27, 1998).

Larsen, Bjorn, and Anwar Shah. 1994. Global Tradeable Carbon Permits, Participation Incentives, and Transfers. *Oxford Economic Papers* 46 (5): 841–56.

Leopold, Evelyn. 1998. European Nations Sign Kyoto Global Warming Treaty. Reuters. http://www.infoseek.com (May 4).

List of Signatories to Kyoto Protocol. 1998. Global Climate Change Coalition. http://www.worldcorp.com (July 11).

Look Who's Trying to Turn Green. 1998. *Time,* November 9, p. 30.

Magada, Dominique. 1998. Focus: Shell Revamps Image. Reuters, April 21. http://www.infoseek.com (May 4).

Manne, Allan S., and Richard G. Richels. 1991. Global CO2 Emission Reductions: The Impacts of Rising Energy Costs. *Energy Journal* 12 (1): 87–107.

National Biodiesel Board. 1998. Agricultural Products Are Key to National Energy Security. February 24. http://biz.yahoo.com (March 2, 1998).

National Corn Growers Association. 1998. National Corn Growers Report. http://www.ncga.com (May 5).

National Impacts. 1998. WEFA National Impacts Link. http://www.api.org (July 27).

Nordhaus, William D. 1991. The Cost of Slowing Climate Change: A Survey. *Energy Journal* 12 (1): 37–65.

Office of Policy and International Affairs. 1997. *Analysis of Carbon Stabilization Cases.* SROIAF/97-01. Washington, D.C.: U.S. Department of Energy.

Ostrom, Elinor, and Edella Schlager. 1996. The Formation of Property Rights. In *Rights to Nature,* edited by Susan S. Hanna, Carl Folke, and Karl-Goran Maler. Washington, D.C.: Island Press.

Pearce, David, and Edward Barbier. 1991. The Greenhouse Effect: A View from Europe. *Energy Journal* 12 (1): 147–61.

Peltzman, Sam. 1976. Toward a More General Theory of Regulation. *Journal of Law and Economics* (August): 211–40.

Pianin, Eric. 1998. Gingrich Halts Move to End Ethanol Subsidy. *Greenville News,* May 7, p. 6D.

Posner, Richard A. 1974. Theories of Economic Regulation. *Bell Journal of Economics and Management Science* (Autumn): 335–58.

Raven, Gerrard. 1998. EU Urges U.S. to Implement Kyoto Global Warming Deal. Reuters. http://www.infoseek.com (May 4).

Salisbury, Laney. 1998a. Anti-Kyoto Group Disappointed by Shell Pullout. Infoseek: The News Channel. http://www.infoseek.com (May 5).

———. 1998b. Enron Exec Wants Clean Energy Tax Break. Reuters, February 26. http://www.yahoo.com (March 2).

Shell Withdraws from Global Climate Coalition. 1998. Friends of the Earth. http://www.foe.co.uk (July 11).

Sinclair, Peter J. N. 1994. On the Optimum Trend of Fossil Fuel Taxation. *Oxford Economic Papers* 46 (5): 869–77.

Singer, S. Fred. 1999. *Hot Talk, Cold Science: Global Warming's Unfinished Debate* (revised second edition). Oakland, California: The Independent Institute.

Sparber, Peter G., and Peter E. O'Rourke. 1998. Understanding the Kyoto Protocol. *Perspectives on Legislation, Regulation, and Litigation* 2 (4). Washington, D.C.: National Legal Center for the Public Interest.

Stark, Craig. 1996. Inventory of Greenhouse Gases. Iowa Energy Bulletin. http://www.state.ia.us/ (May 5, 1998).

Stigler, George J. 1971. The Economic Theory of Regulation. *Bell Journal of Economics and Management Science* (Autumn): 3–21.

Sutherland, Ronald J. 1998. *Achieving the Kyoto Protocol: An Analysis of Policy Options.* Washington, D.C.: American Petroleum Institute.

United Nations Framework Convention on Climate Change. 1998. The Kyoto Protocol. http://www.unfec.de/fccc/conv/file01.htm (October 19, 1998).

U.S. Department of Energy. 1998. *Impacts of the Kyoto Protocol on U.S. Energy Markets and the*

U.S. Economy. Energy Information Agency, Washington, D.C.: U.S. Government Printing Office.

Whaley, John, and Randall Wigle. 1991. Cutting CO_2 Emissions: The Effects of Alternative Policy Approaches. *Energy Journal* 12 (1): 109–24.

World Bank. 1998. *World Resources: 1998–99.* Washington, D.C.: World Bank.

Yandle, Bruce. 1983. Bootleggers and Baptists: The Education of a Regulatory Economist. *Regulation* (May–June): 12–16.

———. 1997. *Common Sense and Common Law for the Environment.* Lanham, Md.: Rowman and Littlefield.

———. 1998. *Bootleggers, Baptists, and Global Warming.* PERC Policy Series PS-14. Bozeman, Mont.: Political Economy Research Center.

Yellen, Janet. 1998. Testimony before the House Commerce Committee, March 4.

Acknowledgments: This chapter draws on "Bootleggers and Baptists in the Afterglow of Kyoto," a paper presented at the Hoover Institution conference "The Greening of U.S. Foreign Policy," October 15–18, 1998.

6

Global Warming and Its Dangers

J. R. CLARK AND DWIGHT R. LEE

We admit at the outset that we know little about the science of global warming. How much, if at all, the earth is warming; whether any warming is a trend or the result of random variations in global weather patterns; and, if a warming trend does exist, how much of it is owing to human activity are questions we cannot answer. Perhaps this ignorance protects us against anxiety attacks when we hear frightening accounts of what lies in store for planet earth and its inhabitants if governments do not immediately take bold and decisive control of the global climate. Our serenity, however, more likely arises from our exposure to public-choice analysis, which convinces us that concern about global warming is being inflamed and inflated as an open-ended rationale for expanding government control over the economy even further. This conviction does not leave us entirely sanguine, however, because we believe a serious danger of this rush to regulate is going largely unnoticed—a danger that might make any actual global warming a far greater problem than it should be.

FIRST, THE BAD NEWS

People are easily frightened, and when they are, governments grow. Fear and crises go hand in hand, and the evidence that government thrives in crises, real or imagined, is overwhelming (Higgs 1987). Claims of impending environmental crisis have proved especially effective in helping to justify an expanded role for government over the past thirty-five years. Widespread famine, acid rain, resource depletion, global cooling (yes, that's right—a big concern in the 1970s), lack of landfills, Alar-laced apples, the spotted owl's possible extinction, and urban sprawl are but a few of the alleged crises used in recent years to justify more reliance on government coercion and less reliance on market incentives. In every case, these alleged crises have proved innocuous or greatly exaggerated and, even when real, have commonly resulted from existing government restrictions on private action. Of course, the government programs put in place to deal with these concerns tend to remain in place, largely hidden from public view, long after public attention has been diverted to a new threat described in even more frightening terms and demanding yet more government programs.

Not surprisingly, the latest episode in this escalating series of crises, global warming, is being described in apocalyptic terms. For example, in *World on Fire: Saving an Endangered Earth,* former Senate leader George Mitchell informs us that global warming, if left unchecked, "would trigger meteorological chaos—raging hurricanes . . . capable of killing millions of people; . . . record-breaking heat waves; and profound drought that could drive Africa and the entire Indian subcontinent over the edge into mass starvation. . . . Unchecked, [global warming] would match nuclear war in its potential for devastation" (qtd. in Moore 1995, 83).[1] If this dire prediction is not frightening enough for you, search the combination of key words *global warming* and *catastrophic* on Google.com, and you will find comments that make Mitchell's account appear sanguine.

We do not want to leave the impression that the global-warming hawks bear only bad news. They invariably soften the threat of doom with the good news that because global warming results from human activity (they ignore what seems to be a warming trend on Mars), we can reverse its destructive effects by changing our behavior. Furthermore, we fortunately have "experts" who know what changes should be made, so our salvation requires only that we give these experts the necessary power and money. This reassuring news does raise a slight problem, however: the experts recommend changes that require government either directly or indirectly to impose controls over almost every aspect of our lives. Greenhouse gas emissions, understood as causes of global warming, now are being defined as pollutants that must be reduced significantly below current levels (as required, at least for developed nations, by the Kyoto Protocol). Carbon dioxide is receiving the most attention, and reducing it as recommended would require lifestyle changes in the developed world, affecting everything from the type of products we consume to the type of occupations we pursue, and would almost surely force the less-developed world to stay that way.[2] The cost of reducing carbon dioxide can be minimized (though remaining huge) by creating global markets for permits to emit carbon dioxide, but the parties whose interests are attached to government control strenuously oppose such markets. Yet even if these markets were created, they would be distorted significantly by direct government controls and by politically influential groups more interested in protecting their interests than in protecting the environment. Absent markets, political attempts to prevent global warming will result in the substitution of government regulations for both private property and market exchange on an enormous scale.

THE BEST SOLUTION: FREEDOM AND PROSPERITY

We admit that without government action, market incentives probably will not reduce greenhouse gas emissions in the short run. However, government regulations that undermine both information flows and adjustments of the market process in an effort to reduce greenhouse gases, even if successful, run the serious risk of increasing the long-run damage of any global warming that does occur.

Two possible, and opposing, approaches to global warming present themselves. The first, and the most familiar one, is the use of government regulations to force greenhouse gas reductions. The second approach is to emphasize arrangements that allow the most efficient response to any changes in the global climate that do occur, without trying to prevent such changes. The latter approach avoids government actions that interfere with the superior ability of markets to provide the information and motivation necessary to adapt quickly and appropriately to changing conditions. Although this approach may not do as much as direct government action to reduce global warming, it results in better responses to any given increase (or decrease) in global temperatures. So, even if warming is greater under the market approach than it is under the government approach, the former may still be preferable. A more efficient response to a worse situation can be better than a less efficient response to a better situation.[3]

Even if the government approach is better than the market approach in reducing greenhouse gases, this success may have little, if any, effect on global temperatures, given the rather minor proportion of total carbon dioxide emissions from human activities.[4] Furthermore, over the long run, the innovation fostered by the disciplined freedom of the market may offer the best hope for reducing reliance on the fossil fuels responsible for most human release of greenhouse gases. Reliance on market forces, with little thought about reducing greenhouse gases, rather than on government regulations specifically aimed at reducing them will do more in the long run to reduce any global-warming problem (and almost everyone agrees that if global warming is a problem, it is a long-run problem) by doing more to promote the economic prosperity and freedom that provide the best foundation for dealing with all problems.

NO FEDERAL GRANT FOR US

With global warming, as with many other issues, a strong political bias favors government coercion rather than market incentives. Most of the benefits from combating global warming with government regulation, if it is successful, will be diffused and delayed, as will the benefits from market responses to any warming that occurs. So there might seem to be no bias favoring political responses resulting from immediate and concentrated benefits. The politically salient considerations favoring government action, however, are not the highly speculative benefits from preventing a small increase in global temperatures many decades in the future. They are the immediate and concentrated benefits from larger bureaucratic budgets and research grants as well as the political advantages created by the appearance of dramatic action to allay a serious threat.[5] Bold and immediate action is certainly much easier to sell to a frightened and rationally ignorant public than an argument for relying on the indirect and little-understood invisible hand of market coordination.

To be sure, government regulations on greenhouse gases impose concentrated costs on business interests that are well organized politically. These interests have prevented the U.S. Senate from ratifying the Kyoto Protocol. How successful will these interests be, though, against a series of small regulations that, in aggregate, seriously constrain the private sector in the name of protecting the public against climate change? Arguments against such regulations are easily depicted as motivated by self-serving disregard for the planet, and they activate considerable amounts of "expressive voting" in support of the regulations.[6] Furthermore, large companies often favor burdensome environmental regulation as a way of hampering competition from smaller rivals.[7] The excessive cost that command-and-control environmental regulation has imposed on business (and on the economy in general) has certainly not prevented this approach from dominating environmental policy.

Market flexibility has received almost no support as a reasonable way of dealing with global warming, and almost everyone assumes that the problem demands government regulation. Lack of support for market solutions, however, should not be taken as evidence that government regulation can do better. Rationally ignorant voters are not solely responsible for giving organized interests the opportunity to force the second-best government approach on reluctant politicians. Politicians themselves often substitute government solutions for superior market solutions. Taking credit for benefits achieved through government action is much easier for politicians than taking credit for benefits achieved through markets, even though the latter benefits are much greater. This difference in credit-taking potential is especially germane in relation to global warming, where the appearance of immediate action may well be the only benefit and certainly the only benefit to be had immediately or within the time horizon of current politicians.

Reinforcing the view of global warming as a major threat to the planet, one that demands immediate government action, is the government's overwhelming dominance as the funding source for research on global climate change. Interestingly, the federal government is almost the only source of funding for such research, and the bias of those who control this funding cannot be doubted. So, as Michaels and Balling point out, "The chance that a finishing graduate student in climatology owes his publications, his dissertation, and therefore his newfound job, to federal global climate change funding is very high. Who among them is going to write a dissertation that global warming is an overblown problem?" (2000, 196). Michaels and Balling do not claim that dissenting views do not get published (after all, their own careers would belie such a claim), but they affirm that those who question the prevailing view on global warming find it more difficult to get their papers through the reviewing process and face more obstacles in achieving successful careers.

CONCLUSION

Once individuals are convinced that global warming is a problem, they automatically assume that it demands a government solution. Although we believe that global warming has been exaggerated by those who stand to gain from larger bureaucratic budgets and from more government controls over private decisions, we cannot render an informed judgment as to how serious a problem it might be. However, based on our understanding of how markets transmit dispersed information on changing conditions and motivate people to coordinate their responses to those conditions in the most appropriate ways, we are convinced that any problem of global warming will be dealt with better through market incentives than with government mandates.

People tend to regard global warming as a problem that should be attacked directly by imposing restrictions on market behavior in an attempt to reduce temperature increases. Almost universally ignored is the argument that a superior solution emphasizes the best response to whatever changes in temperature occur rather than attempts to prevent those changes. Further, even if people acknowledge the importance of appropriate responses to temperature changes, few appreciate the ability of market incentives to inform and motivate such responses; nor do they recognize that government mandates undermine those responses by distorting market incentives or by rendering them completely inoperable.

Even if government mandates were to be more effective than market incentives in reducing global warming, market incentives probably would still be more effective in reducing the harm of global warming by motivating a better response to a worse situation. Moreover, in the long run, reliance on the informed flexibility of the market will do more than government controls to reduce greenhouse gases. No sensible person denies that markets do a far better job than governments in promoting and utilizing technological advances that constitute our best hope for reducing dependence on fossil fuels and for creating the wealth that increases both our demand and our ability to deal with a wide range of environmental problems, including global warming.

The "problem" with the market is not that it is inferior to government in dealing with global warming, but that it handles issues effectively without requiring large government funding of organized groups and without concentrating power in the hands of a few experts. Instead, the market mobilizes the actions of millions of people to solve the problem indirectly by means of marginal and for the most part mundane adjustments for which, no matter how effective they may be, politicians and bureaucrats can take no credit.

NOTES

1. Moore points out that the earth has experienced times, including some in the past few hundred years, when the weather was substantially warmer than it is currently and that those times have been associated with bursts of human progress and improvements in living standards, whereas periods of cooler weather have been periods of stagnation and worse (1995, 83).
2. The human role in carbon dioxide discharges is modest compared to nature's. According to Easterbrook, "naturally occurring carbon emissions outnumber human-caused emissions roughly 29 to one" (1995, 312). Interestingly, some scientists believe that methane may contribute as much to global warming as carbon dioxide does because, though less prevalent, it is far more effective in trapping heat. Moreover, methane reduction would be much less costly. See Easterbrook (1995, 298–300) for the advantages of focusing on methane and for some of the special-interest opposition to doing so.
3. In this regard, we might consider seriously Nordhaus's observation that "perhaps we should conclude that the major concern lies in the uncertainties and imponderable impacts of climate change rather than in the smooth changes foreseen by the global models" (1993, 23). Nordhaus himself, however, uses this observation to emphasize the importance of flexible policy approaches rather than to recommend market adjustments to unforeseen conditions.
4. This argument is stronger for carbon dioxide than for methane emissions, which, as noted previously, are more easily reduced and may be as responsible for global warming.
5. If these benefits were not important in global-warming politics, it would be difficult to explain the exaggerated and frightening scenarios that those who benefit from political action are constantly putting before the public. For example, in a fit of candor Stephen Schneider, a

major activist in the fight against global warming (who in the 1970s warned of global cooling), told the *Boston Globe* in the early 1990s, "It is journalistically irresponsible to present both sides [of the global warming issue] as though it were a question of balance. . . . I don't set very much store by looking at the direct evidence. . . . To avert the risk we need to get some broad-based support, to capture public imagination. That, of course, means getting loads of media coverage. So we have to offer up some scary scenarios, make some simplified dramatic statements and little mention of any doubts one might have. . . . Each of us has to decide what the right balance is between being effective and being honest" (qtd. in Bandow 1998, 35).

6. See Brennan and Lomasky 1993 for a discussion of expressive voting and its distorting effects on political decisions.
7. See Malone and McCormick 1982 and Parshigian 1984 on business support for inefficient environmental regulations.

REFERENCES

Bandow, Doug. 1998. Global Politics, Political Warming. *The Freeman* 48, no. 1 (January): 34–36.

Brennan, Geoffrey, and Loren Lomasky. 1993. *Democracy and Decision: The Pure Theory of Electoral Preference.* Cambridge, U.K.: Cambridge University Press.

Easterbrook, Gregg. 1995. *A Moment on the Earth: The Coming Age of Environmental Optimism.* New York: Penguin.

Higgs, Robert. 1987. *Crisis and Leviathan: Critical Episodes in the Growth of American Government.* New York: Oxford University Press.

Malone, M. T., and R. E. McCormick. 1982. A Positive Theory of Environmental Quality Regulation. *Journal of Law and Economics* 25: 99–123.

Michaels, Patrick J., and Robert C. Balling Jr. 2000. *The Satanic Gases: Clearing the Air about Global Warming.* Washington, D.C.: Cato Institute.

Moore, Thomas G. 1995. Why Global Warming Would Be Good for You. *The Public Interest,* no. 118 (winter): 83–99.

Nordhaus, William D. 1993. Reflections on the Economics of Climate Change. *Journal of Economic Perspectives* 7, no. 4 (fall): 11–25.

Parshigian, Peter B. 1984. The Effects of Environmental Regulation on Optimal Plant and Factor Shares. *Journal of Law and Economics* 27: 1–28.

PART III
Endangered Species

7

The Endangered Species Act

Who's Saving What?

RANDY T. SIMMONS

Environmental policy in general and the Endangered Species Act (ESA) in particular are far removed from our roots in limited government, individual freedom, and personal responsibility. At their core are increasing coercion, expanding government, and shifting responsibility from individuals to society while shifting social costs to individuals. As the government takes charge, however, politicians, bureaucrats, and judges do not necessarily create appropriate laws and regulations. The objectives of some laws and regulations even defy common sense—save every species, zero discharge, zero risk, and natural regulation, to name but a few. The actual content of policy is animated more by emotions than by analysis as Congress makes moral statements rather than establishing functional policies and processes. Major environmental policies such as those that purportedly protect endangered species are implemented without regard to costs or results.

The Endangered Species Act is the federal law dedicated to preserving biological resources. It operates by assigning infinite value to every species and declaring that each must be saved. In 1978, in the Tellico Dam decision (*TVA v. Hill*, 437 U.S. 187), the U.S. Supreme Court declared that the ESA defines "the value of endangered species as incalculable," and requires that species losses must be stopped "whatever the cost" (184).

One of the ESA's most eloquent supporters, Harvard biologist E. O. Wilson (1992), declares that "every scrap of biological diversity is priceless, to be learned and cherished, and never to be surrendered without a struggle" (32). Wilson's statement may be viewed as a modern version of Aldo Leopold's claim that "if the [living world], in the course of eons has built something we like but do not understand, then who but a fool would discard seemingly useless parts?

To keep every cog in the wheel is the first precaution of intelligent tinkering" (1949, 177).

Such high-minded claims may be emotionally satisfying, but by themselves they provide little guide to policy. The question remains: What to do next? As biologist Garrett Hardin put it: "*And then what?* As act becomes policy, as event gives way to cycle of events, what are the consequences? Good intentions are not enough; the mechanism of our policy must produce good results in this time tied world where consequences become causes" (1982, 5). If policy and court decisions declare species to have infinite value, then what? The ESA was created with the intention of saving species, but what have been the results? Are species being preserved? Are they recovering from threats? What are the costs? Who bears them?

As soon as we move from emotive statements about the value of protecting species, it becomes clear that saving every cog or species is costly. It is also clear that the ESA does not allow for reasonable comparisons of costs and benefits. In fact, the act expressly forbids that the secretary of the interior, the administrator charged with implementing it, consider economic effects when he acts to protect a species. Those effects can range from mild irritations to loss of almost all economic value of the land in question. For example, the ESA empowers the salt marsh mouse to hold back bulldozers and the inch-long Delta smelt to reduce freshwater pumping for cities and farms.

Congress does not appropriate enough funds to save every cog. Although state and federal spending on endangered species has averaged $529 billion per year during the 1990s (House Committee on Resources 1997), the Fish and Wildlife Service is so short of funds that it is unable to write recovery plans for more than 400 listed species, to identify critical habitat for more than 750 listed species, to list nearly 180 species that meet the legal definitions of "threatened" or "endangered," or to adequately review nearly 4,000 species that may be declining (O'Toole 1996, 32).

Despite high-minded intentions, the ESA does not achieve its goals. Moreover, it puts obstacles in the way of public and private preservation efforts and creates disincentives for landowners to protect species on their own property. The ESA is, in fact, dishonest legislation. Species are listed but not recovered, and the costs of carrying out the act's public purposes are disproportionately borne by private landowners.

DOES THE ESA WORK?

The ESA authorizes the Fish and Wildlife Service (FWS) to create "critical habitat designations" and requires the development of recovery plans for species on both the threatened and the endangered lists. The purpose is to identify species in trouble, protect them initially, and develop plans and programs to restore the designated populations to viable levels. Although recovery plans have been designated for 653 of the 1,138 U.S. species listed as threatened or endangered, critical habitat has been designated for just 124 (U.S. Fish and Wildlife Service [USFWS] 1997).

A 1994 Fish and Wildlife Service report identified twenty species as having been delisted, eight because they had become extinct and eight others because the original data used to justify their listing were in error (USFWS 1994b, 41–42). According to the report, only four species had been removed from the endangered species list because their populations had "recovered" (41–42). The report also identified the gray whale, the brown pelican, and the American alligator as delisted but remaining in a special listing category (41).

However, the Endangered Species Act helped none of these species. The first four species were delisted because more were discovered than had originally been thought to exist. The alligator is also a case of data error. Officials of the Florida Fresh Water Fish and Game Commission believe that the alligator was originally listed because its population dynamics were misunderstood. A chapter in the National Wildlife Federation's magazine claimed that the "familiar and gratifying" recovery story about the alligator was "mostly wrong" (Lewis, 1987). In addition, alligator farming has greatly increased alligator numbers in the wild by reducing the incentives for poaching. The principal cause of the pelican's decline was reproductive failure due to the pesticide DDT, and its recovery had far more to do with banning DDT than with the Endangered Species Act. (DDT was banned in 1972, and the ESA was passed in 1973.) When the gray whale was listed, its numbers were growing, and it was delisted as its numbers continued to grow. Since the 1994 FWS report, the Arctic peregrine falcon has also been delisted, but, like the brown pelican, the peregrine recovered in large part because of the ban on DDT. The remoteness of its nesting habitat in northern Alaska is also considered a major contributing factor (Competitive Enterprise Institute 1995, 1–3; Gordon, Lacy, and Streeter 1997). Thus, the few species listed under the ESA that are now considered recovered are not ESA success stories at all, even though Secretary of the Interior Bruce Babbitt called the act "the most innovative, wide-reaching and successful environmental law that has been passed in the past quarter century" (1994, 55).

Another reason the ESA does not live up to its publicity is that, besides keeping species from going extinct, its purpose is to return them to viability. The ESA successfully *lists* species, but it brings few back from the edge of extinction. Over the past decade an average of 38 species have been added to the list each year, but of all listed species only 10 percent are improving, and the backlog of species waiting to be listed keeps growing (Reid 1994, 5–6).

DATA QUALITY

Nothing in the ESA regulates the quality of the data admissible for the purpose of listing species. Data are not subjected to a scientific review process, they are not field-checked, and the Secretary of the Interior is not required to identify any gaps in the data that have been collected. One result is that species are sometimes listed needlessly (Gordon, Lacy, and Streeter 1997). Eight of the nineteen species that have been removed from the Endangered Species List are in this category. Two examples are the Mexican duck and the tuamoc globeberry, species that were listed and then delisted once new data became available.

During the 1970s the Mexican duck was listed as endangered, but it was later removed upon discovery that there is no such thing as a "Mexican" duck. What biologists initially thought was a distinct species turned out to be a blue-eyed version of a mallard that was not genetically different from regular mallards (USFWS 1978).

In 1986 the tuamoc globeberry was listed, and plans were implemented to save it from extinction. From 1989 to 1991, the Bureau of Land Management, the Department of Defense, the Army Corps of Engineers, and the Bureau of Reclamation spent $1.5 million to protect the globeberry. But the FWS eventually discovered that the tuamoc globeberry existed in far greater numbers and in more places than initially thought, so it was delisted.

One might argue that the tuamoc globeberry and the Mexican duck are successes of the act, not failures. After all, new data were allowed into consideration and, based on those data, the listing decision was reversed. Moreover, spending a few million dollars on a mistake seems virtually harmless when viewed amid the bigger picture of government spending and pork-barrel politics. Of course, a process that allows wrong decisions to be reversed is good, but in addition to the possibility of reversal, a fair policy would require that initial listing decisions rest on sound, relatively complete data. Such a requirement is especially important given the disruptive effects the resulting decisions often have on people's lives.

COSTS OF THE ENDANGERED SPECIES ACT

Although court decisions regarding endangered-species protection use terms such as "incalculable," "the highest of priorities," and "whatever the cost," Congress does not appropriate funds as if species had incalculable value or as if species protection were the highest of priorities. Claims about expenditures on endangered-species protection vary by as much as a factor of ten, but the actual government expenditures appear to be quite modest. The Environmental Defense Fund's fact sheet on endangered species claims, "The core budget of the U.S. Fish and Wildlife Service's endangered species program is $58 million, or about equal to what residents of the metropolitan Washington, D.C., area spend each year on pizza delivery from just one major chain" (Environmental Defense Fund 1998). The House of Representatives Committee on Resources claims that total state and federal spending from 1991 to 1997 was $3.706 billion, or an average of $529 million per year (House Committee on Resources 1997, 23). Even this amount does not approach infinity; nor does it suggest that Congress places "the highest of priorities" on species preservation.

The relatively small amount of funding appropriated for endangered-species programs suggests that Congress is unwilling to carry out the policy implications of legislation the Court interprets to mean that Congress considers the worth of endangered species to be incalculable. Others are bolder than Congress in identifying the policy implications. For example, Dr. Reed Noss (1992, 1994), editor of the prestigious scientific journal *Conservation Biology*, argues that huge amounts of land must be set aside. For a species to maintain genetic variation sufficient to cope with environmental uncertainty and to guard against nature's catastrophes, an interbreeding population of at least 1,500 to 2,000 individuals must be maintained. Therefore, Noss claims, 484 million acres are required to support a minimum viable population of 2,000 grizzlies, 400 million acres for 2,000 wolverines, and 200 million acres for 2,000 wolves. Other estimates are somewhat smaller. Bader (1992), for instance, estimated that a minimum of 32 million acres is required to support a population of 2,000 grizzly bears. Even this lower estimate represents an area equal to one-third of Montana.

As of March 31, 1997, there were 1,082 species listed as threatened or endangered (USFWS 1997). In addition, more than 3,000 others are waiting to be listed (Carroll et al. 1996), and a few thousand more are waiting to be "discovered." These estimates may be low: others claim that more than 9,000 plants and animals are biologically threatened in the United States (Halverson 1995). Saving all the habitat required to protect minimum viable populations of all these endangered species will require depopulating large areas of the United States, hardly a credible policy.

An alternative to depopulating large sections of the country was suggested by ecologist Paul Ehrlich and biologist E. O. Wilson. In a chapter in *Science,* they claimed that biodiversity is in such danger that the United States must "cease developing any more relatively undeveloped land" (Ehrlich and Wilson 1991, 758). Only then, they argue, can the intentions of the ESA be met.

A less costly plan would be simply to pay the costs of listing, recovering, and delisting species currently listed or proposed for listing. But estimates of those costs range from $7.63 billion to $13.56 billion (National Wilderness Institute 1994, 38–39), and Congress does not come close to providing funding at those levels. The National Wilderness Institute reviewed the costs estimated by 306 recovery plans for 388 of the 853 species then listed under the ESA and found the total cost would be $884,164,000 in 1994 dollars (Gordon and Streeter 1994, 1). The National Marine Fisheries Service's proposed recovery plan for the Snake River salmon makes cost projections ranging from $166 million to $338 million per year over the seven-year life of the plan (House Committee on Resources 1997, 18).

Other costs, however, are often not immediately clear to analysts. Half of the economic activity around Bruneau, Idaho, for example, was threatened when the Fish and Wildlife Service began cutting off water rights to fifty-nine farms and ranches in order to protect the Bruneau Hot Springs snail.[1] In Oregon and Washington, millions of acres of productive timberlands are off-limits for timber harvest, as are similar amounts of loblolly pine stands on private and public lands in the South. Western rangelands are subject to use prohibitions because of the desert tortoise and numerous desert plant species, and the mining of phosphates and other minerals has been banned in much of northern Florida. Property owners in the Texas hill country face criminal prosecution if they clear brush along their fencerows. Water rights of the Ute and Navajo tribes are being expropriated by the federal government to protect endangered species. Highways are being rerouted and airport expansions prohibited across the country. The kangaroo rat regulations in Riverside County, California, make the rat effectively the largest landowner in the county. Current plans to save the Sacramento Delta smelt are likely to cost billions. According to Ike Sugg, the executive director of the Exotic Wildlife Association, "conservative estimates of the act's costs are in the tens of billions of dollars" (1993, A15).

Proponents claim, however, that the ESA causes very few economic disruptions, and they point to what are called "agency consultations" as evidence. Any federal agency proposing an activity that may affect an endangered species must consult with the Fish and Wildlife Service or with the National Marine Fisheries Service to ensure that the activity produces no more than minimal harm to protected species. After a biological assessment determines that the project is likely to adversely affect the species or its critical habitat, the Fish and Wildlife Service must issue a biological opinion, known as a jeopardy opinion, that officially declares whether the action is likely to jeopardize the continued existence of a listed species or to result in the destruction or adverse modification of its critical habitat. If the Fish and Wildlife Service or the National Marine Fisheries Service finds that the proposed action will jeopardize members of an endangered species or adversely modify an endangered species' critical habitat, the agency issues a jeopardy opinion that suggests reasonable and prudent alternative actions.

A 1992 World Wildlife Fund (WWF) analysis of 71,560 federal agency consultations from 1987 to 1991, for example, found only 18 projects that were stopped because they would endanger a species. Many of the agency consultations, though, were simply telephone calls to the FWS requesting a list of endangered species. Of the 2,000 contacts that became formal consultations, jeopardy opinions were issued for 350. Eighteen activities were blocked, and 35 were left unresolved. The WWF study uses these data to show that, depending on the outcomes of the unresolved cases, from 0.9 to 2.65 percent of the 2,000 formal consultations and from 4.1 to 15.1 percent of the jeopardy opinions stopped proposed actions.

Looking carefully at the WWF numbers, however, one reaches different conclusions. Of the 350 jeopardy opinions, 169 were separate opinions for a single EPA pesticide program. If those 169 opinions are counted as one, the percentage of actions blocked rises to 24 percent. The WWF study's results also change dramatically if one includes the 44 timber sales adversely affected by the spotted-owl decisions (National Wilderness Institute 1994, 18–19).

The WWF analysis is restricted to actions between government agencies and does not consider how many actions on private lands are blocked by the ESA. Private landowners who wish to work around an endangered species on their land can face huge costs as projects are dropped or delayed, some indefinitely.

Another California illustration of the ESA's hidden costs is provided by the pocket mouse, a rodent rediscovered in 1993 after not having been seen since 1971. The mice were found during an environmental survey for a proposed resort, to be built on the Headlands, 121 acres overlooking Dana Point Harbor. The survey found thirty-nine mice on four acres of the site. The Fish and Wildlife Service moved quickly to place the mouse on the emergency endangered-species list, thereby delaying construction of the four-hundred-room hotel complex and 394 homes. The *Los Angeles Times* quoted a spokesman for the local chapter of the Audubon Society who welcomed the listing because, even though it might not stop the project, it would cause delay (Haldane and Hall 1994, B1).

Projects such as the Dana Point project are being delayed all across the country by the ESA. The Fish and Wildlife Service's *Report to Congress on the Endangered and Threatened Species Recovery Program* (1993), which claimed to be the "first comprehensive accounting" of conservation and recovery efforts since 1967, estimated that "approximately 25% of all listed species have conflicts with development projects or other forms of economic activity." A content analysis by the National Wilderness Institute (1994, 39) of 306 recovery plans shows the expected conflicts detailed in table 1. With more than 3,000 species waiting to be reviewed and then possibly listed, the amount of economic conflict will expand exponentially.

PRIVATE VERSUS PUBLIC RIGHTS

The agency consultations mentioned earlier are required by Section 7 of the ESA. That section requires federal agencies to ensure that their actions, broadly defined as activities authorized, funded, or carried out by an agency, including issuing permits and licenses, do not "jeopardize the continued existence of any endangered or threatened species or result in the destruction or adverse modification of habitat." Building dams, roads, canals, highways, or housing; granting grazing permits or hunting licenses; and allowing access to federal lands are all activities of the federal government that might alter habitat. If the agency determines that its action would jeopardize a species, it must either recommend reasonable and prudent alternatives that would avoid harming the species (*U.S. Code*, vol. 16, sec. 1536) or apply to the Endangered Species Committee for an exemption from the requirements of the act. This committee, known popularly as the "God Squad," consists of seven political appointees. It was created by Congress in 1978 to serve as a political relief valve during the controversy over the Tellico Dam and the snail darter.

When the ESA was first passed, it did not include a provision for private citizens to have the same right to consult and receive recommendations for "reasonable and prudent" alternatives. There was no mechanism for property owners to request a written opinion about whether activities on their land adversely affected a listed species; nor could they apply for an exemption from the requirements of the Act. In 1982, Congress remedied that inequity with an amendment to section 10 of the Act, allowing the Fish and Wildlife Service and the National Marine Fisheries Service to issue incidental take permits to private parties meeting certain requirements. Those requirements, as stipulated in the regulations adopted in 1984 by the Fish and Wildlife Service (see

Table 1: Conflicting Activities and the Number of Recovery Plans in Which Each Is Mentioned

Agriculture	153
Cattle	100
Collecting	117
Development	245
Grazing	128
Habitat manipulation	199
Hunting/fishing	83
Irrigation	43
Mining	121
Off-road vehicles	63
Outdoor recreation	146
Pesticides	150
Water development	147
Wetlands degradation	21

Source: National Wilderness Institute 1994, 39

50 C.F.R. secs. 17.22b, 17.32b) and in 1990 by the National Marine Fisheries Service (see 50 C.F.R. secs. 222.22), require applicants for an incidental take permit to submit a Habitat Conservation Plan (HCP). A proposed HCP must include an assessment of impacts likely to result from the proposed taking; measures the permit applicant will undertake to monitor, minimize, and mitigate such impacts; a listing of alternative actions and reasons for not adopting them; and additional measures the permitting agency may require as "necessary and appropriate."

One problem for property owners granted some of the early HCPs was the continuing threat that the federal agency could come back and claim that the agreement was not sufficient, that something new was required once new and better information had become available. In response, the Clinton administration adopted a new policy in August 1994 entitled "No Surprises: Assuring Certainty for Private Landowners in Endangered Species Act Habitat Conservation Planning" (the "No Surprises" policy). As Secretary Babbitt (1997a) explained in a July 17, 1997, speech at the National Press Club,

> We thought about [future changes to HCPs] and fought about it, and said, you know, if we're going to make this act work on the ground in the real world, and ask timber companies and developers to make those kinds of concessions to find that balance, we've got to establish one simple com-

> monsense principle, and that is one bite at the apple—take a good one—thrash it out, but then say to the developer, "Okay, a deal's a deal—there aren't going to be any surprises." And if it turns out that we need a little more habitat or a few more adjustments, well then the obligation should be on the public, the participating public agencies, including the federal government, to put up the resources to rebalance the plan.[2]

By adopting the "No Surprises" policy, the Clinton administration officially recognized that previous arrangements had been deals that the private parties, but not the government, were required to live by. "No Surprises" was an attempt to correct that problem.

Secretary Babbitt's comments highlight the cost issue. Even under the HCP process, the property owner or developer has to mitigate his incidental taking of endangered species by providing land, money, procedures, or whatever is necessary. Babbitt says the agencies will take as large a bite of the private "apple" as they deem necessary, without compensating the owner, and then the landowner gets the rest of the apple.

One cynical but not unrealistic way of looking at HCPs and "No Surprises" is to view them as analogous to protection schemes.[3] The property owner occupies much the same position as a small merchant on Chicago's South Side during the 1920s. The merchant was offered protection from unspecified future problems if he would purchase insurance. His decision to purchase was, of course, "voluntary," and if he went along with the plan, he would be able to continue to operate his store without "surprises." The modern property owner with endangered species on his property has the same deal. If he gives up a portion of his property or some of its potential uses "voluntarily," he will be able to use the rest of his property, with no "surprises."

TAKING PRIVATE PROPERTY UNDER THE ESA

The ESA limits activities on public and private lands designated as endangered-species habitat. Given that two-thirds of U.S. lands belong to private owners, endangered-species policy entails federal control over much private property. Ninety-five percent of private lands are classified as "rural and non-developed," and those lands harbor the broadest diversity of species, including endangered species. According to data published by the U.S. General Accounting Office (1994), more than 75 percent of listed species in the United States depend on private land for all or part of their habitat requirements. Lynn Dwyer, Dennis Murphy, and Paul Ehrlich (1995, 736) found that fully 50 percent of federally protected species are found exclusively on private land. Although the distribution across private and public lands of the roughly 3,000 species being considered for listing under the ESA is not known (Carroll 1996, 5), a reasonable assumption is that they are distributed similarly.

Texas provides an illustration of the impact the ESA is having on private property owners. Ninety-five percent of the land in Texas is private, and that land supports 82 federally listed species and 305 candidate species. The result is that the costs of protecting and conserving endangered species in Texas fall on rural landowners, who are primarily farmers and ranchers. In most cases, they must bear those costs without compensation or assistance (McKinney 1993, 63–65).

Endangered-species policy makes it unlawful for any private citizen to interfere in any way with an endangered species or its habitat, and it imposes severe penalties on those who do. Farmers violate the ESA if they plow their land and in some cases even if they allow grazing on a pasture when an endangered species is present. In many cases, property owners are prohibited from cutting trees, clearing brush, using pesticides, planting crops, building homes, protecting livestock or even themselves from predators, and building roads. They are often required to set aside numerous acres for no purpose other than aiding the endangered species.

One illustration of how government agents try to use the ESA is provided by the case of *U.S. Fish and Wildlife Service v. Shuler*. In March 1994, a Department of the Interior administrative law judge fined Montana rancher John Shuler $4,000 for killing a grizzly bear. The bear was one of four the rancher found outside his home one summer night in 1989. The bears had previously killed $1,200 worth of sheep, and the rancher, upon seeing one from his living-room window, ran outside with his rifle. About thirty feet from his house, three bears were running through his flock. He fired over them, and a fourth bear came out of the dark and turned as if to attack him. The rancher shot the bear and assumed it was dead. The next day he found it was still alive. When it reared up on its haunches and came after him, he shot the bear twice more, finally killing it.

The Department of the Interior initially assessed a $7,000 penalty for "taking" the bear in violation of the ESA. Shuler appealed under the Act's self-defense clause, but Interior's judge ruled the rancher wasn't defending his life, just his property, and that he was "blameworthy to some degree in bringing about the occasion for the need to use deadly force." He could have watched from the safety of his living room while the bears devoured his sheep. But he "purposefully place[d] himself in the zone of imminent danger of a bear attack." Even so, the judge reduced the fine from the original $7,000 to $4,000

(Sugg 1993, A15).[4] Shuler then appealed to Interior's Ad Hoc Appeals Board, which upheld the initial ruling and increased the fine to $5,000. The board rejected Shuler's claim of self-defense and argued that his dog, which had gone on point when the bear reared, had "provoked the bear." Shuler's final step was to appeal to the U.S. District Court for the District of Montana. On March 17, 1998, nearly nine years after he had confronted the bears, his conviction and fine were finally vacated.[5]

The dollar costs imposed by the ESA can be substantial. Property values in Travis County, Texas, dropped $359 million after the golden-cheeked warbler and black-capped vireo were listed as endangered (Wilkinson 1993, 15). One Texan saw the appraised value of her land fall from $830,000 to $38,000 (Adler 1995, 3), and the state of Texas is losing $2 million in taxes each year because of property-tax assessment declines caused by the ESA (Wilkinson 1993, 15). People whose investments are tied up in land have had their retirement plans nearly wiped out by the listing of the warbler and the vireo, because their property now has little value and is nearly impossible to sell.

In North Carolina, Ben Cone had 1,560 acres of his 7,000 acres of land taken by the ESA to provide habitat for the red-cockaded woodpecker. He still owns the land, of course, but no timber may be harvested within a half-mile radius of each colony of woodpeckers. In addition, each colony must be protected from the controlled burning Cone does to improve wildlife habitat on his property, so now he has to rake and bushhog around each nest tree before burning because a burned tree would constitute a taking under the ESA.

In addition to increased management costs, Cone has suffered a loss of land value. He had the 1,560 acres appraised in 1993 and found that, absent the woodpeckers, the estimated value of the hunting leases, timber, and pine straw was $2,230,000. With the woodpeckers, the land was worth just $86,500, or 96 percent less. The twenty-nine birds therefore cost Cone $73,914 each, a cost he alone has borne.[6]

In order to protect the rest of his land from the same loss of value, Cone began clear-cutting more acres annually than he had previously, in order to decrease the amount of habitat attractive to red-cockaded woodpeckers. He also sent a letter to adjacent property owners informing them of their possible liabilities if the woodpecker nested on their lands. They immediately began clear-cutting their timber adjacent to Cone's property (Welch 1994, 151–97).

Cone's story generated a great deal of interest in the press and in Congress. In response, Defenders of Wildlife listed his story as one of the "top ten lies about the ESA." They posted the following claim on their Web site (Defenders of Wildlife, 1996):

> Ben Cone has been legally using his property for years in a variety of ways, including timber sales. Mr. Cone's land had been managed as a quail plantation, which provides good habitat for red-cockaded woodpeckers. USFWS offered to enter into a Habitat Conservation Plan which would allow Mr. Cone to harvest the timber and would provide for the incidental take of future woodpeckers that might occupy any thinned areas. Mr. Cone declined to enter into an HCP, but did submit a management plan, which was approved. He continues to sell pine straw from his property, operate a hunting lease on it, and harvest timber on it.

This version of the story is technically correct but misleading. Cone was offered a Habitat Conservation Plan that would have allowed him to log all of his land except for the acres of woodpecker habitat. Thus, by signing the agreement, he would have relinquished his right to receive compensation for the timber on those acres, which was valued at $1,425,000. Even without signing the plan he already had the right to cut timber on the portion of his property not occupied by woodpeckers. Cone does continue to rake pine straw on his land and to sell hunting leases. The total income from these activities on all 7,000 acres is less than $20,000 per year—hardly comparable to the $1.4 million in forgone timber revenues.

Although there are many stories similar to Cone's, reducing property values by listing a species under the ESA is not considered a "taking" under the U.S. Constitution. A constitutional "taking" refers to the phrase in the Fifth Amendment that states "nor shall private property be taken for public use, without just compensation." (It is not to be confused with the "taking" of an endangered species. Under the 1973 legislation, to take an endangered species is "to harass, harm, pursue, hunt, shoot, wound, kill, trap, capture, or collect or to attempt to engage in such conduct.") Under a strict interpretation of the Fifth Amendment, an owner whose property has been devalued because of government regulations could file a claim and be compensated, but no court has yet adopted this strict definition.

Until 1922, the U.S. Supreme Court required that a Fifth Amendment taking involve actual government appropriation or physical invasion of private property. However, in *Pennsylvania Coal v. Mahon* (260 U.S. 393 [1922]), the Court decided that the regulation of property use can constitute a taking and required compensation in that particular case. Opponents of the ESA's regulatory powers argue that ESA designations do, in fact, constitute a taking of private property without just compensation (Welch 1994, 186). But proponents of the ESA's existing broad powers argue that the ESA's provisions constitute a kind of zoning, and courts have not considered zoning laws to be a constitutional taking (Dwyer, Murphy, and Ehrlich 1995). Flagstaff, Arizona, for example, uses its zoning powers to prohibit landowners from cutting down pine trees within the city limits except to make room for improvements authorized under the city's planning and building codes. Similarly, Palmdale, California, prohibits its landowners from cutting Joshua trees, a species of yucca, on their land. If the ESA is interpreted as simply zoning, then Secretary Babbitt may have been correct to say that "the Fish and Wildlife Service has never come close to a constitutional taking" (1994, 55).

Although a 1995 Supreme Court decision, *Babbitt v. Sweet Home Chapters of Communities for a Greater Oregon* (515 U.S. 687), has been interpreted by some as strengthening Babbitt's position, the Court actually avoided the takings issue. The question before the Court was whether the definition of "harm" to an endangered species included habitat modification. The Court ruled that Congress meant to include habitat modification in the definition of harm and that the Fish and Wildlife Service's application of regulations prohibiting certain uses of private lands was consistent with congressional intent.

The Court's failure to decide whether the regulations constituted a taking of private property without just compensation is striking in light of two previous decisions that suggested the Court was moving in a direction that would allow FWS actions to be defined as takings. The first case was brought by a South Carolina resident, Mr. Lucas, who bought two residential lots on a South Carolina barrier island with the intention of building single-family homes. After he purchased the property, the state enacted the Beachfront Management Act, which barred him from erecting any permanent, habitable structures on his land. He sued the state agency, asserting that the ban on construction deprived him of the "economically viable use" of his property and was therefore a taking. The Court ruled in his favor and stated that government regulation of land that completely eliminates the economic use is per se a taking (*Lucas v. South Carolina Coastal Council*, 505 U.S. 1003 [1992]).

The second case was filed after the city of Tigard, Oregon, denied Florence Dolan a building permit for her property unless she dedicated 7,000 square feet of land for storm-water management and a park. The Supreme Court ruled in her favor, declaring that the city was attempting to force Dolan "to bear public burdens which, in all fairness and justice, should be borne by the public as a whole." The majority opinion further stated that the city should not try to avoid "the constitutional way" of paying for what it wants and that the Fifth Amendment should not be relegated to the status of "poor relation" compared with other amendments in the Bill of Rights (*Dolan v. City of Tigard*, 515 U.S. 374 [1994]).

When property values are being reduced from $900,000 to $38,000, as in the case of the Texas woman, or from $2,230,000 to $86,500 as in Ben

Cone's case, it is at least plausible that the Court might declare that ESA regulations amount to an attempt to avoid "the constitutional way" of paying for what Congress wants and are therefore unconstitutional takings. Clearly, Justice Scalia seems ready to make such an argument. In his dissenting opinion to the *Sweet Home* decision, he wrote that "the Court's holding that the hunting and killing prohibition incidentally preserves habitat on private lands imposes unfairness to the point of financial ruin—not just upon the rich, but upon the simplest farmer who finds his land conscripted to national zoological use." If Scalia's position became the majority opinion in a case that claimed FWS regulations were taking property without just compensation, protecting species would require far more innovation than heretofore seen.

Environmentalists greatly fear such a decision by the Court, because they believe it would make the costs of protecting species prohibitive, and they are fearful of congressional actions that might have similar effects (Dwyer, Murphy, and Ehrlich 1995). Tim Eichenberg, program counsel for the Center for Marine Conservation, reflected that fear in his comments regarding the *Sweet Home* decision. He called the Court's decision "a welcome dose of sanity in what has become an irrational and shortsighted rush in Congress to undo one of our nation's most important environmental laws" (quoted in Sturges 1995, 126). Some Republican members of Congress responded to *Sweet Home* by calling for immediate action to protect the rights of property owners (Pombo 1995). Secretary Babbitt called one property-rights bill introduced in Congress a "proposed raid on the public treasury" (1994, 55). If that bill had passed, it would have required federal agencies to compensate property owners for any diminution in value caused by any regulatory action taken under the authority of any environmental laws, including the Endangered Species Act.[7]

WARM-HEARTED DISHONESTY

Endangered-species policy had a warm-hearted beginning. Most members of Congress thought they were protecting charismatic species—grizzlies, whales, manatees, whooping cranes, bald eagles, and the like. But the laws they wrote protect far more species. Charismatic species are relatively few—birds, mammals, and other vertebrates as a whole constitute only about twenty percent of the 1.4 million known species. The rest are primarily insects and plants (Wilson 1992, 131–37), most of which are not charismatic.

Lawmakers also did not expect the ESA to be used as a land-use-planning tool, but if any one of the birds, mammals, plants, insects or fungi located on a person's land is listed as endangered, the ESA controls many of the uses of that land. Restrictions can result in anything from mild irritations to the loss of almost all land value. Congress required that, in listing species, the secretary of the interior not consider economic effects. This law is the basis for court decisions and agency regulations that claim every species has infinite value. But policy makers have not considered that protecting subjects with infinite value might require infinite funds; nor have they acted in accord with the assertion of infinite value.

Proponents of the ESA claim that it saves species, but they offer little evidence beyond the number of threatened and endangered species listed each year, as if listing instead of delisting were the real purpose of the law.

Endangered-species policy has become an exercise in environmental theology and political dishonesty. Few politicians, judges, bureaucrats, or endangered-species activists intend to be dishonest. They simply fail to recognize the incentives created by the ESA or to consider less emotionally satisfying but more practical means of protecting biodiversity. It is time to stop preaching species rights and promoting ineffective policy. Instead, policy makers need to ask: Why do the lists of endangered and threatened species keep getting ever longer? Why have only a few species ever been taken off the list? Where will the money come from to pay for protecting the species on the lists as well as the thousands waiting to be listed?

NOTES

1. A federal judge threw out the listing of the snail under the ESA, citing inadequate data, and stopped the FWS from reducing the flows of water.
2. This quotation is from the transcript provided by the National Press Club the day after Babbitt's speech. The transcript available from the Department of the Interior (Babbitt 1997b) does not contain the phrase "And if it turns out that we need a little more habitat or a few more adjustments, well then the obligation should be on the public, the participating public agencies, including the federal government, to put up the resources to rebalance the plan." The Department of the Interior's version of that portion of the speech reads: "No surprises" is a policy that was worked out in the intense give and take that went into the Southern California NCCP process. This solution basically says that once the government and a landowner agree as to what, where, and how much shall be done to minimize and mitigate damage by a development project to a listed species, both sides must then stick to that package. The government cannot come back to the landowner pleading for more. It boils down to four words: "A deal's a deal." I am absolutely convinced that it is fair and that the idea of closure that it embodies is essential to bringing the private sector into the conservation process.
3. Ike Sugg of the Exotic Wildlife Association suggested this analogy.
4. Defenders of Wildlife tell the story somewhat differently. The following is posted on their Web site available at: http://web.archive.org/web/19980612174607/http://www.defenders.org/esatop.html:

John Shuler did not shoot the bear in self-defense, as he claimed, and could have taken a legal route to have the bear removed or killed. The regulations regarding grizzlies clearly allow people to shoot bears in self-defense when there is a real threat. On a few occasions in 1989, a few bears entered John Shuler's sheep bedding area and killed a number of sheep. Wildlife officials began trapping in the area and prepared to dart bears in order to remove them from the area. They also offered to finance the installation of an electric fence to protect the corral but Mr. Shuler refused the offer. Before officials could remove or kill the problem bear, Mr. Shuler shot it in his corral. He did not notify USFWS. The next morning he found it lying about 400 yards from his house and he shot it again although it was already mortally wounded. Mr.Shuler was fined by the court because there was no evidence to suggest that Mr. Schuler was defending himself. After the incident, Defenders of Wildlife and the Great Bear Foundation jointly paid for the construction of a large electric fence for Mr. Shuler's corral. Since the construction of the fence, no livestock have been killed by bears inside the corral.

5. See U.S. District Court for the District of Montana, Case No. 96-110-GF-PGH.
6. Cleaves et al. (1994) provide cost data against which Cone's may be compared. They compared the cost of private red-cockaded woodpecker (RCW) conservation relative to timber stocking, ownership parcel size, and cluster size. "Given the smallest RCW cluster size, [RCW protection on] 60-acre parcels showed costs of $1,200 per acre, or 28 percent of maximum present value; 300-acre parcel costs were $143 per acre, or 3 percent of maximum present value. The total cost of providing for one RCW cluster on a 60-acre parcel was $72,000 compared with $42,900 for a 300-acre parcel."
7. Contrary to Secretary Babbitt's claim, I believe it would be good for endangered-species policy if the courts or Congress decided that regulatory takings were unconstitutional. It would immediately make the costs of preserving species more apparent and would, therefore, spur the search for a broad range of innovative approaches. The net effect would be a diversity of policy instruments that would operate in the political economy much as biodiversity works in the ecology—some would fail, but the adaptive, flexible ones would succeed.

REFERENCES

Adler, Jonathan. 1995. Testimony before the Committee on Environment and Public Works, U.S. Senate, July 12.

Babbitt, Bruce. 1994. The Triumph of the Blind Texas Salamander and Other Tales from the Endangered Species Act. *E Magazine* 5 (2), March–April, 54–55.

———. 1997a. Address by Bruce Babbitt, Secretary of the Interior, to National Press Club luncheon, July 18, 1997. Washington, D.C.: Federal News Service.

———. 1997b. To reauthorize the Endangered Species Act: Why, where, and how we should translate our success stories into law. Remarks of Secretary of the Interior Bruce Babbitt to the National Press Club, Washington, D.C., July 17, 1997. Retrieved 4 December 1997 from http://www.doi.gov/alcove/esapress.html.

Bader, Mike. 1992. A Northern Rockies Proposal for Congress. *Wild Earth*, Special Issue, 61–64.

Carroll, Ronald, et al. 1996. Strengthening the Use of Science in Achieving the Goals of the Endangered Species Act: An Assessment by the Ecological Society of America. *Ecological Applications* 6(1): 1–11.

Cleaves, Dave, Rod Busby, Brian Doherty, John Martel. 1994. Costs of Protecting Red-cockaded Woodpecker Habitat: The Interaction of Parcel and Cluster Size. In *Forest Economics on the Edge*, edited by D. H. Newman and M. E. Aronow, pp. 276–93. Proceedings of the 24th

annual Southern Forest Economics Workers Meeting, Savannah, Ga., March 1994.
Competitive Enterprise Institute. 1995. *Delisted Endangered and Threatened Species*. Washington, D.C.: Competitive Enterprise Institute.
Corn, Lynne M. 1995. Endangered Species: Continuing Controversy. Congressional Research Service Issue Brief IB95003.
Defenders of Wildlife. 1996. Top 10 Lies about the ESA. Available at: http://web.archive.org/web/19980612174607/http://www.defenders.org/esatop.html.
Dwyer, Lynn E., Dennis D. Murphy, and Paul R. Ehrlich. 1995. Property Rights Case Law and the Challenge to the Endangered Species Act. *Conservation Biology* 9 (4): 725–41.
Ehrlich, Paul R., and Edward O. Wilson. 1991. Biodiversity Studies: Science and Policy. *Science* 253 (August 16): 758–62.
Environmental Defense Fund. 1998. The Endangered Species Act: Fact vs. Myths. Retrieved July 30, 1998, from the World Wide Web at http://www.edf.org/pubs/FactSheets/c_ESAFact.html.
Gordon, Robert E., James K. Lacy, and James R. Streeter. 1997. Conservation under the Endangered Species Act. *Environment International* 23 (3): 359–419.
Gordon, Robert E., and James Streeter. 1994. *Going Broke? Costs of the Endangered Species Act as Revealed in Endangered Species Recovery Plans*. Washington, D.C.: National Wilderness Institute.
Haldane, David, and Len Hall. 1994. Pocket Mouse Added to Endangered Species List, PuttingResort on Hold. *Los Angeles Times*, February 2, p. B1.
Halverson, Anders. 1995. A Full-Court Press to Save Ecosystems. *High Country News* 27 (9): 9–10.
Hardin, Garrett. 1982. Sentiment, Guilt, and Reason in the Management of Wild Herds. *Cato Journal* 3 (Winter): 823–33.
House of Representatives, Committee on Resources. 1996. *Funding of the Endangered Species Act: Are We Adequately Funding the Act?* Office of the Chief Counsel, 1320 Longworth
House Office Building, Washington, D.C. 20515.
Leopold, Aldo. 1949. *A Sand County Almanac*. Oxford: Oxford University Press.
Lewis, Thomas A. 1987. Searching for Truth in Alligator Country. *National Wildlife*, September–November, 14–18.
McKinney, Larry. 1993. *Reauthorizing the Endangered Species Act—Incentives for Rural Landowners, in Building Economic Incentives into the Endangered Species Act: A Special Report from Defenders of Wildlife*. Edited by Hank Fischer, project director, and Wendy E. Hudson. Washington, D.C.: Defenders of Wildlife.
National Wilderness Institute. 1994. Endangered Species Blueprint. *NWI Resource* 5 (1), Fall: 1–43.
Noss, Reed F. 1992. The Wildlands Project: Land Conservation Strategy. *Wild Earth*, Special Issue: 10–25.
———. 1994. Building a Wilderness Recovery Network. *George Wright Forum* 11 (4): 17–40.
O'Toole, Randal. 1996. Nine Reasons Why the Act Doesn't Work. *Different Drummer* 3 (1): 32.
Pombo, Richard. 1995. Press release from the office of U.S. Representative Richard Pombo, June 29.
Reid, Walter. 1994. Status and Trends of U.S. Biodiversity. *Different Drummer* 1 (3): 5–6.
Sturges, Peyton M. 1995. Supreme Court Strengthens Protection of Endangered. *Daily Report for Executives*, The Bureau of National Affairs, Inc., June 30, A126.
Sugg, Ike. 1993. If a Grizzly Attacks, Drop Your Gun. *Wall Street Journal*, 13 November, p. A15.
U.S. Fish and Wildlife Service. 1978. Deregulation of the Mexican Duck. *Federal Register* 43 (143), July 25: 32258–61.
———. 1993. *Report to Congress on the Endangered and Threatened Species Recovery Program*. Washington, D.C.: U.S. Government Printing Office.
———. 1994a. The Reintroduction of Gray Wolves to Yellowstone National Park and Central

Idaho: Final Environmental Impact Statement. Helena, Mont.: U.S. Fish and Wildlife Service.

———. 1994b. *Endangered and Threatened Wildlife and Plants.* Document no. 1994—380-789/20165. Washington, D.C.: U.S. Government Printing Office.

———. 1997. Endangered Species General Statistics. Retrieved June 26, 1997, from the World Wide Web at http://www.fws.gov/~r9endspp/esastats.html.

U.S. General Accounting Office. 1994. *Endangered Species Act: Information on Species Protection on Non-Federal Lands, Report to Congressional Requestors.* GAO/RCED-95-16, December 20.

Welch, Lee Ann. 1994. Property Rights Conflicts under the Endangered Species Act: Protection of the Red-Cockaded Woodpecker. In *Land Rights: The 1990s Property Rights Rebellion,* edited by B. Yandle. Lanham, Md.: Rowman and Littlefield.

Wilkinson, Cynthia M. 1993. Endangered Species or Endangered Rights. Paper presented at the Third Annual Texas Water Law Conference, Austin, Tex., October 28.

Wilson, E. O. 1992. *The Diversity of Life.* Cambridge, Mass.: Belknap Press of Harvard University Press.

8

Fixing the Endangered Species Act

RANDY T. SIMMONS

In a 1934 essay by Aldo Leopold, titled "Conservation Economics" (Flader and Callicott 1991, 193–202), we can find some direction for improving on the command-and-control approach embodied in the Endangered Species Act (ESA). Leopold's insights, as usual, are telling. He began the essay by noting that in his day the accepted theory of the birth of the moon was that a large planet had passed near enough to pull a large piece of the earth into space, creating a new heavenly body. He compared the birth of conservation programs to that process:

> Conservation, I think, was "born" in somewhat the same manner in the year A.D. 1933. A mighty force, consisting of pent-up desires and frustrated dreams of two generations of conservationists, passed near the national money-bags whilst opened wide for post-depression relief. Something large and heavy was lifted off and hurled forth into the galaxy of the alphabets. It is still moving too fast for us to be sure how big it is, or what cosmic forces may rein in its career. . . .
>
> [Conservation's] history in America may be compressed into two sentences: We tried to get conservation by buying land, by subsidizing desirable changes in land use, and by passing restrictive laws. The last method largely failed; the other two have produced some small samples of success.
>
> The "New Deal" expenditures are the natural consequence of this experience. Public ownership or subsidy having given us the only taste of conservation we have ever enjoyed, the public money-bags being open, and private land being a drug on the market, we have suddenly decided to buy us a real mouthful, if not indeed, a square meal.

> Is this good logic? Will we get a square meal? These are the questions of the hour. (Flader and Callicott 1991, 193–194)

These are still the questions of the hour. To extend Leopold's analogy, beginning in 1970 conservation was hurled into a higher orbit with even greater infusions of government cash and regulation. To the "galaxy of the alphabets" were added the EPA, ESA, CIRCLA, RPA/NFMA, and a host of others. The big difference in the years since 1970, as compared to the years from 1933 to 1970, is that government-sponsored conservation rediscovered a new and stronger drug—direct command-and-control regulation, despite Leopold's claim that that method had largely failed. It continues to fail today.

To overcome the failure of endangered-species policy, I propose eight guiding principles, four of them political and four ecological. They are natural extensions of the lessons learned since 1933 and, in fact, reaffirm many of the principles Leopold promoted as he tried to direct the development of a positive political ecology. Adherence to these principles would dramatically alter existing management systems, and the ESA would be replaced with pragmatic, effective, intellectually honest policy.

The biological principles are as follows:

- Preserving habitat is a more important and achievable goal than saving all species.
- Global extinctions are more serious than local extinctions, which are more serious than local population extinctions.
- Preventing ecological wrecks is more feasible and efficient than rescuing them.
- Managing nature protects biological integrity better than does "natural regulation."

The political principles are these:

- Conserving habitat and species requires enlisting private-property owners on the side of conservation.
- Positive incentives are more effective than penalties, if only because penalties are ex post facto.
- Decentralizing biodiversity activities is more effective than centralizing them. That is, twenty competing answers are better than one, especially inasmuch as no one knows which one is right.
- Depoliticizing biodiversity changes incentives for private individuals, public officials, and interest-group representatives and thereby improves the chances of spending funds effectively and creates more private support for conservation.

The foregoing principles lead away from the idealism and moralism of much of the endangered-species debate and inject pragmatism into the discussion. Although they are consistent with noble goals, they suggest policies that allow for experimentation and creativity.

These principles cannot be adopted under existing endangered-species legislation because they require a more decentralized framework in order to operate effectively. If Congress will allow more decentralization and innovation, if politicians, interest groups, and agency personnel will move beyond what Leopold called "unending insistence on grooves of thought" (Flader and Callicott 1991, 151), then effective policies can be crafted.

BIOLOGICAL STRATEGIES

Include Habitat in the Species Equation

Suzanne Winckler (1992) was correct when she wrote in the *Atlantic Monthly*, "It makes little sense to rescue a handful of near-extinct species. A more effective strategy would focus on protecting ecosystems that support maximum biological diversity" (74). This exhortation goes beyond the obvious point that protecting more habitat is preferable to protecting less. It implies, at least, that before public funds are spent on protecting a particular species an assessment of the appropriate and available habitat should be made. Such an approach would indicate, for example, that it makes sense to spend money to rescue the whooping crane because appropriate and possibly adequate summer and winter habitat exists and within that habitat are the species on which the crane preys. It would *not* make sense, however, to make heroic efforts to save the California condor in the wild, because its habitat requirements are not likely to be met again. Similarly, efforts to singularly protect the lynx or wolverine in the northern Rockies would be judged bad policy inasmuch as those species were historically rare precisely because the habitat is not well suited to them.

This approach is not opposed to the targeted, private actions of organizations such as the Nature Conservancy that purchase small parcels of land in attempts to protect microclimates that are home to species particularly adapted to those microclimates. Those efforts are laudable and possibly important but seem better suited to private rather than public action. Microclimates are vulnerable even to small changes in climate or weather patterns. Limited public funds should be spent where they are most likely to have lasting effects.

Rank Global, Local, and Population Extinctions

The principle of ranking global, local, and population extinctions represents simply a recognition that resources are scarce and that the nation cannot afford to indulge the noble impulse to save every population, every subspecies, or even every species. Policy makers must make tough choices, and in doing so they should rank their priorities. Charles Mann and Mark Plummer (1992) quote Gardner Brown, a University of Washington economist, on this issue:

> We can't save every species out there, but we can save a lot of them if we want to, and save them in ways that make sense economically and scientifically. To do that, we have to make some choices about which species we are going to preserve. And nobody wants to do that! Nobody! (66)

Mann and Plummer asked if nobody wanted to make choices "because they are dismayed by the prospect of playing God?" Brown responded, "Oh, sure. But in this case God is just sitting on his hands, which is a pretty dangerous thing for him to do."

As the nation sits on its collective hands, species will continue to disappear at a more rapid rate than necessary, and they will continue to do so as long as the legislative mandate remains "save everything." A much more realistic if less emotionally satisfying rule dictates that policy makers establish priorities and make trade-offs (Easter-Pilcher 1996; Czech and Krausman 1998). In fact, despite the rhetoric about the incalculable value of every species, priorities are being established and trade-offs made under existing policy. The U.S. Fish and Wildlife Service (FWS) has chosen to spend its money on a few species—particularly visible, charismatic ones.

Favor Prevention over Rescue

Almost everyone who writes about endangered-species policy calls for earlier conservation efforts than those adopted under the ESA. Under the current process, populations often fall to nearly irreversible lows before they are nominated for listing as endangered. Immense biological and ecological problems attend the effort to recover a species that is nearly gone. The management problems are also intensified when dealing with a species approaching extinction. Tim Clark, Richard Reading, and Alice Clarke (1994) explain:

> As a species continues to decline and approach extinction, management options narrow, costs rise sharply, and the sense of urgency grows nerve-rackingly high. Fear of failure can become paralyzing; flexibility for experimentation approaches nil. As a result, the context of the recovery

> program deteriorates into a politically charged and conflict-laden mess with little room for maneuvering. Simply starting conservation *before* a species is severely endangered would alleviate much of the pressure, keep more options open, and reduce the costs. (424–25)

More options at less cost ought to be the motto of species preservation. Choosing such a path would require tough choices, but it makes little sense to spend large sums of public money on a species or habitat that is nearly gone if the opportunity cost is to allow other species to slide into a steeper decline.

Manage Nature

In his book about "reinventing nature," William Cronon (1995) explains that "many popular ideas about the environment are premised on the conviction that nature is a stable, holistic, homeostatic community capable of preserving its natural balance more or less indefinitely if only humans can avoid 'disturbing it'" (24). This assumption, which he calls "problematic," descends from the work of botanist Frederic Edward Clements, for whom the "landscape is a balance of nature, a steady-state condition maintained so long as every species remains in place" (Barbour 1995, 235).

Central to this belief is the presumption that nature is highly structured, ordered, and regulated, and that disturbed ecosystems will return to their original state once the disturbance is removed. This view of nature is an integral part of successional theory, in which species are seen as replacing one another in an ordered procession, culminating in climax communities. Such thinking continues to animate many modern activists. Bioregionalist Stephanie Mills (1995), for example, writes of "our species beginning finally to take an interest in attending to what the land itself has wanted to bring forth, its creation of self-regulating communities of organisms, climax ecosystems" (3).

Today, however, the "balance of nature" idea is widely rejected by ecologists (Botkin 1990, 1991, 1992; Pielou 1991; Johnson and Mayeux 1992; Pickett, Parker, and Fiedler 1992; DeGraaf and Healy 1993; Tausch, Wigand, and Burkhardt 1993) and by many in the environmental community (Worster 1995; Lewin 1986; Foreman 1995–96).[1] Yet, as ecologist Norman Christensen, dean of Duke University's Nicholas School of the Environment and chair of the Ecological Society of America's panel that reviewed the 1988 Yellowstone National Park fires, argues, "everything from the Endangered Species Act to the Clean Water Act has implicit in it the notion of an equilibrium ecology, the idea that systems tend toward these stable end points and that they are regulated by complex feedbacks—a sort of balance of nature that is almost Aristotelian" (Basgall 1996, 39).

Paradoxically, however, this view is neither modern, progressive, nor scientific. The "balance of nature" idea is in reality an old, conservative, religious view of the natural world that dates to the dawn of written history (Botkin 1990). Instead of constancy and stability, disturbance and change have been the norm throughout the evolutionary history of the earth. Glaciers that covered large portions of North America advanced and retreated repeatedly over the last 3 million years. Not only has the climate fluctuated widely, but what we in the United States view as "normal"—what we have experienced during our lifetimes or since the birth of our nation—is, when viewed in the perspective of the last several hundred thousand years, an abnormally warm, dry period. The longer-run "normal" climate for most of Canada, for instance, is associated with several thousand feet of ice, not with the landscape we see today (Pielou 1991). As one might expect, the distributions of plants and animals have also contracted and expanded over time. Local extinctions are a fact of life, and so is the extinction of entire species. Disturbance and change are the only true constants of ecosystems.

Christensen suggests that the widespread misconception of nature bodes ill for the ESA, which "assumes that we can know what a minimum viable population of a plant or animal is in a very predictable way." He says, "The nonequilibrium approach to ecology suggests that species populations fluctuate constantly. Species may go locally extinct in a given area. They may appear and reappear. That's very frustrating for managers of endangered species and for a public that expects much more deterministic answers from science" (Basgall 1996, 41). The implication is that human beings have a strong role to play in managing the "natural" world. Indeed, current ecological processes have already been structured by human actions, and removing human effects, including fire, will substantially change existing habitats and processes, not necessarily for the better (Budiansky 1995; Kay 1995). Simply setting lands aside as wilderness or preserves and then letting nature take its course will not necessarily save species or protect ecological integrity (Botkin 1990).

POLITICAL STRATEGIES

Enlist Property Owners

The single most important step a new Endangered Species Act should take is to remove the power of government agencies to take private property without compensation. Private property rights should get the same protection as the rights to free speech, a free press, or free assembly. Speaking at the Smithsonian Earth Day Conference on Biodiversity, Randal O'Toole (1995) explained the relationship:

> Imagine that freedom of the press meant that the government could censor "only" 20 percent of *U.S. News and World Report* or the *Wall Street Journal*. Or imagine that freedom of assembly meant that the government could forbid "only" 20 percent of all public or private meetings. Anyone would argue that such freedoms would be meaningless under these conditions.

O'Toole then explained that the way in which the U.S. Fish and Wildlife Service sometimes exercises its power under the ESA makes private property rights vacuous.

The intellectual justification for diminishing private property rights rests on an economic argument that the members of the species are not private goods; they are public goods, and their public-good nature justifies taking some portion of people's property in attempts to preserve endangered species.

Public goods are those for which provision is nonexclusive and consumption is nonrivalrous. The mainstream economic presumption is that government must provide public goods. Technically, most species are more correctly defined as common-pool resources: their provision is nonexclusive but their consumption is rivalrous. Certain aspects of endangered species are public goods. In fact, each of the most commonly cited justifications for preserving species focuses on their character as public goods: (1) they serve ecological functions; (2) they are sources of knowledge that can be turned to consumptive and nonconsumptive human uses; (3) they are sources of scientific information, models, and theory; (4) all species have rights worthy of respect; and (5) a species-rich world is esthetically superior to a species-poor world. The first three justifications qualify on public-goods grounds—benefits flow to everyone, and no one's consumption of those benefits diminishes anyone else's consumption. The esthetic justification is also a public-good argument to the effect that if we allow species loss to continue, each of us is impoverished, or each person's soul is diminished.

The public-good nature of species may help to explain why opponents of the Fifth-Amendment-takings argument revolt at the proposal that the government compensate the owner when regulations reduce the economic value of private land. They view the landowner as a polluter who should be fined for his actions or stopped altogether. After all, he is reducing an endowment the earth provides to all—and by "all" they mean not just humans but all species. In fact, they argue that the landowners are acting immorally and unfairly. John Humback, a property-rights expert at Pace University, argues, "The whole idea that government needs to pay people not to do bad things is ridiculous. The reason the government exists in the first place is to define what is for the common good and what's not" (Harbrecht 1994, 6).

Secretary of the Interior Bruce Babbitt voiced a similar objection. He was especially concerned that some groups were using takings arguments to resist the protection of endangered species. In a speech, he argued against bills such as House Bill 1388, titled the "Just Compensation Act of 1993." That bill, if enacted, would have required federal agencies to compensate property owners "for any diminution in value" caused by environmental regulations. His response:

> Let's examine the implications of this proposed raid on the public treasury. The Kesterson National Wildlife Refuge in California is one of the great migratory bird stops on the Pacific flyway. But a few years ago, the waterfowl were dying, and they were deformed at birth. It turned out to be selenium poisoning running off into the refuge from nearby farm irrigation wastewater. Under the Endangered Species Act, I tell the farmers: Clean up the pollution or we'll sue you. But under this new proposal, I am undeniably causing a "diminution in value" of a property right—it will cost those farmers money to clean up. They'll comply, but then they'll send me the bill! The old legal maxim, "make the polluter pay," would be replaced by a new legal rule: "It pays to pollute; the government will reimburse your costs." (Babbitt 1994, 55)

Secretary Babbitt's example is important, but his analysis is wrong. His claims are representative of the efforts to mischaracterize property-rights arguments and of a fundamental misunderstanding of the nature of property rights and the proper role of government in protecting them.

As any introductory economics textbook explains, one of the justifications of government action is to prevent one party from harming others and their properties, and, failing that, to punish those who transgress against the rights of others. Controlling pollution has the same justification, namely, preventing the unwanted imposition of wastes or toxins by one party on another. Pollution is a form of "trespass" or "nuisance" under the principles of common law, and those trespassed against can properly call on the power of government to gain restitution.

Babbitt's hypothetical scenario, in which corporations could claim to have been harmed by a prohibition from injecting toxins into the groundwater, does not fit within our tradition of property rights or our system of law. Polluters, being trespassers on others' property, can be stopped; and they are not entitled to compensation. The polluter-pays principle requires that the polluter bear legal responsibility for his actions. [2] In the Kesterson example the story is complicated because the selenium-laden water entering Kesterson arrived in a publicly funded and managed drainage system built by the Bureau of Reclamation. Without that system, the water would never have reached the refuge, so a question remains: Is the polluter the farmers or the bureau?

Still, the principle Secretary Babbitt tried to assert is important. He wants to establish that reducing habitat or harming an endangered species on private property is pollution—an externality landowners are forcing on the rest of the world. Because they are creating costs for others, they should pay to fix the problem. If proponents of that approach can make the externality or pollution argument stick, they will prevail over the constitutional takings argument. If polluters ought to pay *and* if reducing the habitat of endangered species constitutes pollution, then landowners must pay.

But Secretary Babbitt's logic is flawed for at least two reasons. First, biodiversity is a slippery concept. It may be true that once the earth dips below a certain level of biodiversity, it is a poorer place biologically, but until we reach that threshold, the marginal loss of species has little effect on our store of biological "wealth." This claim applies with particular force when species are defined so narrowly that distinct subpopulations are treated as if they were single species. Saving every species and every population of each species, no matter how biologically insignificant, makes no biological or economic sense. It requires an enormous stretch of the imagination, for example, to suppose that the extinction of the Bruneau Hot Springs snail or the Colorado humpbacked chub would impoverish their former habitats or diminish biodiversity in any scientifically significant way.

This insignificance argument should not be pushed too far, however. Many tiny reductions in biodiversity can and sometimes do add up to significant reductions in biodiversity, just as many nonpoint sources of pollution can add up to serious pollution problems. Such is the essence of the tragedy-of-the-commons argument: each individual loss is insignificant, but all of them together amount to a tragedy. So, although saving every species is a silly goal, consciously sacrificing a species or a population is not necessarily trivial.

Reevaluating the nature of the situation provides a second and better reason why the public-goods and polluter-pays arguments are wrongly applied to private producers of species habitat. Humanity[3] may "own" the rights to biodiversity, but landowners own the habitat that individual members of species occupy. Thus, the landowners are being asked to produce a public benefit upon which all of humanity can free-ride. In the case of Kesterson or other examples of pollution, the landowners ought to bear responsibility for the costs they impose on others. In the case of biodiversity, the landowners are producing a *benefit*, and if members of society value the biodiversity the landowners produce, ways should be found to encourage the landowners to continue to produce that positive externality.

The difference here is between public costs (negative externalities) and public benefits (positive externalities). The pollution-producing landowner passes *costs* on to others who have not contracted to bear them. The biodiversity-producing landowner passes *benefits* on to others who have not contracted to receive them. The policy responses should differ: punish those who create costs but reward those who create benefits. Notice, however, that the ESA punishes those who produce benefits—a perverse policy indeed!

The polluter-pays principle has been misapplied in endangered-species policy. The assumption has been that owners who wish to develop their land are "polluters" of the species found on that land and must be either stopped from undertaking activities that would "pollute" the species or made to pay for any effects their actions have on the species. California property owners who are being denied permission to create firebreaks on their property or to build additions to their homes without paying mitigation fines exemplify the issues involved. Landowners required to stop timber harvest on nearly 30 percent of their property provide another example.

But what if property rights were recognized so that the owners' use of their property could not be curtailed without compensation even if a listed species resided on the property? Then, in effect the landowners would own the species on their property even though the title to the species still rested nominally with the state. The species could then be treated as private goods rather than common-pool goods, and those who wished to protect or save the species would operate within markets as opposed to the political realm as they do now. No longer treated as polluters, property owners would be treated as producers of something of value to others.

Defenders of Wildlife is already using such methods to protect wolves, because they consider hearing wolves howling again in Montana to be a public good for which they are willing to take responsibility (Anderson 1994). Ranchers, however, view wolves as polluters: wolves kill an occasional cow, calf, or sheep and thereby impose uncompensated costs. Solving wolf "pollution" problems under the polluter-pays principle requires that the polluter be identified and fined. In this case, Defenders of Wildlife claims responsibility and stands willing to pay. But instead of calling the payment a fine, they call it a reward.

In the spring of 1994, a rancher near August, Montana, collected a $5,000 reward from Defenders for having three wolf pups successfully raised on his property. In anticipation of the reward, the rancher followed advice from state and federal biologists about how to minimize human disturbance, and he managed to leave the wolves alone. A rancher's usual response to wolves on his property is to "shoot, shovel, and shut up," because wolves threaten to impose costs. In this case, Defenders of Wildlife paid the costs.

Defenders have been paying the costs of wolves since 1987, when they created a Wolf Compensation Fund to pay for livestock killed by wolves. So far, the Fund has paid $12,000 to about a dozen ranchers. One problem of this approach is that the landowner cannot decide the price of the tolls. Another is that, although Defenders' compensation-insurance program may cover the costs of a replacement cow, it does not pay the rancher for the time spent proving the cow was killed by wolves, arranging for the replacement cow, or organizing transportation. Clearly the system is not perfect, but no system is. And because the ranchers' normal means of excluding wolves is to quietly kill them, a system that compensates for use—employing the user-pays principle —increases the chances of Defenders' members being able to hear wild wolves howl again. By adding a reward for allowing wolves to use one's land, Defenders of Wildlife have turned the liability of being the provider of a public good into an asset and turned ranchers' incentives in a new direction. What is more, the payments are relatively small and are paid by private parties, not out of the public treasury.

Another organization pursuing the same kinds of innovative policies is the Delta Waterfowl Foundation, a private nonprofit organization dedicated to reversing the downward trend in North American duck populations by stopping the loss of habitat. The foundation conducts research and provides education and economic incentives to farmers. It is supported from tax-deductible contributions.

One of the foundation's programs is Adopt-a-Pothole. Funds are raised from contributors all over North America, each of whom receives an aerial photograph of the adopted pothole, a quarterly report on its status, and an annual estimate of duck production. The farmer receives $7 per acre to maintain pothole habitat and $30 per acre to restore pothole habitat. In addition, farmers are offered production contracts that pay based on the actual numbers of ducks produced. The production contracts encourage farmers to improve and protect nesting habitat.

Results have been impressive. After just two years of operation, contributions totaled nearly $1 million from more than a thousand individuals and organizations, and eighteen thousand pothole sites were enrolled. Nest density is twice as great for adopted sites as for unadopted sites, and nesting success averages 51 percent for adopted sites, compared to 10 to 15 percent for unadopted ones. The program has even developed a special nesting box that protects birds from predators, and potholes using the device have nesting success rates of 90 percent (Delta Waterfowl 1994).

An important difference exists between the Defenders of Wildlife wolf program and the Delta Waterfowl program. The wolf is listed under the ESA, so ranchers who attract it to their property run the risk of having the uses of their property regulated by the FWS. None of the mallards, canvasbacks, shovelers, blue-winged teal, green-winged teal, gadwalls, redheads, or pintails that nest in the prairie potholes are endangered. Farmers know they can attract the ducks without having to worry that the value of their property will be reduced because they protect and develop duck habitat. Thus, they are pleased to be paid to attract ducks by improving habitat and changing farming practices. They win and the ducks win. If the threat of ESA regulations were removed from wolves, western ranchers would be more interested in attracting them than they are now.

Employ Positive Incentives

The importance of understanding and using incentives may seem obvious. Yet Congress and the implementing agencies ignore the incentives their laws and rules create. Randal O'Toole (1995) argues that our existing laws and most proposals to change them "are based on the same assumptions upon which the Soviet Union based its entire economy—assumptions that people will do what they are told or follow some moral principle even if their incentives run in the opposite direction" (1). A large public-choice literature (Mitchell and Simmons 1994) shows that not only do citizens respond to perverse incentives; bureaucrats do, too. Policy makers must carefully consider, therefore, the incentives a new Endangered Species Act will create, or they will be disappointed by its consequences.

A host of private individuals and groups such as Defenders of Wildlife and Delta Waterfowl are using positive incentives to promote species and habitat preservation. They provide private benefits to landowners who produce a public good. Many other opportunities exist to let the species pay their own way; and many endangered species can pay their own way if allowed to do so. South Africa, for example, decided that private entrepreneurs were the best agents to save several species of endangered vultures —birds as big as California condors. Tourists who wish to view and photograph these endangered raptors pay to see them at "vulture restaurants," where carrion is provided for the birds. Local Boy Scouts gain service hours by hammering carrion bones into fragments small enough for the birds to swallow. The bone fragments are a necessary source of protein and were once broken up by hyenas, which are now extinct in the vulture's breeding range (Reiger 1993, 14).

In the United States, a broad range of ventures, from exotic game ranches in Texas to greenhouses producing cacti for the supermarket trade, allow species to pay their own way. In 1979 the FWS revised its regulations to allow commercial foreign trade in American alligators. Alligator farming has become so successful that wild populations have exploded, and universities in the South now offer courses in alligator farming. One entrepreneur received permission to raise alligators in a warm-springs area in southern Idaho, a location far outside the alligator's normal range. He expects to feed them dead cows from nearby dairy farms, thereby solving a difficult disposal problem for the farmers and gaining a free source of food for his alligators.

The Costa Rican government allows rain forests to pay their way by means of the discovery and patenting of genetic resources. It is often argued that great human benefits will accrue from species we have not yet even studied. One congressional report asked, "Who knows, or can say what potential cures for cancer and other scourges, present or future, may lie locked up in the structures of plants which may yet be undiscovered, much less analyzed?" However, in most countries natural genetic resources cannot be patented, and so private companies have little incentive to prospect for them. Costa Rica's formal response to this problem was to allow the Instituto Nacional de Biodiversidad (INBio) to contract with pharmaceutical companies to prospect for and develop indigenous genetic resources. In the first agreement, Merck, Inc., a major international firm, agreed to make an initial payment of $1 million over two years. In addition, the company will make a royalty payment to INBio from commercial sales of products developed (Sedjo and Simpson 1995, 175). Considering that one-quarter of today's cancer drugs derive from random testing of organisms, the potential for significant royalties is clear.

In speaking of "letting species pay," I emphasize *letting*. Many environmentalists are appalled by commercialization of wildlife, especially endangered species. They strongly oppose letting the species pay their own way. The U.S. Fish and Wildlife Service, for example, in 1983 rejected a proposal to allow commercial use of captive-bred green sea turtles. The current controversy over the private Grizzly Discovery Center in West Yellowstone, Montana, is another example. The owner wanted to create a park where the viewing public could see grizzly bears. He proposed stocking the park with nuisance bears from Yellowstone and other parks. The alternative is to kill the bears, as the Park Service does routinely. But the Park Service would not allow him to have any of its nuisance bears. One member of the Greater Yellowstone Coalition, a local environmental group, was quoted in *Newsweek* as saying that she would rather see the bears killed than put into the "artificial" habitat of the private park. She refuses to let the bears pay.

Roger Beattie, a New Zealand conservationist, tells of a similar conversation with an official from the Mount Bruce Endangered Species Unit. The official took Beattie on a tour of the complex, describing each endangered species and its management. When they came to a species of kakariki, the official explained that the female birds were in one aviary and the males in another. Beattie (1994) asked why, prompting the following exchange:

> OFFICIAL: We do not want them to breed any more.
> ROGER: Do you mean to say that you have birds in an endangered species unit that you are deliberately not breeding?
> OFFICIAL: Yes.
> ROGER: Why?
> OFFICIAL: We do not know what to do with the extra young birds.
> ROGER: Have you thought of selling them?
> OFFICIAL: Oh no! You couldn't do that! (6)

Selling the extra young birds would let the birds pay at least part of the costs of their preservation. That approach has worked well for alligators, provides income from the Costa Rican rain forest, and protects South African vultures. It should be used more widely.

Attitudes toward letting the species pay may be changing, and the change may be hastened just by rewording the slogan. Instead of saying "let the species pay," we can say "make the users pay." Thus, the polluter-pays principle can be evoked to protect habitat. Mill Creek Canyon, east of Salt Lake City, Utah, is an example. This publicly owned canyon was frequently visited but received just $3,000 per year from the Forest Service to be managed for human use. Salt Lake County built a toll booth just outside the canyon and now collects a modest toll from all who enter. The money is given to the local Forest Service Office under condition that it be spent on the canyon. The toll booth is now generating more than $125,000 per year to be spent on riparian restoration, control and disposal of human and pet waste, and protection of the canyon's fragile watershed (Smart 1994).

The same approach can be used in other areas of critical habitat, but politicians must resist the temptation to dictate how the fees are spent; otherwise they will restrict the managers' entrepreneurial capacities. An ongoing study at the Political Economy Research Center in Bozeman, Montana, is finding that for state parks, restrictions on the spending of fee income reduce creativity and even the incentive to collect the fees (Leal and Fretwell 1997).

Decentralize Conservation

Environmentalists tend to push for centralized policies. If a preferred solution is imposed nationally, lobbyists do not have to deal with fifty state legislatures and the local interest groups of each state. Some issues do require a national policy. Some species, such as the Colorado squawfish, cross state boundaries and require attention from a regional if not a national body. But most endangered-species problems are local.

Decentralization can entail turning responsibility for endangered species over to states, but it can also allow private groups and individuals, as well as local, state, and national officials, to participate in protecting endangered species. Some private groups will help species without compensation, and governments should allow them to do so. One model is the restoration of the peregrine falcon, an accomplishment of Dr. Tom Cade of Cornell University and the Peregrine Fund. Using techniques developed by falconers over centuries, they raised birds in captivity and then released them into the wild. The birds nested in very surprising places—on bridges into New York and other cities and on urban skyscrapers. One-way glass and television monitors have been installed so people can watch the falcons nest, raise their young, and devour pigeons. Today more pairs of peregrine falcons are nesting in New Jersey than at any other time for which records exist.

The wood duck provides another example. Early in this century people expected it to follow the passenger pigeon and the Carolina parakeet into extinction because its wetland habitat and the dead trees in which it nested were disappearing. But a massive, national, voluntary campaign to build and place artificial nesting boxes reversed the trend. The wood duck is now the second most common duck species in North America, and wildlife agencies are encouraging hunters to take more of them and fewer of other species (Seasholes 1995, 8). Imagine Audubon Society members, Boy Scouts, hunters, and other interested citizens—the same groups that organized to save the wood duck—approaching owners of timberlands and asking permission to put up spotted owl nesting boxes today. Very few would give permission. Had the ESA been in place when the wood duck was endangered, few landowners would have been willing to allow the nesting boxes on their property, and the wood duck, in all likelihood, would now be extinct.

Electric companies attract bald eagles to their property because of the wetlands and ponds created by their cooling operations and because the fenced, patrolled property keeps people from disturbing the birds. But under the ESA, companies become financially responsible for any eagles they inadvertently attract. The ESA does allow the issuance of incidental-take permits that absolve the companies of some responsibility, but the process for getting the permits is cumbersome and expensive. If the rules were changed to allow the FWS and private organizations to act as Delta Waterfowl officials do, electric companies might behave very differently. Instead of inadvertently attracting eagles, they might actively encourage them. Think of the difference it could make if, rather than being fined when a young eagle on its first flight was killed by flying into power lines, the company was rewarded whenever a young eagle migrated from its property. Consider, too, the many other species that could benefit from a company's actions to protect an endangered species and its habitat.

Some people enjoy creating preserves on their property and will bear substantial expense to protect species there. Roger Beattie and his wife created a preserve on their New Zealand farm. One block of the farm contains a stand of native forest that is home to several native bird species as well as nonnative predators. They fenced 20 hectares of that forest as a nature preserve, erecting a six-foot-high predator-proof fence, a combination of deer, rabbit, and bird netting and electric wiring. Inside the fence, they spent two months setting bait stations and traps to rid the preserve of rats, stoats, cats, and dogs. Their goal was to reintroduce the eastern buff weka, a rare bird species, onto the preserve. This species is no longer found in Canterbury but was introduced on the Chathams during the last century and flourishes there. Getting the permits to transport and release the weka on the reserve took longer than eradicating the predators, but in April 1994 the birds were released. Beattie said,

> Conservationists have spent much time and effort deliberating over how they would reintroduce the weka to Canterbury. We just set out and did it. . . . The success of our private reserve shows there is a better option for nature conservation than preying on taxpayers. That success comes from immediate and effective action. (1994, 7)

The Beatties plan to capitalize on their success by selling information, assistance, and services to other predator-proof nature preserves.

Private groups can also involve themselves in preserving species if they are allowed to bid for uses of public land just as they now bid for private land. On private land, you may buy timber and let it stand, pay for a grazing right and leave the grass for wildlife, or hold a mineral right but leave the underground minerals undisturbed. If the same system were applied to public lands, anyone could bid on commodity sales and then use the land for conservation purposes instead of extraction. Allowing public land managers to make such sales would introduce a new dynamic to public lands management, because private parties could preserve habitat they consider significant. One can even envision private groups purchasing conservation easements to public lands.

Some environmental groups already have purchased rights to lands they consider important. In 1998 a coalition of five environmental groups in the Pacific Northwest purchased the equivalent of a conservation easement in Loomis State Forest in north-central Washington. They purchased the logging rights to 24,000 acres, the only roadless area in the 2-million-acre forest. The

Loomis is a high-elevation lodge-pole pine forest that harbors grizzlies, fishers, and Canadian lynx. Washington's Common School Trust, which uses the income from timber harvest for the state's public school system, manages the Loomis. The purchasers will pay the trust what it would have received had the site been logged and pay for an additional harvestable tract elsewhere in the state. In 1996, the Forest Guardians of New Mexico outbid a rancher for a 644-acre degraded, riparian-area grazing lease. Instead of managing for cattle, they managed for wildlife by planting willows and other vegetation, and they removed the livestock (Brown and Shaw 1998).

An Idaho environmentalist, Jon Marvel, attempted to purchase grazing leases on state lands in Idaho. He believes the state lands are overgrazed and hopes to lease the lands and exclude cattle. So far he has failed because, although he outbid the ranchers, the ranchers convinced the Idaho Land Board to reject his bid. The 1995 Idaho legislature even passed a law to make it very difficult for a nonrancher to bid on state grazing lands. But Marvel has taken his case to the Idaho Supreme Court and expects to win. He points to an Oklahoma case in which the court, relying on trust language in the state constitution, ruled that the state could not offer leases to ranchers at below-market rates. Expired Oklahoma leases are now advertised on television (Stuebner 1995).

Land leases for a variety of nontraditional uses should soon become prevalent. Once the political power of traditional users is broken through court decisions or by increased public attention, whole new patterns of endangered-species protection will emerge, including easements and purchases. Under such arrangements, groups that currently oppose grazing might even use it as a management option to benefit endangered species. Many private and public wildlife refuges graze cattle, for example, to manage the vegetation. Domestic livestock are not necessarily harmful (Kay and Walker 1997). What is harmful is a system of political management that suppresses creativity and adaptation.

Depoliticize Biodiversity

Enlisting property owners, providing positive incentives, and decentralizing biodiversity protection cannot be accomplished until the process is depoliticized. In part because of the need to generate continued political support for endangered-species protection, the FWS spends most of its money on charismatic species (Dwyer, Murphy, and Ehrlich 1995, 738–39). In addition, managers pursue strategies that make little ecological sense but a lot of political sense. The core of a strategy to depoliticize the process pertains to the budget. Simply put, the FWS must pay its own way.

For that to happen, the funding mechanism for endangered-species protection must be changed. Currently, the FWS receives a budget from Congress, but in reality it has a virtually unlimited budget because it can list species and thereby control private landowners' actions without having to pay the costs associated with the listing. If, however, the FWS could not "take" property without compensation, it would have to make tough choices, especially if it had a fixed budget.

Insisting that the FWS make tough choices does not imply that biodiversity would be less protected, but rather that the mix of policies would change and that managers would have incentives to innovate. As Richard Stroup, an economist at the Political Economy Research Center, points out in public lectures about the ESA, bureaucrats must make choices in order to spur the search for alternative policies. He notes how aluminum-can manufacturers reduced the amount of aluminum in each can by more than 60 percent as they sought to use the metal more efficiently. He suggests that bureaucrats would search for the same kinds of efficiencies if they were given a mandate to maximize biological integrity with a fixed budget.

One proposal is to create a biodiversity trust fund that collects a fixed share of public-land user fees each year (O'Toole 1995, 3). A user fee of $6 per day would generate about $4.6 billion per year in addition to the commodity user fees of roughly $3.4 billion. A biodiversity trust fund that received just 10 percent of those fees would have $800 million, a great deal more than the 1994 FWS endangered-species appropriation of $6.7 million (Corn 1995, 8). Another possible funding source was suggested by a former Department of the Interior economist, who proposed earmarking a portion of royalty and bonus payments from the development of oil reserves on the Arctic National Wildlife Reserve (Nelson 1995, 122).

Having a trust fund charged with promoting biodiversity and funded from user fees would relieve many of the political pressures that now drive endangered-species policy. Funding determined as a percentage of user fees would replace funding determined by congressional whim and interest-group politics. Because the trust's income would come from fees for the use of federal lands, the trustees would have an incentive to make sure the federal land agencies charged market value for the consumptive and nonconsumptive resources located there. The trustees would also have a vested interest in providing information to Congress when resources were being provided at less than their value. O'Toole suggests having the biodiversity trust fund managed by a board of trustees made up of conservation biologists or ecologists appointed by cabinet officials such as the secretaries of agriculture and interior or the director of the Smithsonian Institution and serving nine-year terms (1995, 2). The model is the federal reserve board. Expenditures from the trust fund would have to be justified to the trustees, who could also accept proposals from state agencies and private groups for protecting biodiversity. Competition for funds would spur innovation and creativity. Proposals to spend millions to restore wolves to Yellowstone, for example, would have to compete with proposals to spend the same amount of money to protect a host of truly endangered species across the nation.

Trustees might discover that it makes sense to allow many species into markets where owners would have an incentive to protect them. Fee hunting, farming, and captive breeding are all market activities that require relatively small funds to oversee and monitor. Because many species cannot be adequately protected in markets, however, to protect those species the trustees would have to consider a range of positive incentives, from awards or other forms of recognition to cash awards to rental of habitat. Like Delta Waterfowl, the innovative conservation officer might pay for output (e.g., hatchlings per pothole) rather than simply setting aside habitat.

Where no such policies can work, it would make sense to rent or purchase habitat. Even then, however, buying easements rather than land might prove sufficient. Landowners might be invited to enroll their lands in a program that required them to manage those properties in a particular way. In return, they would receive payments or other forms of compensation. Such a program resembles the "conservation reserve" or "wetlands reserve" programs of the federal government, in that landowners enrolling lands in those programs commit themselves to certain land management practices and receive payments. For bureaucrats with fixed budgets and a broad range of possibilities for preserving habitat, a better approach would be to tie compensation to actual increase in numbers or the attraction of particular species rather than to the simple adoption of certain management practices.

The underlying dynamic is that agency personnel or trust fund officers would have a mandate and a fixed budget. Required to carry out the mandate and subject to their budget, they would have a strong incentive to innovate and to rank choices about which species to concentrate on and which policies would best meet the conservation objectives. They would have to ask whether a particular species or habitat can be protected most efficiently if treated as a private, a public, or a toll good. Thus, endangered-species policy would become dynamic, innovative, and purposeful.

A simple way to depoliticize conservation is to establish conservation rental contracts (CRCs). Renting most habitat would be far cheaper than purchasing it and could be customized to meet the needs of the species and the landowner. Thomas Bourland and Richard Stroup (1996), the originators of the CRC concept, suggest that rental contracts are especially important in the southeastern United States, where about two-thirds of the commercial forest belongs to private owners. Attempting to put together large-scale programs under that ownership pattern is difficult, but targeted rental efforts could operate successfully.

CRCs could be used for spotted owls and desert tortoises. Every *X* years a census could be taken to determine how many owls or tortoises lived on someone's land, and a payment made to reward the owner for having maintained the population. An additional reward might be offered for having attracted new owls or turtles. In this way, the owls and turtles become quasi-private property of the landowners. The endangered species generate income rather than expenses. By protecting the species and their habitat, the landowners make themselves better off. Such a system could work for a broad range of animal and plant species.

If endangered-species proponents had to produce rewards for landowners instead of legislation penalizing landowners, new political dynamics would emerge. Federal agents calling on landowners in Utah's Garfield and Iron counties, which harbor Utah prairie dogs, would receive the same welcome as a realtor. Both the agent and the realtor would have something positive to offer the landowner and, because they would not be limited by the strict rules of the ESA, they could in many cases work out arrangements that benefited everyone.

It is impossible to estimate how much money is "needed" to protect endangered species. In fact, need statements make little sense apart from the question, "at what price?" Given scarce resources, we must recognize that, although all species may be important, some are more important than others. If inexpensive ways are found to protect species, more will be protected. Decentralizing species protection into the many systems we have suggested provides ways to discover efficiencies and thereby to protect more species. But until the endangered-species budget is fixed and the power to shift costs to others is removed from the FWS, the search for innovation and efficiencies will be stifled.

Stop Subsidizing Ecosystem Disruption on Private Lands

As part of a move to depoliticize and decentralize conservation, federal and state subsidies to private landowners must be ended. Leopold wrote of laws and programs that "frequently clash, or at best, fail to dovetail with each other" (Flader and Callicott 1991, 199). Today we see subsidies for private lands that certainly do not mesh with endangered-species protection and often work against it (Losos et al. 1995). The Washington, D.C., promoters of those programs seldom see how they encourage development on fragile lands that would otherwise never be economically feasible. O'Toole (1995) claims that "subsidies are the biggest threats to rare species on both public and private land. They include everything from below-cost timber sales to animal damage control programs to import tariffs on sugar that encourage sugar production near the Everglades" (10). Loan guarantees and loans below market rates of interest have serious ecological consequences. A major conference center was built in a canyon near my home using loans subsidized by the nation's taxpayers. The development might eventually have occurred without the loans, but they were a factor in its happening when it did. Similarly, federal crop insurance, flood insurance, and disaster relief all contribute to overdevelopment of private lands.

Floods in the Midwest and coastal fires in California are particularly important examples. Given the propensity of rivers to flood, private bankers are not likely to grant loans for building on a floodplain unless the loan applicant has federal flood insurance that assures the banker the federal government will cover the bank's losses in case of flooding. Thus, federal disaster relief creates an additional incentive to build on the floodplain. The rules of the game in California's dry hills are much the same. The homebuilder is told, in effect, "Build in areas that have a history of fires, and if the fires come the state and federal governments will provide disaster relief as well as firefighters and equipment at no additional cost to you."

Ending those subsidies will not stop the disruption of all remaining habitats, but it will stop some of them, because developers, farmers, and homeowners will have to bear more of the costs of their own actions. It is no secret that as individuals are made less responsible for their actions—that is, required to pay less of the cost—their behavior becomes more reckless. Removing subsidies reduces the recklessness.

Stop Subsidizing Ecosystem Disruption on Public Lands

A better name for the public lands is "the political lands." Their use is determined not by what people want but by what the organized interests want. These political lands contain many habitats for endangered species, because federal and state governments own more than a third of the onshore land in the United States. Subsidized private users, however, unnecessarily disrupt those habitats. More than half of Forest Service timber sales are below-cost sales. The recreation subsidy is even worse in terms of the dollars lost to the treasury; most recreation opportunities are simply given away. As Terry Anderson suggested in a telephone conversation, "If environmentalists were serious about having the federal government stop giving away resources, alongside their 'Cattle Free in '93' bumper stickers they would have had 'Hike No More in '94.'"

As with subsidies to private landowners, subsidies to consumptive and nonconsumptive uses of the public lands encourage overuse and reckless use. Doing away with those subsidies is a major political problem; it will not be solved by simply outlawing them. Instead, structural changes need to be made to the agencies that administer the lands. Depoliticizing the political lands poses a daunting challenge. For examples of market-oriented approaches, see *Reforming the Forest Service*, by Randal O'Toole (1988), and *Public Lands and Private Rights: The Failure of Scientific Management*, by Robert H. Nelson (1995).

Move Away from Coercive Incentives

Analysts who recognize the value of incentives but prefer to have them handled more by government than by markets propose policies based on what they call "market incentives" but are more properly termed "coercive incentives." These incentives are innovative and appear to be more effective at protecting biodiversity than is the ESA. But they rest on different premises than the ones I suggest. I begin with the premise that landowners actually own their land and that public and private organizations that want particular biological outputs of that land should pay for them. Coercive incentives start with the premise that property rights are vested in government, which grants use rights to private parties. For example, Larry McKinney (1993) proposes the following set of policy tools, which contains a mix of positive and market incentives:

- Property tax credits for habitat maintenance
- Tax credits for habitat improvements
- Partial tax credits for ESA compliance expenditures
- Income-tax deductions for farm lands managed to support endangered species
- Tax penalties for habitat conversion
- Prohibition on the use of federal subsidies and tax benefits for activities causing habitat loss or degradation
- Creating a market for development rights on important biological habitat

He offers the following caveat:

> The opportunity costs of not extracting marketable resources or converting land to commercial or residential uses can be substantial. These tax incentives do not seek to bridge the considerable distance between status quo, land-based revenues, and unrealized opportunity costs. They are intended as motivating incentives and economic signals, *not* as compensation for the effects of lawful and appropriate government regulation. (McKinney 1993, 4)

His proposals are less expensive for landowners than the absolute taking that occurs under the ESA, but McKinney emphasizes that he does not propose to offer full compensation. Landowners are "granted" development rights that are insufficient to develop their own property. In effect, this "grant" of development rights is actually a reduction of the owners' rights to develop their property. In order to develop, they must purchase development rights from other landowners. Thus, landowners, developers, and customers bear the financial burden of conservation on the property. They are forced to privately fund public and common-pool goods.

One example is the system of tradable development rights (TDRs) implemented in Montgomery County, Maryland, and the New Jersey Pinelands. In each case, trading development rights in one area in order to exceed density restrictions in another has preserved open space. In Maryland thousands of TDRs have been transferred, and in the New Jersey Pinelands 10,000 acres have been protected (Goldstein and Heintz 1994, 53).

Elizabeth Kennedy, Ralph Costa, and Webb Smathers (1996) suggest a similar mechanism they call "marketable Transferable Endangered Species Certificates (TESC)" for red-cockaded woodpeckers. They model their proposal on TDRs and pollution-rights certificates and suggest limiting the TESCs to landowners within a statewide habitat conservation plan. Landowners who wanted to alter habitat on their property could purchase TESCs from other landowners or by translocating juvenile birds from their property to protected properties. Those who owned certificates would be allowed to alter habitat in ways normally not allowed under the ESA. The price of TESCs would be determined by the costs of engaging in more costly mitigation or management practices. The authors note that under their proposal, "Costs (i.e., the cost of accumulating or purchasing the necessary number of certificates) would be more than those for forestry operations without the ESA obligation, but landowners could potentially cut timber sooner, decreasing opportunity costs" (Kennedy, Costa, and Smathers 1996, 25). They cite estimates of current opportunity costs on highly stocked stands that range from $143 to $1,486 per acre (23).

Again, notice that the costs of all these proposals fall on the landowners. Nor do the TDRs, TESCs, and tax credits require agency personnel to rank conservation priorities, make careful choices, or search for innovation.

CONCLUSION

People will try to conserve species if doing so serves their interest. Unfortunately, many of the people whose involvement is critical to saving species and habitat do not find preservation in their interest. Public and private land managers pursue private agendas, which may or may not be in the public interest, depending on the costs and benefits they face in that pursuit. Thus, National Park Service personnel with a vested interest in a paradigm they helped to develop justify their actions regardless of what the data tell them. Agents of the Fish and Wildlife Service, anxious to retain or increase their congressional funding, spend most of the endangered-species budget on charismatic species such as wolves and grizzly bears while they do little to aid truly endangered species with less emotional appeal. Organized interests use elephants to raise money and promote Western eco-imperialism, even though their actions reduce incentives for those who live among the elephants to protect them. Private timber owners accelerate harvesting if they fear that their land harbors endangered red-cockaded woodpeckers. Alternatively, they may cut down trees with holes in them, cut them into sections, and dispose of the section with the hole. All cavity nesters lose, including other species of woodpeckers, flickers, screech owls, and flying squirrels.

Positive incentives would cause owners, land users, activists, and policy makers to recognize the costs of their actions and would bring human management to the fore. Adopting the proposals presented here would produce a multitude of approaches to habitat and species preservation instead of the limited, prescriptive policies imposed by the ESA. Specifically, government can

- Let species pay their own way by allowing private groups and individuals to profit from protecting species and by allowing public agencies to charge user fees.
- Decentralize biodiversity protection efforts to states and private organizations. Many competing answers are better than one, especially inasmuch as no one knows what the right answer is.
- Depoliticize the protection process by creating a biodiversity trust fund that would change incentives for private individuals, public officials, and interest-group representatives and thereby improve the chances of spending funds effectively while creating private support for conservation.
- Stop subsidizing ecosystem disruption on private lands to reduce development on fragile lands.
- Charge user fees on public lands so that users bear some of the costs of preservation as well as paying the costs of their own actions.

Adopting these proposals will not save all species or keep people from making mistakes. They will result in actions that rely on trial and error, learning from mistakes, and using that new knowledge to respond to new challenges and problems. This approach is, therefore, emphatically not advanced as a set of "solutions." I do not know what solutions are best. I believe, however, that the most effective policies would allow resilience and flexibility in a dynamic, changing world. My approach will allow society to harness the creativity of people everywhere to innovate, experiment, take new risks, and produce new knowledge; it is, in fact, adaptive management (Walters 1986). Such a process would produce responses that no one can predict. Just as those who predicted timber shortages could not foresee chipboard and particle board or the technology that turned previously worthless trees into usable building materials, thereby creating timber surpluses, we cannot foresee the exact responses to many of the present—and certainly not the future—species controversies. The best we hope for is that policy makers will set in motion processes that allow innovation and experimentation to produce those responses. Aldo Leopold summarized this argument well in the conclusion of his essay "Conservation Economics":

> This paper forecasts that conservation will ultimately boil down to rewarding the private landowner who conserves the public interest. It asserts the new premise that if he fails to do so, his neighbors must ultimately pay the bill. It pleads that our jurists and economists anticipate the need for workable vehicles to carry that reward. It challenges the efficacy of single-track laws, and the economy of buying wrecks instead of preventing them. It advances all these things, not with any illusion that they are truth, but out of a profound conviction that the public is at last ready to do something about the land problem, and that we are offering it twenty competing answers instead of one. Perhaps the cerebration induced by a blanket challenge may still enable us to grasp our opportunity. (Flader and Callicott 1991, 202)

NOTES

1. A 1998 ballot measure in Oregon (Measure no. 64) attempted to codify the balance of nature. If passed it would have outlawed harvesting trees in excess of 30 inches diameter at breast height and would have required leaving at least seventy well-distributed trees per acre harvested. Although some environmentalists opposed such attempts to maintain stasis in forests, the proposal was supported by the Oregon chapter of the Sierra Club, the Native Forest Council, Forest Guardians, and other local environmental groups.
2. The polluter-pays principle pertains to one of the major economic concerns of environmentalism— negative externalities. Negative externalities are costs of one person's actions that are passed on to others without their consent (e.g., water pollution from pesticide residues).

3. The logic does not change if we accept the anti-"humanist" argument that "ownership" is broader than just humanity and should include all species.

REFERENCES

Anderson, Terry L. 1994. Home on the Range for Wolves. *Christian Science Monitor*, April 14, p. 18.

Babbitt, Bruce. 1994. The Triumph of the Blind Texas Salamander and Other Tales from the Endangered Species Act. *E Magazine* 5 (2): 54–55.

Barbour, Michael G. 1995. Ecological Fragmentation in the Fifties. In *Uncommon Ground: Re-Thinking the Human Place in Nature*, edited by William Cronin. New York: Norton.

Basgall, Monte. 1996. Defining a New Ecology. *Duke Magazine*, May–June, pp. 39–41.

Beattie, Roger. 1994. Free Market Conservation: Protecting Native Plants and Birds to Death. *Free Radical*, August, pp. 6–7.

Botkin, Daniel B. 1990. *Discordant Harmonies: A New Ecology for the Twenty-first Century*. New York: Oxford University Press.

———. 1991. A New Balance of Nature. *Wilson Quarterly* 15 (2): 61–72.

———. 1992. A Natural Myth. *Nature Conservancy Magazine* 42 (3): 38.

Bourland, Thomas R., and Richard L. Stroup. 1996. Rent Payments as Incentives: Making Endangered Species Welcome on Private Lands. *Journal of Forestry* 94 (4): 18–21.

Brown, Mathew, and Jane S. Shaw. 1998. Paying to Prevent Logging Is Breakthrough for Environmentalists. *News Tribune*, Tacoma, Wash., August 13.

Budiansky, Stephan. 1995. *Nature's Keepers: The New Science of Nature Management*. New York: Free Press.

Clark, Tim W., Richard P. Reading, and Alice L. Clarke, eds. 1994. *Endangered Species Recovery: Finding the Lessons, Improving the Process*. Washington, D.C.: Island Press.

Corn, Lynne M. 1995. Endangered Species: Continuing Controversy. *Congressional Research Service Issue Brief* IB95003.

Cronon, William, ed. 1995. *Uncommon Ground: Toward Reinventing Nature*. New York: Norton.

Czech, Brian, and Paul R. Krausman. 1998. The Species Concept, Species Prioritization, and the Technical Legitimacy of the Endangered Species Act. *Renewable Resources Journal* 16 (1): 17–21.

DeGraaf, Richard M., and William H. Healy. 1993. The Myth of Nature's Constancy: Preservation, Protection, and Ecosystem Management. *Transactions of North American Wildlife and Natural Resource Conference* 58: 17–28.

Delta Waterfowl Foundation. 1994. *Delta Waterfowl Report*. Deerfield, Ill.: Delta Waterfowl Foundation.

Dwyer, Lynn E., Dennis D. Murphy, and Paul R. Ehrlich. 1995. Property Rights Case Law and the Challenge to the Endangered Species Act. *Conservation Biology* 9: 725–41.

Easter-Pilcher, Andrea. 1996. Implementing the Endangered Species Act: Assessing the Listing of Species as Endangered or Threatened. *Bioscience* 46: 355–63.

Flader, Susan L., and J. Baird Callicott, eds. 1991. *The River of the Mother of God and Other Essays by Aldo Leopold*. Madison: University of Wisconsin Press.

Foreman, Dave. 1995–96. Wilderness: From Scenery to Nature. *Wildearth* (Winter): 11.

Goldstein, Jon H., and H. Theodore Heintz, Jr. 1994. Incentives for Private Conservation of Species and Habitat: An Economic Perspective. In *Building Economic Incentives into the Endangered Species Act*, edited by Wendy E. Hudson. Portland, Ore.: Defenders of Wildlife.

Harbrecht, Doug. 1994. A Question of Property Rights and Wrongs. *National Wildlife* 32 (6): 4–11.

Johnson, Hyrum B., and Herman S. Mayeux. 1992. Viewpoint: A View on Species Additions and

Deletions and the Balance of Nature. *Journal of Range Management* 45: 322–33.
Kay, Charles E. 1995. Aboriginal Overkill and Native Burning: Implications for Modern Ecosystem Management. *Western Journal of Applied Forestry* 10: 121–26.
Kay, Charles E., and John W. Walker. 1997. A Comparison of Sheep and Wildlife Grazed Willow Communities in the Greater Yellowstone Ecosystem. *Sheep Research Journal* 13: 6–14.
Kennedy, Elizabeth L., Ralph Costa, and Webb M. Smathers, Jr. 1996. Economic Incentives: New Directions for Red-Cockaded Woodpecker Habitat Conservation. *Journal of Forestry* 94 (4): 22–26.
Leal, Donald R., and Holly L. Fretwell. 1997. Back to the Future to Save Our Parks. *PERC Policy Series*, June. Bozeman, Mont.: Political Economy Research Center.
Lewin, Roger. 1986. In Ecology, Change Brings Stability. *Science* 234: 1071–73.
Losos, Elizabeth, Justin Hayes, Ali Phillips, David Wilcove, and Carolyn Alkire. 1995. Taxpayer-Subsidized Resource Extraction Harms Species: Double Jeopardy. *Bioscience* 45(7): 446–55.
Mann, Charles C., and Mark L. Plummer. 1992. The Butterfly Problem. *Atlantic Monthly*, January, pp. 47–70.
McKinney, Larry D. 1993. Reauthorizing the Endangered Species Act: Incentives for Rural Landowners. In *Building Economic Incentives into the Endangered Species Act: A Special Report from Defenders of Wildlife*, edited by Hank Fischer and Wendy E. Hudson. Washington, D.C.: Defenders of Wildlife.
Mills, Stephanie. 1995. *In Service of the Wild*. Boston, Mass.: Beacon Press.
Mitchell, William, and Randy T. Simmons. 1994. *Beyond Politics: Markets, Welfare, and the Failure of Bureaucracy*. Boulder, Colo.: Westview Press for The Independent Institute.
Nelson, Robert H. 1995. *Public Lands and Private Rights: The Failure of Scientific Management*. Lanham, Md.: Rowman and Littlefield.
O'Toole, Randal. 1988. *Reforming the Forest Service*. Washington, D.C.: Island Press.
———. 1995. Incentives and Biodiversity. A talk given at the Smithsonian Earth Day Conference on Biodiversity, Washington, D.C., April. Transcript available from the Thoreau Institute, Oak Grove, Oregon.
Pickett, Steward T. A., V. Thomas Parker, and Peggy L. Fiedler. 1992. The New Paradigm in Ecology: Implications for Conservation Biology above the Species Level. In *Conservation Biology*, edited by Peggy L. Fiedler and Subodh K. Jain. New York: Chapman and Hall.
Pielou, E. C. 1991. *After the Ice Age: The Return of Life to Glaciated North America*. Chicago: University of Chicago Press.
Reiger, George. 1993. Footing the Bill. *Field and Stream* 98 (May): 14.
Seasholes, Brian. 1995. Species Protection and the Free Market: Mutually Compatible. *Endangered Species Update* 12 (4–5): 7–9.
Sedjo, Roger A., and R. David Simpson. 1995. Property Rights Contracting and the Commercialization of Biodiversity. In *Wildlife in the Marketplace*, edited by Terry L. Anderson and Peter J. Hill. Lanham, Md.: Rowman and Littlefield.
Smart, Angela. 1994. Recreational User Fees on Public Lands. Master's thesis, Utah State University, Logan, Utah.
Stuebner, Steve. 1995. Bidding War for State Rangelands. *Different Drummer* 2 (3): 57.
Tausch, Robin J., Peter E. Wigand, and J. Wayne Burkhardt. 1993. Viewpoint: Plant Community Thresholds, Multiple Steady States, and Multiple Successional Pathways: Legacy of the Quaternary? *Journal of Range Management* 46: 439–47.
Walters, Carl J. 1986. *Adaptive Management of Renewable Resources*. New York: Macmillan.
Winckler, Suzanne. 1992. Stopgap Measures. *Atlantic Monthly* 269 (January): 74–78.
Worster, Donald. 1995. Nature and the Disorder of History. In *Reinventing Nature*, edited by Michael E. Soule and Gary Lease. Washington, D.C.: Island Press.

9

Environmental Colonialism

"Saving" Africa from Africans

ROBERT H. NELSON

Religious ideals have always been a central element in the interaction of the Western world with African society. Religion was a major motivation for the original fifteenth-century Portuguese explorations to discover the coast of Africa. Henry the Navigator was seeking to reunite European Christianity with the Christian kingdom of "Prester John," known to have survived in isolation for approximately a thousand years in Ethiopia. In the nineteenth century, David Livingstone opened up the interior of Africa in hopes of bringing Christianity to these domains. Yet the results of these religious missions have not always been very "Christian." Indeed, the spread of slavery and other forms of exploitation of ordinary Africans frequently followed in their wake.

The greatest current efforts to "save" Africa are associated with contemporary environmentalism. The results have not been as devastating as the experience of slavery, yet they have often served Western interests and goals much more than the interests of ordinary Africans. In some cases, local populations have been displaced and impoverished in order to create national parks and to serve other conservation objectives. Under the banner of saving the African environment, Africans in the last half century have been subjected to a new form of "environmental colonialism."

Many informed observers have held the view, although not well known to the general public in Europe and the United States, that environmental activism exhibits a neocolonial character in Africa. Raymond Bonner, for example, came newly to the African scene in the early 1990s from a long career in investigative journalism. He found that most of his preconceptions about African wildlife management—typical of popular attitudes shaped by conservation organizations and an uncritical Western media—were wrong. Indeed, he would write that "the longer I stayed in Africa, . . . the more I realized that the issues weren't so simple. . . . I realized that the way I, a Westerner, looked at wildlife wasn't necessarily the way Africans did" (1993, 7). As Africans achieve greater political maturity, however, Bonner thought, they will no longer "allow themselves to be dominated by Europe and the United States" (286). "They threw off colonialism" once, and Bonner now predicted that "one day they will throw off eco-colonialism" (286) in the management of their wildlife and other aspects of the environment.

In further exploring the neocolonial character of Western environmentalism in the African setting, I draw here on an impressive body of recent scholarly research. Many of these studies are by people who would be placed on the traditional left of the political spectrum. As seen from their perspective, it is no longer businessmen who are today most likely to be exploiting Africans for their own gain (most current capitalists are actually almost entirely indifferent to Africa, preferring to put their money elsewhere, where the returns are higher and more predictable), but rather the activities of the environmental movement.

I am not suggesting that the problems of environmental colonialism have gone entirely unnoticed until now; some observers, even some within important components of the environmental community, have noticed it. Indeed, for at least a decade international conservationists based for the most part in southern and eastern Africa have led a strong movement for community-based natural resource management (CBNRM) (Hulme and Murphree 2001; Western, Wright, and Strum 1994). The CBNRM advocates have argued that successful wildlife conservation requires the assistance of local African populations (Child 1995; Murombedzi 1992) and have emphasized the importance of local economic benefits in order to create positive incentives for the protection of wildlife.

The efforts of such African conservationists, however, have often been undermined by their European and American counterparts (Hutton and Dickson 2000). Financial contributors and other environmental supporters in Europe and the United States have found the myths of Africa more attractive than the realities. The international commitment to CBNRM so far has been more rhetorical than real. Although environmental colonialism is no longer as overt as it was in the original colonial era prior to the 1960s, it has continued in practice in the policies of many current African governments.

A number of the leading examples of past and present environmental colonialism are found in Tanzania, even though this nation is often considered to be among the most enlightened of African countries—a place where the corruption of government, the exploitation of ordinary people, the divisions of tribalism, and other African ills have been less severe. Yet in Tanzania, too, the creation of national parks and other game preserves has been and still is being accomplished only with the displacement of native tribal groups from their historic homelands, leaving them worse off economically and in some cases in dire poverty today.

Like Christianity historically, current environmentalism is possessed of a strong missionary spirit. In this respect and others, the rise of environmental colonialism is not unrelated to Christianity in defending forms of colonialism. Hence, a brief digression on the religious character of modern environmentalism and its relationship to the Christian religious heritage of the West helps to set the stage for the subsequent parts of this chapter (see also Nelson 1990, 1993, 1997).

ENVIRONMENTAL FUNDAMENTALISM

The environmental movement is best understood as a complex—and ultimately confused—reaction to some of the more disturbing elements of "the modern project." For many people, scientific knowledge undermined traditional Christian belief. In Europe today, the Christian churches are more museum pieces than vital institutions of society. The modern age often substituted a belief in economic progress: in Marxism, socialism, and other secular faiths, hopes for a new heaven on earth replaced expectations of salvation in the hereafter (Nelson 1991, 2001). Yet the history of the twentieth century put the lie to the conviction that material progress would transform human nature, that a "new man" (and woman) would emerge from the enjoyment of far greater material abundance than had ever been known previously to human experience. Germany, Japan, Russia, and other nations did experience unprecedented economic growth and development, but that "progress" seemed more to disorient their world than to save them from sinful actions.

Thus, as the end of the twentieth century approached, a religious vacuum in Western society existed. Neither traditional Christianity nor the secular twentieth century religions of progress could fulfill the spiritual aspirations of large numbers of people. In this circumstance, the contemporary environmental movement emerged as one way to fill the vacuum. Environmentalism has some very prosaic goals, such as making the air less polluted and reducing the incidence of cancer. Within the wide scope of the environmental movement, some card-carrying environmentalists are committed merely to making the world a more pleasant place in which to live. One might say that their goal is to carry on in the progressive tradition, taking some further steps toward the improvement—and perhaps ultimately the perfection—of the human condition.

Yet much of the crusading energy of current environmentalism derives from a much different source. For many of its followers today, environmentalism has been a substitute for fading mainline Christian and progressive faiths—its religious quality obvious to any close observer of its workings. Its language is often overtly religious: "saving" the earth from rape and pillage; building "cathedrals" in the wilderness; creating a new "Noah's Ark" with laws such as the Endangered Species Act; pursuing a new "calling" to preserve the remaining wild areas; and taking steps to protect what is left of "the Creation" on earth. At the heart of the environmental message is a new story of the fall of mankind from a previous, happier, and more natural and innocent time—a secular vision of the biblical fall from the Garden of Eden.

Before the corrupting influence of modern civilization (and perhaps the true date of the fall can be traced as far back as the rise of agriculture, almost ten thousand years ago), human beings lived in genuine harmony with nature. Since the fall, the rise of acquisitive urges and the destructive powers of modern science and industrial production have defiled the innocence of nature almost everywhere. Environmentalism therefore seeks to protect the vestiges of the original natural order and perhaps in some places to restore a "true" nature—the original Creation, as it is in effect regarded within the movement—that has been lost.

The idealism of this vision is not in question; neither is the need for a religious faith to give meaning to the events of the world. However, contemporary environmentalism has lacked any well-developed body of thought to explicate its theology in a careful fashion. There has been no Thomas Aquinas of environmentalism to work out the core precepts of the faith and to ensure that they are logically ordered and rigorously defensible and that they meet other requirements of intellectual and theological coherence. For example, writing of the movement to protect nature in Great Britain, W. M. Adams notes that "conservation in the UK has grown up without a coherent philosophy, a cultural and scientific rag-bag of passion, insight and good intentions." There is a familiar "set of practical concerns (rare species, characteristic habitats, beautiful landscapes) and a set of recognized and institutionalized activities (particularly the complex pattern of British protected areas) [but] underneath this established pattern, conservation floats on a maelstrom of diverse ideas" (1996, 99).

Despite its modern appearance, environmentalism is closer to an old-fashioned form of religious fundamentalism, perhaps reflecting the fact that John Muir, Dave Foreman, and other leading sources of inspiration for the environmental movement were spawned by forms of Protestant fundamentalism. For modern environmentalism, as for the classical Protestantism of old, the core religious understanding is not achieved by a process of rational understanding (in the Protestantism of Martin Luther and John Calvin, it was by "faith alone"). Indeed, although modern environmentalism sometimes has put science to use for its purposes, at heart it has been deeply skeptical of science. The rise of modern science has been the decisive factor in the destruction of the natural order. Martin Luther, it might be noted, similarly regarded Thomas Aquinas's more rational and intellectual exposition of the truths of Christianity as a negation of true religion, as the actual death of any possibilities for a valid faith. Modern environmentalism in some cases has been in direct conflict with scientific facts and knowledge.

For people who have regarded the environmental movement as a branch of science, the level of indifference to scientific knowledge and fact often comes as a surprise. Ecologist Daniel Botkin, for example, explains his motive for writing the book *Discordant Harmonies,* wherein he seeks to provide a more scientific basis for environmental policy:

> In the mid-1970s, I confronted several curious contradictions that I attempted to explain: decisions about managing nature were based on ideas that were clearly contradicted by facts; in my own field of ecology, those same ideas dominated, yet the facts that contradicted them were gathered by ecologists. We repeatedly failed to deal successfully with our environment, and we seemed to ignore the very facts that could most help us. . . . The search for an explanation led down many paths and required peeling back layer after layer of impression and observation. At the surface were the activities of our society: scientists doing research; legislators signing bills; government officials dealing with policies. Underneath these was a layer of belief, myth and assumption, of symbol and metaphor. . . . [At issue was] the character of nature undisturbed. . . . What is the proper role for human beings in nature? At this level, the solution to the paradox lies with a shift of perception, the change in metaphor, myth, and assumption. (1991, vii)

Environmentalism was in fact answering questions similar to those answered by Genesis in the Bible. Indeed, despite the secular overlay of environmental language, the environmental answers were often remarkably similar to the biblical answers, and in some cases the tension with modern scientific realities was equally as great, although this tension was not sufficient to dissuade true believers in either case.

Religion can be a double-edged sword: though necessary for human existence, it has the potential to create havoc in society as well. Following the rise of the new Christian fundamentalism of the Protestant Reformation, Europe was plunged into 150 years of terrible religious warfare. Hence, it is not enough that the environmentalists of today are religiously well meaning, that they genuinely desire to save the world, and that some of them are willing to make significant personal sacrifices in the service of their highest ideals. The ideas of environmentalism must be subject to a searching critical analysis and scrutiny. The environmentalists' efforts must be judged by their real consequences, not by their intended outcomes. When the members of the environmental movement are (largely) unwilling to undertake such a searching inquiry, it falls to others to perform the task.

In the remainder of this chapter, I review the rise of a new environmental colonialism in Africa and draw many of my examples from the Tanzanian experience, which has been especially well documented.

HOLLYWOOD AFRICA

In *The Myth of Wild Africa* (1996), Jonathan Adams and Thomas McShane describe how images of nature in Africa have been crafted to appeal to European preconceptions. The image of the "noble savage" has had an enduring attraction for many Western minds. Even though Scottish missionary and explorer David Livingstone knew better, he wrote for the English public in the mid–nineteenth century that "to one who has observed the hard toil of the poor in civilized countries, the state in which the inhabitants [of Africa] live is one of glorious ease. . . . Food abounds and very little effort is needed for its cultivation; the soil is so rich that no manure is required" (qtd. in Adams and McShane 1996, 15). This account exemplified a common depiction of Africa as a virtual Garden of Eden, innocent of the ills of modern civilization. As Roderick Neumann has observed, "The identity myth of a colonizing society returning to or discovering an earthly Eden is deeply implicated in the establishment of national parks [in Africa]" (1998, 18). (This romantic image, to be sure, often conflicted with another common view of Africa as a land of wild savages whom Christian religion and modern ways of living must rescue from a barbaric condition.)

Today, Africa is still being presented in such Edenic terms. However, the Eden myth with Africans present has been supplanted by images of an Edenic wilderness in which current Africans, as well as non-Africans (except tourists), must be excluded. Neumann aptly comments, "national parks in Tanzania could accommodate the presence of the noble savage for only a brief time" (1998, 18). In Western eyes, the original innocence of nature is now found in places where modern Africans themselves are kept out because they have acquired the technical power to subject nature to human domination. Thus, the

African Eden now survives only in protected places such as the national parks created in many African countries.

These parks are marketed to Western tourists as places where they can see nature in its "true" form. A recent promotional brochure of a South African safari operator, for example, explains that:

> Tanzania, the land of Kilimanjaro and undoubtedly one of the most beautiful countries of Africa, boasts some of the most sensational wildlife refuges in the world. Tanzania has long been considered the finest safari destination in all of Africa and within its borders lie legendary game reserves and game areas that combine incredible concentrations of wildlife. The Ngorongoro crater and the Serengeti National Park contain almost two million animals. The Selous Game Reserve [in southeastern Tanzania] is the largest wildlife reserve in all of Africa, much of it totally unexplored. Here the lion remains "king of beasts" over large populations of buffalo. It is remote and peaceful, but more importantly, it is the true Africa, undamaged and unspoilt. (Mafigeni Safari and Tours 2002)

Such imagery boosts tourist interest in the Selous Game Reserve and serves both the interests of safari operators and the revenue goals of the Tanzanian national government, but it has almost nothing to do with the reality of Selous history. Until the end of World War I, Germany controlled the colony of Tanganyika. In 1905, the native Africans living in the Selous area revolted against their colonial masters. Finding it difficult to defeat the small guerrilla groups of Africans by direct military means, the Germans adopted a strategy of deliberately starving the local populations. As John Reader explains in his magisterial history of the African continent, "three columns advanced through the region, pursuing a scorched earth policy—creating famine. People were forced from their homes, villages were burned to the ground; food crops that could not be taken way or given to loyal groups were destroyed" (1999, 600). By some estimates, as many as three hundred thousand Africans died, perhaps a third of the total population in the area.

In this fashion, the groundwork was laid for the eventual creation of the Selous Game Reserve, advertised today as "the true Africa, undamaged and unspoilt." Reader can hardly contain his sarcasm in describing the ironies of the situation:

> Paraphrasing Tacitus' verdict on the Roman warfare in Germany, a commentator wrote that "the Germans in East Africa made a solitude and called it a peace." The *Maji-Maji* districts were at peace again, but it was the peace of the wilderness. Survivors attempting to re-establish themselves in the region found it transformed, with forest encroaching on

> village sites and game reoccupying previously cultivated land. More ominously, the tsetse fly was there too. . . . For agriculturalists in the southern regions of German East Africa, . . . vast areas of their homeland were uninhabitable; from its midst the British colonial administration [which had replaced the Germans after World War I] carved out the world's largest game park—the Selous. (600)

To be sure, in an area as large as the Selous, some African populations survived. A population of forty thousand scattered through the Liwale District had to be removed in order to create the Selous Reserve. In his 1977 classic *Ecology Control and Economic Development in East African History,* Helge Kjekshus (former lecturer at the University of Dar es Salaam) reports that "the man in charge of the operations, Rooke Johnston, . . . held that development [of the Selous] depended on the eradication of all human rights and interests in the areas" (178). Johnston would write that in this pursuit "I went all out to achieve what I had conceived in 1931 to be the betterment of Liwale District and its people, namely its elimination." If the Selous appears today to be "wild Africa," it is really the product of the extermination and removal of its peoples by deliberate European strategy in the twentieth century. Earlier, thriving populations of Africans had actively engaged in the manipulation of the Selous environment for their benefit.

THE RINDERPEST PLAGUE

Most of the national park areas of Africa were not depopulated by military means and administrative actions, however. Instead, as in North and South America a few centuries earlier, diseases introduced by Europeans wiped out native populations. Many Africans died from smallpox, which was introduced from Europe and to which Africans had no natural immunity. Unlike in the Americas, the new diseases in Africa had the greatest impact on its animal populations. Reader describes the rinderpest epidemic of the late nineteenth century as "the greatest natural calamity ever to befall the African continent, a calamity which has no natural parallel elsewhere" (1999, 589). Between 1889 and the early 1900s, the rinderpest plague killed 90 to 95 percent of all the cattle in Africa. The rinderpest first appeared in Somaliland and spread rapidly to engulf the entire continent, reaching as far as Cape Town in South Africa. Goats and sheep also were affected. For the many African tribes that depended on livestock, their economic means of support was decimated. Whole areas where livestock raising had traditionally taken place were depopulated. By one estimate, two-thirds of the Maasai population in Tanzania died as a result of the rinderpest plague (Reader 1999, 590).

The rinderpest affected wildlife as well. Over wide areas of Africa, the existing populations of buffalo, giraffe, eland, most small antelopes, and warthogs were virtually wiped out. Thus, ordinary Africans suffered the loss of this traditional source of sustenance as well. The ecological balance that had kept the tsetse fly under control was in turn disrupted. Cattle grazing traditionally had kept the grasslands from growing into dense fields and thickets. With cattle removed and much of the wildlife gone also, these grasslands could grow without any check from the clipping and thinning of animal foraging. The new habitat that grew up was much better suited to the tsetse fly. In Uganda, an estimated two hundred thousand people died between 1902 and 1906 from sleeping sickness spread by new hordes of tsetse flies that spread across the landscape.

The native wildlife populations of Africa had long been exposed to tsetse flies and were immune to the sleeping sickness the flies spread by their bites. Domestic cattle were a more recent arrival to the continent and were susceptible, along with humans, to the disease. With most of Africa's cattle now dead, large areas of habitat were newly available for wildlife without the traditional competition from livestock. Even though the rinderpest had decimated wild animals as well, they rebounded rapidly. Thus, in the early twentieth century, free of traditional cattle grazing and other human impacts, large areas of Africa had newly abundant wildlife populations. For European conservationists, typically ignorant of the recent ecological history of the continent, this landscape appeared to be the "true Africa" of wild game.

That African conservationists and game park managers lack basic scientific knowledge is an observation of long standing. In 1973, A. D. Graham would declare, based on his long experience in Kenya and other parts of Africa, that "to the scientist it was their abysmal professional ignorance that was so disappointing. Simple facts about the animals and the wilderness were evidently quite unknown to the conservationists. Yet, almost without exception, the preservationists themselves claimed a profound knowledge of those very facts; claimed them in fact with such authority that the uninitiated accepted their distinction without demur" (27).

The conservationists actively sought to set aside preserves in natural parks in order to prevent the recurrence of human impacts. Reader again describes these events—the source of much of the national parkland of current eastern and southern Africa—with dripping sarcasm:

> The overall effect of the rinderpest plague, compounded by initial depopulation and the subsequent migration of people away from the bite of the

> tsetse fly, was to shift the ecological balance of the trypanosome [sleeping sickness] cycle heavily in favour of wild-animal populations. In East Africa in particular, areas which had once supported large and relatively prosperous populations of herders and farmers were transformed into tsetse-infested bush and woodland inhabited only by wild animals. Influential colonists during the colonial period assumed that these regions were precious examples of African environments which had existed since time began. Believing that the plains and woodlands packed with animals were a manifestation of "natural" perfection, untouched by humanity, they declared that they should be preserved from human depredation forevermore. Most are now tsetse-infected game parks: Serengeti, the Masai Mara, Tsavo, Selous, Ruaha, Luangwa, Kafue, Wankie, Okayango, Kruger. (1999, 592)

Kjekshus similarly reports that long before Europeans arrived in the late nineteenth century, ordinary Africans had established "a relationship between man and his environment which had grown out of centuries of clearing the ground, introducing managed vegetations, and controlling the fauna" ([1977] 1996, 181). A main goal of this active management of the environment was to limit the harmful influence of the tsetse fly, and, indeed, for centuries such management had succeeded in making the tsetse fly "a largely irrelevant consideration for economic prosperity" (181). This happy world was destroyed at the beginning of the twentieth century when an "eruption of tsetse-borne sleeping sickness epidemics" produced a "sudden human and cattle depopulation and the attendant loss of control over the environment" (181). In the larger scope of East African history, Kjekshus considers that the social impacts of these ecological developments exceeded even "earlier events like slave-raiding and intertribal warfare, to which historians have given so much attention" (181).

The creation of a national park in eastern and southern Africa thus typically served to prevent ordinary Africans from reoccupying areas from which they had been expelled by European military force and disease in the previous half century. The "true Africa" seen by tourists visiting the parks was the product of the decimation of traditional African life in the aftermath of the arrival of European settlement.

Ironically, the creation of a park area would also serve to change the behavior of the wild animals. Lions previously had never allowed humans to approach within a few feet, as is now possible in vehicles in park areas. The national parks of Africa increasingly are taking on the character of large open-air zoos. The tourists love the experience because they can see animals that in earlier times would have taken care to stay far removed from any human presence. If today someone wants to see the behavior of a "natural" lion, he can find the closest approximations in the areas *outside* park boundaries, where Africans continue to hunt lions.

THE "PARADISE" OF SERENGETI

Kruger National Park in South Africa was created in 1926. The Convention for the Protection of African Flora and Fauna met in London in 1933. Of this meeting, Jonathan Adams and Thomas McShane write that "the age of Africa's national parks truly began with the international agreement of 1933" (1996, 47). The events of World War II intervened, but soon afterward the colonial administrations were creating war national parks across Africa (MacKenzie 1988). In 1951, the British administrators of Tanganyika created the Serengeti National Park (which had already been protected under a less-formal status), today perhaps the most famous national park in the world and widely (but of course wrongly) regarded as a surviving remnant of "original Africa." The park is 5,600 square miles, about the size of Connecticut.

For several centuries, the Maasai people, themselves invaders from the north, had occupied the Serengeti area. In the mid–nineteenth century, before the arrival of Europeans, some fifty thousand Maasai occupied large areas of what is now Kenya and north-central Tanzania. The Maasai lifestyle was based on the raising of cattle, and therefore the Maasai were among the tribes decimated by the rinderpest. The Maasai also lost large parts of their land as a result of colonial policies that evicted them in order to make way for European agricultural settlement. Yet another large part of the Maasai land was taken away for the creation of national parks—not only Serengeti but Tarangire and Lake Manyara National Parks in Tanzania, as well as the Nairobi, Amboseli, and Tsavo National Parks in Kenya (along with the Masai Mara National Reserve, also in Kenya).

As a result of the ecological consequences of the rinderpest, in 1951 "woodlands covered the northern reaches of the Serengeti, though less than half a century earlier the area had been open, grassy plains, inhabited by people and their animals" (Adams and McShane 1996, 48). Leading conservationists persisted in calling Serengeti "a glimpse into Africa as it was before the white man ever crossed its shores" (from a 1955 report by the Royal National Parks Department, qtd. in Adams and McShane 1996, 48), but the truth was closer to the opposite. The European arrival in the late nineteenth century had massively altered both the human and the wildlife circumstances of the Serengeti. The Europeans who saw wild nature in the Serengeti were actually seeing the product of their own recent manipulations. The Serengeti was now a large "garden," some parts the (unintended) equivalent of wild weeds, other parts more like (intended) domestic plants.

Before the manipulations of the past 125 years,

> Tsetse had long inhabited the no-man's-land between African settlements, such as the ungrazed areas that separated one Maasai settlement from another in and around the Serengeti Plain. Africans knew of these focal points of infection and avoided them, while Maasai cattle ate young sprouts, preventing them from maturing into tough, thorny scrub, and thus kept the tsetse in check. The hunting practices of tribes other than the Maasai also helped deter the spread of tsetse by regulating wildlife populations that could provide hosts for the flies. Africans . . . had thus established "a mobile ecological equilibrium" with wildlife and their associated diseases.
>
> The equilibrium collapsed when Africans and their cattle began dying in large numbers from diseases brought by Europeans. On the Serengeti and elsewhere, a vicious cycle began: the bush returned because cattle no longer kept the bush down, the flies multiplied, further lowering both human and cattle populations, leading to more habitat for tsetse, and so on. (Adams and McShane 1996, 49)

Yet in areas outside the protected national parks, new forces in the 1950s and 1960s would again radically alter the ecological order. The new availability of modern medicines led to the recovery and significant increase of Maasai populations. In the late 1950s, a new campaign was waged against the tsetse fly, using modern insecticides and traps. As the Maasai experienced greater health and vitality, they returned to older burning practices on the plains that further reduced the habitat suited to tsetse flies. Maasai cattle numbers again grew rapidly, pressing against the capacity of the grazing resource to support these numbers, especially in light of the large areas of traditional Maasai lands now converted to agriculture and set aside in national parks and other reserves.

The Maasai looked to return to their former grazing lands in the national parks, and in the mid-1950s the Tanganyika legislature voted to cut the size of Serengeti park in half in order to allow them to reoccupy the central plain. This change would have provided the Maasai with much-needed flexibility as they moved their cattle from area to area according to traditional practice. It also would have posed little hazard to wildlife; populations of livestock and wildlife had previously coexisted in these areas for centuries. Moreover, the Maasai—unusual among African tribes—have religious prohibitions against the routine killing and eating of wild animals in times when there was no shortage of food.

Led by famed international conservationist Bernhard Grzimek, president of the Frankford Zoological Society, the world conservation movement mobilized to block any such Maasai aspirations, however. Because the British still made the final decisions in colonial Tanganyika, the plan to reduce Serengeti was soon shelved. The center of controversy then shifted to the Ngorongoro Crater, where the Maasai had begun grazing as a result of the loss of their traditional grazing areas in Serengeti and elsewhere. Grzimek led a campaign to evict the Maasai from this area as well in order to create yet another area "free of human impact." The Ngorongoro Conservation Area (NCA) was set aside in 1959, and Maasai cattle grazing was banned within the area of the crater.

In a compromise reflecting a greater recognition of the needs of native Africans, grazing was allowed in the 1960s in some areas outside the Ngorongoro crater but within the NCA. Agriculture was also permitted within these areas, allowing the Maasai to support cattle populations in part through raising grain. However, in 1975 the Tanzanian government banned agriculture altogether in the NCA, thereby essentially eliminating any feasibility of sustainable livestock raising there by the Maasai.

The modern environmental movement is no gentle society of aristocrats. The many triumphs of environmentalism over the past half century have been won by hardball practitioners of politics and media relations. As John McPhee (1971) memorably describes in *Conversations with the Archdruid,* Sierra Club leader David Brower pioneered these methods in the United States in the 1950s and 1960s in battles over Dinosaur National Monument, the Grand Canyon, and other park areas. For Brower and many other environmental activists to come, factual accuracy would have to take a back seat to practical results when the fate of the earth was at stake.

While Brower was working away in North America, Grzimek was applying the same kinds of hardball tactics on the African environmental scene. He produced an internationally acclaimed book and film of the same name, *Serengeti Must Not Die.* Although enormously influential, this book was also, as Adams and McShane report in hindsight, "another of Grzimek's propaganda tools, filled with misleading, often falsified data." The overall image, immensely appealing in its own way to European and American audiences, was "that Africa is dying and . . . what little remains must be saved from mankind" (1996, 53)—that is to say, saved from the Maasai use of the land as it had been taking place for several centuries. Adams and McShane summarize this infamous episode in conservation history:

> In 1959, Tanganyika seemed poised to take the crucial step of allowing local people to share their land with wild animals in and around a pro-

> tected area. Bernhard Grzimek, however, was horrified at the thought of people wandering around in "his" national park, so he fought the NCA as he fought all the battles over wildlife conservation, with any weapon at his disposal; "First by soft line, then by hard line, next by bribery, and if necessary by outright blackmail," according to one journalist. . . . Grzimek once described himself as "a showman of pity." Indeed, his campaigns to save wild animals were based on manipulating the emotions and expectations of both the general public in Europe and politicians in Africa.
>
> The NCA is today just another park or preserve, and a poorly managed one at that. The goals set for the NCA in 1959 . . . have never been realized. The harmonious existence of people, livestock, and wild animals has not been achieved, and the rights and needs of the local Maasai community are often ignored. (1996, 53)

Reflecting similar outcomes across many parts of Africa, Barnabas Dickson commented recently that "when the effect of past conservation policies on indigenous people is properly recognized, the record is a shameful one" (2000, 176). He describes the "colonial approach to conservation"—often carried forward by new African governments even in the aftermath of the colonial era—as both a practical failure and "unjust":

> [This] approach involved the state assuming ownership of wildlife and instigating widespread restrictions on the use of wildlife. . . . [But] it did not work because the rural people living closest to wildlife had little incentive to conserve wildlife. Since they had no legal claim on that wildlife they saw little long-term gain from it. On the contrary, it was often a threat to their livelihoods (when wild animals destroyed their crops) and sometimes to their lives. They had no reason not to acquiesce in poaching and positive reason to engage in the practice themselves. In these circumstances, it should not have been surprising that state attempts to protect wildlife often ended in failure. The colonial approach was condemned as unjust because the colonial authorities had deprived indigenous people of a valuable resource that, prior to colonialization, they had regarded as their own. In addition, the state typically sought to protect wildlife under its nominal ownership by the use of extremely harsh methods, including the extrajudicial execution of suspected poachers. (176)

MKOMAZI GAME RESERVE, 1988

One might have thought that the end of the colonial era in Africa would have brought the end of environmental colonialism as well. The forms of European influence on environmental policy did indeed shift; it was no longer possible simply to issue an administrative edict from London or Paris. However, African nations and governments survived in a condition of great dependence on outside donor agencies. For the Africans who were fortunate enough to be able to live a Western lifestyle, the money to pay for it typically came from these agencies and from foreign tourists and other foreign sources. Continuation of the flow of money depended in significant part on a deep respect for the wishes of Europeans and Americans, including prominently international environmental organizations and their constituencies.

The European conquest of Africa often exploited the deep divisions among Africans themselves. Much earlier, the practice of slavery had depended on the Africans' willingness to capture, transport, and sell slaves to European (and Arab) slave traders. There have always been Africans who have found that serving outside needs and demands was the easiest route to their own prosperity and well-being. Their advantage, admittedly, was frequently derived from the suffering of other Africans. The bonds among different tribes and different regions of Africa have never been strong. In the most recent illustration of this phenomenon, African government administrators of protected park and wildlife areas have sought actively to please European and American donors and clients even as ordinary Africans suffered from their actions. In Kenya in 2002, the Constitution of Kenya Review Commission, created by the government, listened to local people throughout the country and reported back that "one of the most common areas of complaint related to the use of land for game parks but to the exclusion of the local people" (Constitution of Kenya Review Commission 2002). The commission heard of "a sense experienced very widely: that local control of resources, and therefore of their lives, had been wrested away" by outsiders.

Sharing a long border, Kenya and Tanzania were similar in this respect. In 1988, the local inhabitants of Mkomazi Game Reserve in northeastern Tanzania were expelled from the area by action of the Tanzanian national government. In *Fortress Conservation: The Preservation of the Mkomazi Game Reserve, Tanzania* (2002), Dan Brockington describes the circumstances that led to this action and the consequences for the people there—yet another story of environmental colonialism. The new twist, however, is that native Tanzanians stepped into the shoes of the old colonial overseers. Similar developments occurred widely over the African continent. In nation after nation, new African governing elites captured old colonial instruments of state control for their own private purposes (Bayart, Ellis, and Hibou 1999).

As Brockington explains, "the rural poor in Africa tend to be weak and marginal to their countries' affairs." In Mkomazi and many other places, "they can be, and often are, ignored by their rulers." By contrast, "conservation receives continual and valuable support from a number of non-governmental organizations (NGOs), which lobby and raise money for conservation causes. They provide valuable funds to African governments" (2002, 10). Foreign tourism also brings in large revenues that can be used to support the Africans who staff government agencies. The NGOs are important not only for the direct infusions of money they contribute but also for the political legitimacy they provide. According to Brockington, "the resources provided by conservation interests, as well as the powerful rhetoric of providing for future generations, may serve to justify the existence of protected areas to government officials" (10) who themselves benefit significantly from the existence of these areas.

The Mkomazi Game Reserve is located in northeastern Tanzania adjacent to the border with Kenya. The reserve was formally established in 1951 in order to protect an area with significant numbers of elephants and other wild animals (the area is part of the broader ecosystem that includes Tsavo National Park in Kenya). In deference to longstanding use, the grazing of cattle was allowed to continue in the eastern part of the reserve; in 1969, the western part also was opened to livestock grazing. In subsequent years, the numbers of domestic animals in the reserve increased rapidly, reaching a total of ninety thousand cattle and thirty thousand sheep and goats in 1984. As part of this grazing use, some settlement also occurred.

Arguing that this livestock use was degrading the reserve and reducing its value for wildlife preservation, international conservation organizations pressed the Tanzanian national government for the expulsion of the livestock from the reserve. They argued that the heaviest grazing users were not indigenous to the area and thus had less moral claim to continued use. Initial efforts to remove the livestock had limited success in the face of local resistance, but in 1986 the Tanzanian Department of Wildlife finally issued an order to remove all livestock and associated settlement. The actual removal took place in large part in 1988. Although some illegal use continued, it is estimated that the number of cattle in the reserve decreased by 75 percent. The people evicted suffered severe and uncompensated economic losses, but, aided by international human rights groups, Maasai and Parakuyo tribesmen eventually brought court cases seeking compensation for those losses.

In reviewing this history, Brockington (2002) finds that the truth is illusive in the midst of numerous claims and counterclaims. In a complex ecological system—both in human and in plant and animal terms—it would require large resources to undertake the scientific studies required to disentangle all the various factors. One of the major uncertainties concerns the impact of grazing. According to one scientific view, "the disturbance caused by grazing and burning does not necessarily cause damage; it is more likely to result in disturbances that foster biodiversity. Livestock do not necessarily exclude wildlife, rather the greatest concentrations of wildlife in East Africa depend on pastures grazed with livestock" (Brockington 2002, 56). Indeed, livestock grazing is not a recent innovation on the continent; it has been a part of the African ecological dynamic for thousands of years. The absence of livestock does not protect "original nature," but rather creates something brand new.

All in all, as Brockington concludes, "there is no clear evidence about the effect of people and their stock on the biodiversity of the Reserve. It remains possible that they enhanced it" (2002, 73). He also refutes the claims about the benefits of the absence of a long-term presence of grazing in the reserve. He portrays systematic misuse of information by world conservation organizations in their enthusiasm to "save" a part of Africa. These organizations and their allies in the Tanzanian government frequently claimed that the Mkomazi Reserve was "one of the richest savannas in Africa and possibly the world in terms of rare and endemic fauna and flora" (Tanzanian Department of Wildlife, *Draft Management Plan,* qtd. in Brockington 2002, 80). The reality is that, "as regards biodiversity, Mkomazi is species-rich for plants and birds, but not outstanding in global or regional terms" (80). There is nothing extraordinary about the mammal populations, although "invertebrates are numerous" (80). Overall, any grand claims for biodiversity in Mkomazi are misleading; the "evaluation of its conservation value awaits better research in similar ecosystems" (80) that may in fact be biologically richer. Yet the international conservation organizations engaged in a powerful campaign in Europe and America in the 1980s to portray Mkomazi as a unique biological resource in Africa, thus justifying the removal of the local African populations from the area.

In Brockington's view, "the international representation of Mkomaze ends up being an almost Orwellian rewriting of the Reserve's, and its people's, histories" (2002, 126). The advocates of exclusion of people were driven by the familiar myths of a "wild Africa" that must be maintained in its "original wilderness" condition. The emotional power of these images for European and American audiences is not in doubt; nor is their usefulness for fund-raising purposes. From a crassly cynical point of view, one might suggest that the spreading of fictions can promote the maximum utility in society: the fictions do make many people feel good. In this sense, although the international conservation organizations belong in the same category as Hollywood producers of illusion, the propagation of their myths may actually enhance the world's total economic product.

Indeed, if social science is truly to be value neutral, as it often claims to be, there may be no grounds to object to the use of falsehoods that make people feel better. It is obvious that Brockington personally is offended by the outcomes he observed in the course of his Mkomazi research. Feeding the emotional needs of Europeans and Americans on the backs of the rural African poor is not a pretty sight. However, Brockington also recognizes an obligation to the canons of the academy that supposedly limit subjective judgments based on strong personal moral convictions. Thus, at some points in *Fortress Conservation,* he attempts to adopt an "objective" posture with respect to the international conservation organizations' obvious illusions and deceptions:

> The case of Mkomazi suggests two reasons for the strength of fortress conservation [that requires the exclusion of people]. The first is that myths work. The Mkomazi myths can bring in much revenue. They result in the enforcement of exclusion and the creation of wilderness in the image desired by the creators. Myths may be wrong, but that is not the point. Myths are powerful. They motivate people; they help them to organize and understand their worlds; they provide structure and meaning; they are the source of beliefs, hopes and plans. (2002, 126)

Beliefs that help people "organize and understand their worlds" are often called religions. In the developed Western world, environmental religion has exerted an extraordinarily strong attraction for many people over the past thirty-five years. In a world of rapid change, where new scientific discoveries are announced every day, many people feel disoriented. The citizens of the developed world seem willing to cling to any rock available, and the environmental movement has offered them hopes that some vestiges of real and permanent "nature" can be found. These last remaining places where human impacts have supposedly not already transformed the natural world must be preserved as "wilderness." Even if very little is historically accurate in all this, environmental religion would not be the first religion to maintain a hold over masses of believers in the face of strong contrary scientific evidence.

As Daniel Botkin observes, "there is no longer any part of the Earth that is untouched by our actions in some way, either directly or indirectly." As a result, "there are no wildernesses in the sense of places completely unaffected by people" (1990, 194). Yet, for many people, the idea of "true nature" unaffected by human action is a necessary benchmark for their dealing with the natural world. Without it, they lack a sense of purpose and direction in the human interaction with nature. It is as though "God is dead" in a modern sense. Botkin is aware of this element and describes the psychological disorientation that many people feel in the face of modern scientific knowledge of the disorderly realities of the natural world:

> To abandon a belief in the constancy of undisturbed nature is psychologically uncomfortable. As long as we could believe that nature undisturbed was constant, we were provided with a simple standard against which to judge our actions, . . . providing us with a sense of continuity and permanence that was comforting. Abandoning those beliefs leaves us in an extreme existential position: we are like small boats without anchors in a sea of time; how we long for safe harbor on a shore. (1990, 188–89)

In the face of this "extreme existential position," many people will prefer to find comfort in myth. They will believe something—whatever it is—before they will believe nothing. That inclination is a major problem facing those who would seek a more scientifically informed environmental debate. Indeed, many fundamentalist Christians today also continue to believe in "the Creation" as literally presented in the Bible, despite strong scientific evidence to the contrary.

ARUSHA NATIONAL PARK

Yet another rendition of the myth of wild Africa in Tanzania involves Arusha National Park, set aside as a reserve in 1953 and made into a national park in 1960. Roderick Neumann examines this case in *Imposing Wilderness: Struggles over Livelihood and Nature Preservation in Africa* (1998). Arusha is Tanzania's third largest city, the jumping-off point for many visitors to Serengeti National Park and the NCA as well as for climbers of nearby Mt. Kilimanjaro. Arusha National Park is not in the same category of world attraction, but it does receive large numbers of visitors, reflecting its close proximity to a major city. Its most prominent physical feature is Mt. Meru, more than sixteen thousand feet high, which looms spectacularly over the city of Arusha.

This story, as Neumann relates it, is yet another in which "the portrayal of the national park as pristine nature symbolically and materially appropriates the landscape of Mount Meru for the consumptive pleasures of foreign tourists while denying its human history" (1998, 13). The more recent "European appropriation of the African landscape for aesthetic consumption" follows directly in the path of an earlier colonial tradition of "appropriation of African land for material production" (9).

Arusha National Park is surrounded by local populations of Meru and other tribes engaged in agriculture. The relationship between park authorities and the surrounding villages has changed little from colonial times, characterized as it is by deep suspicions on both sides. The villagers must respect the superior coercive power available to park managers, but they attempt to subvert park management through poaching of animals, capture of plants, and other forms of illegal activity that are difficult to detect and prevent. Neumann found that "much like their colonial predecessors, state authorities [now] present an implied and often explicit image of villagers as either backward peasants or as criminals" (192). For their part, the villagers make "serious accusations of abuse" with respect to the park guards. In one example, "a villager said that after the right-of-way was closed, the guards would beat people and rape the women that they caught inside [the park]" (189).

For the villagers living today in proximity to Arusha National Park, there are clear "parallels between the park [management] and colonialism." The people living near the park still experience at present "a humiliation and deprivation that . . . cannot do other than resurrect memories of the worst injustices of the colonial government" (Neumann 1998, 194). For one thing, the park was largely formed from lands that had been taken from local Africans in order to make them available for German and then British settlement. After the colonists left, local Africans had hoped to recover their lands, but that recovery was not to be. Now additional lands are being taken over for the park with no more regard for local feelings than existed during colonial times. "As a local villager whose family farm was partly taken over by the park expressed bitterly, 'Do you think we have independence [*uhuru*]? Isn't this like colonialism [*kama ukolini*]?'" (194).

Ordinary Africans' experience of the management of Arusha National Park, as Neumann explicitly characterizes it, amounts to "the new colonialism" (194). Tanzanian park authorities and others in the Tanzanian government justify the park as a boost to tourism and thus as a source of large revenues generated for the support of state institutions at the national level. The tourists are attracted in part because of beliefs they have about the history and purposes of the national parks of Tanzania, however fictional the basis for those beliefs may be. Such beliefs also benefit international environmental organizations for revenue-raising purposes and serve to legitimize the neocolonial practices of the current Tanzanian park authorities. In terms similar to those employed by other recent scholars, Neumann describes the situation as follows:

> The European settlers are now gone. Significant portions of their former estates lie not in the hands of [indigenous] Meru farmers, but behind the boundaries of the national park. The land has taken on new meanings

> derived from European representations of Africa. . . . The late poet and author Evelyn Ames was much taken by Arusha National Park, describing her experience there as . . . like being "alone in Eden." In her account of leaving the park we can hear many of the themes of nature that African national parks were meant to embody for Europeans: the park is primordial, undisturbed, unchanging, and pure in the absence of humans. . . . The representation of Arusha as a prehuman remnant providing refuge from society is also developed in another popular depiction, where the park provides "a sense of complete withdrawal from the world of man and of immersion in the peace of unspoilt nature."
>
> Tanzania's independent government has accepted the national park model based on these Western ideals of pristine nature. Arusha National Park remains principally an attraction for tourists to experience "primeval Africa." (177)

Neumann recognizes that the allusions to Eden are more than a metaphor. Western conservation efforts in Africa are infused with a missionary spirit; at the famous Arusha conference in 1961, "conservationists were encouraged to 'work among the masses with missionary zeal' and 'to awaken African public opinion to the economic and cultural values of their unique heritage of wildlife'" (141). It is easy to see in such efforts "striking parallels with the efforts of early Christian missionaries, particularly their ideas about Africans as 'natural Christians.' Likewise it appears that Africans were [now] regarded as 'natural conservationists'" (141). The Christian religion, unlike many other faiths, has always assumed that its values are universal, in the end meant to spread across the entire world.

As related in Genesis, God created the world. To see nature unaltered by human hand, to enter into nature "undisturbed" and "unspoilt," is to encounter a direct product of the divine handiwork. God is not literally in nature—such a supposition would be the heresy of pantheism—but the experience of "original nature" comes close to putting a person in the very presence of God. The tourists who flock today to Africa's national parks are a modern version of the pilgrims who have long flooded Rome or descended on Lourdes in southern France. As the visitors to "original nature" in Africa have received spiritual nourishment and replenishment, accommodating their needs has proved good business for many Africans.

At present, serving the needs of wildlife pilgrims is the most rapidly growing area of the economy of African nations such as Tanzania, which lacks any base of manufacturing or other industry. The Africans need not share the spiritual motive—Neumann comments that "of all the inherited colonial institutions, wildlife conservation was least understood within African culture" (1998, 141)—but they can well appreciate the economic gains that tourism brings.

In some parts of Africa, to be sure, the economic benefits have not been as great, and the motives or capacities of African national governments have been insufficient for the protection of wildlife even in park areas. The bushmeat trade has decimated wildlife populations over parts of West Africa including the parks. John Oates (1999) argues that the old colonial approach—protected areas with local Africans excluded by direct coercive means—may be the only workable solution to protect the wildlife in such cases. He criticizes environmental leaders for their unwillingness to confront the real world, as they pretend that local "community-based" approaches to conservation can succeed everywhere. Although the themes are now altered, even the community-based style of international environmentalism remains a political crusade to save the world. This newer form of environmental thinking also includes a greater element of guilt about the past. Formal appearances are changed, but the old colonial attitudes are still manifested, and efforts on the ground to protect wildlife or to help the African poor commonly fail. According to Oates, many

> international conservation planners [now] stress the need to "empower" local people. This form of paternalism seems to be an entrenched feature of Third World development and humanitarian aid projects, which are typically planned and implemented by highly educated middle-class Westerners. The project planners and managers generally maintain (or improve) their own lifestyles, while displaying attitudes that seem to be colored both by colonial-style paternalism toward people they regard as the benighted peasants of the Third World, and by guilt for the perceived wrongdoing of their colonial antecedents. This pursuit of a mixture of material and sociopolitical aims has become endemic in Third World conservation projects initiated by Westerners and, as I have argued, has its roots in the liaison that developed in the 1970s between international conservation and development organizations. (1999, 234)

CONCLUSION

The national parks of Tanzania and other African countries have today become grist for the scriptwriters of environmental fantasies. A cynic might say that this "Disneyland management" of Africa's park areas is their actual highest and best use. Fantasy sells, and millions of people in Europe and the United States, living in London, New York, and other urban centers, enjoy images of the Garden of

Eden, whether in Africa or elsewhere in the world. By contrast, the rural people in these areas who are directly affected by the setting aside of surrounding park lands constitute a small and less-moneyed minority that has less political influence both with their own national governments and in international arenas.

Yet a critical problem with the use of rural Africa as a playland for romantic fantasies is the potential for contrary images to arise. Many religious prophets have lost their following by specifying an actual date for the end of the world. When Hollywood filmed the life of John Nash in *A Beautiful Mind,* it took large dramatic license. That portrayal was acceptable for a movie, but no such license is granted the scriptwriters for Africa's national parks. If the current Hollywood imagery and management practices are exposed as such, the viewer pleasures will be greatly diminished. Large European and American commitments of funds—and other large costs borne by the local people who live in close proximity to the African park areas—will have gone for naught. Rather than moral heroes, many American and European environmentalists may come to be seen as the Elmer Gantrys of our time.

REFERENCES

Adams, Jonathan S., and Thomas O. McShane. 1996. *The Myth of Wild Africa: Conservation Without Illusion.* Berkeley and Los Angeles: University of California Press.

Adams, W. M. 1996. *Future Nature: A Vision for Conservation.* London: Earthscan.

Bayart, , Stephen Ellis, and Beatrice Hibou. 1999. *The Criminalization of the State in Africa.* Bloomington: Indiana University Press.

Bonner, Raymond. 1993. *At the Hand of Man: Peril and Hope for Africa's Wildlife.* New York: Alfred A. Knopf.

Botkin, Daniel B. 1990. *Discordant Harmonies: A New Ecology for the Twenty-first Century.* New York: Oxford University Press.

Brockington, Dan. 2002. *Fortress Conservation: The Preservation of the Mkomazi Game Reserve, Tanzania.* Bloomington: Indiana University Press.

Child, Graham. 1995. *Wildlife and People: The Zimbabwean Success.* Harare, Zimbabwe: Wisdom Foundation.

Constitution of Kenya Review Commission. 2002. *The People's Choice: A Draft Constitution.* September 22. Nairobi, Kenya: Constitution of Kenya Review Commission.

Dickson, Barnabas. 2000. Global Regulation and Communal Management. In *Endangered Species, Endangered Convention: The Past, Present, and Future of CITES,* edited by Jon Hutton and Barnabas Dickson, 161–77. London: Earthscan.

Graham, A. D. 1973. *The Gardeners of Eden.* London: George Allen and Unwin.

Hulme, David, and Marshall W. Murphree. 2001. *African Wildlife and Livelihoods: The Promise and Performance of Community Conservation.* London: Heinemann.

Hutton, Jon, and Barnabas Dickson, eds. 2000. *Endangered Species, Threatened Convention: The Past, Present, and Future of CITES.* London: Earthscan.

Kjekshus, Helge. [1977] 1996. *Ecology Control and Economic Development in East African History: The Case of Tanganyika, 1850–1950.* London: James Curry.

MacKenzie, John M. 1988. *The Empire of Nature: Hunting, Conservation, and British Imperialism.*
Manchester, U.K.: Manchester University Press.
Mafigeni Safari and Tours. 2002. Promotional advertisement, Duiwelskloof, South Africa. Available at: http://www.mafigeni.co.za/more_info_tanzania.htm. Visited August 14.
McPhee, John. 1971. *Conversations with the Archdruid.* New York: Farrar, Giroux and Straus.
Murombedzi, James. 1992. *Decentralization or Recentralization: Implementing CAMPFIRE in the Omay Communal Lands of the Nyaminyami District.* CASS NRM Working Paper no. 2. Harare, Zimbabwe: Center for Applied Social Sciences, University of Zimbabwe.
Nelson, Robert H. 1990. Unoriginal Sin: The Judeo-Christian Roots of Ecotheology. *Policy Review* (summer): 52–59.
———. 1991. *Reaching for Heaven on Earth: The Theological Meaning of Economics.* Lanham, Md.: Rowman and Littlefield.
———. 1993. Environmental Calvinism: The Judeo-Christian Roots of Environmental Theology. In *Taking the Environment Seriously,* edited by Roger E. Meiners and Bruce Yandle, 233–55. Lanham, Md.: Rowman and Littlefield.
———. 1997. Does "Existence Value" Exist? An Essay on Religions, Old and New. *The Independent Review* 1 (spring): 499–521.
———. 2001. *Economics as Religion: From Samuelson to Chicago and Beyond.* University Park: Penn State University Press.
Neumann, Roderick P. 1998. *Imposing Wilderness: Struggles over Livelihood and Nature Preservation in Africa.* Berkeley and Los Angeles: University of California Press.
Oates, John F. 1999. *Myth and Reality in the Rain Forest: How Conservation Strategies Are Failing in West Africa.* Berkeley and Los Angeles: University of California Press.
Reader, John. 1999. *Africa: A Biography of the Continent.* New York: Vintage.
Western, David, R. Michael Wright, and Shirley C. Strum. 1994. *Natural Connections: Perspectives in Community-Based Conservation.* Washington, D.C.: Island.

10

The Ivory Bandwagon

International Transmission of Interest-Group Politics

WILLIAM H. KAEMPFER AND ANTON D. LOWENBERG

In October 1989 the Convention on International Trade in Endangered Species of Wild Fauna and Flora (CITES), meeting in Lausanne, Switzerland, voted to classify the African elephant as an endangered species and to make trade in ivory illegal. That policy decision displeased many conservation experts in Africa and elsewhere, who argued that a ban on the ivory trade would prove disastrous for the African elephant. Opponents of the ban favored a free-market approach that would offer rural Africans tangible benefits as an incentive to preserve their elephant herds and would help to compensate for the costs of coexisting with elephants.

The free-market approach presupposes consumptive utilization of elephants, which in turn requires that Africans have access to a market for ivory.[1] The 1989 CITES treaty abolished the legal ivory market, overriding the protests of the scientists and economists who advocated consumptive utilization. The blanket ban on ivory trading prevailed until June 1997, when it was partially lifted, at the request of southern African countries, to allow limited sales of existing ivory stockpiles. Widespread ivory trading remains illegal, however, under the international regime established by the CITES treaty.[2]

How could such a policy have been implemented despite considerable evidence that a ban on ivory trade would have the perverse effect of exacerbating the decline of the African elephant? We try to answer this question by examining the interest-group pressures that led to the CITES decision. We show how lobbying and publicity efforts by a small group of animal-rights activists and preservationists, whose views did not coincide with those of mainstream conservationists, ultimately succeeded in generating broad-based public support for an ivory-trade ban. Our analysis provides a case study of the spread of a political position, publicly articulated, from one interest group to another and even from one country to another. Starting from a very small base of support, a policy with little scientific respectability can easily snowball into a national and international program with unstoppable momentum, and thus tiny special-interest groups can wield an enormously disproportionate degree of influence in the political process.

ELEPHANTS: MENACE OR TREASURE?

Elephants can do a considerable amount of damage to livestock and crops, and they are extremely dangerous to humans, more so than most other wildlife. Elephants spend sixteen hours a day eating. An adult bull consumes 300 pounds of trees and 50 gallons of water per day and can weigh more than 5,000 kg. An elephant clan's home range can cover up to three thousand square kilometers, depending on the availability of water and forage (Kreuter and Simmons 1995, 148). "A herd of elephants goes through an area like a slow tornado, snapping off branches and uprooting trees, leaving devastation behind" (Bonner 1993, 101). No wonder that the prevailing sentiment toward elephants among rural African villagers is one of fear (Bonner 1993, 28, 221–23).[3]

Africans who compete with wild animals for land and food have strong incentives to kill them (Kreuter and Simmons 1995, 148). Indeed, rural Africans have traditionally hunted elephants and other game. Hides were used for clothes, shields, and containers, and ivory and rhino horn were carved into ornaments and jewelry. European and Arab traders in East Africa purchased ivory and rhino horn for sale abroad. By the 1950s many East Africans were engaged in widespread commercial hunting to supply meat and skins to growing urban populations (Bonner 1993, 43, 45). Conservation was viewed as a less-than-honorable profession by many Africans, for whom wild animals were a potentially life-threatening liability, and such anti-conservation sentiments led to an escalation of hunting in post-independence Africa (Kreuter and Simmons 1995, 149).

The attitude of rural Africans toward elephant conservation contrasts starkly with that of many Westerners. Africans living among the elephants incur all of the costs of allowing elephants to exist in the wild, but the benefits accrue largely to Westerners, who view elephants as "an important conservation symbol with high aesthetic and emotional appeal" (Kreuter and Simmons 1995, 149). Westerners obtain "existence value" from the elephants, a benefit deriving from the knowledge that elephants continue to exist in the wild even if the Westerners in question will never personally have any contact with them (Kreuter and Simmons 1995, 149; 't Sas-Rolfes 1998, 17).[4]

In 1933 the European powers that had African colonies held a conference in London to discuss conservation. Although some conservationists at the time were motivated by a desire to preserve biological diversity, others were big-game hunters who wanted to ensure a steady supply of wildlife for hunting safaris (Bonner 1993, 39–46). The 1933 conference called for the creation of a system of national parks in which licenses would be required for access to the game, whether for photography or for hunting. The philosophy underlying the parks was one of separating wildlife from indigenous people, thus preserving the wildlife in a pristine state as a part of the country's national heritage and primarily for the benefit of foreign visitors.[5] Human populations were relocated to areas outside the parks, often resulting in disruption of their agricultural economies and consequent impoverishment.[6]

But the notion that wildlife and rural Africans can be separated is fundamentally flawed. Elephants in particular are notoriously difficult to confine, and they have broken through virtually every type of barrier, including electric fences, which they tear down with their tusks. Any barrier that might successfully keep elephants within a large perimeter would require a technology far too expensive for rural Africa (Bonner 1993, 215–16). By roaming around in agricultural areas adjacent to the parks, the elephants became pests to African farmers and ranchers, who were then even more inclined to shoot them. This issue of animal damage to agriculture is important for the survival of wildlife species, especially inasmuch as 80 percent of Kenya's wildlife live *outside* parks; for elephants in Africa as a whole the figure is 50 percent (Kreuter and Simmons 1995, 158; Bonner 1993, 223).

Another problem with the separation approach is that the parks require game wardens or rangers to enforce laws against hunting and poaching. But African countries are among the poorest in the world, and expenditures on maintaining and protecting the parks are low.[7] The consensus among conservationists is that, ideally, $400 per square kilometer must be spent annually to protect elephants and rhinoceroses from poachers and that at least one ranger per fifty square kilometers is required. Yet in the late 1980s Kenya, one of the wealthier countries in sub-Saharan Africa, was spending $10 per square kilometer in Tsavo National Park, and Zambia had only one warden per four hundred square kilometers of parks. At $400 per square kilometer, Tanzania would require $48 million annually to adequately protect its parks; that country's actual wildlife expenditure in 1991 was less than $5 million. Tanzania was paying its game rangers a salary of about $30 per month, and Zambia less than $20 per month (Bonner 1993, 93–94, 195).

With such low wages and minimal resources for law enforcement, it is not surprising that corruption is rampant among game wardens and other officials in Africa's parks. In East Africa in particular, corruption in game parks is so entrenched that it has severely undermined conservation efforts. Game wardens are easily bribed to ignore, and even assist, poachers.[8] Poorly paid rangers poach out of desperation for food and to support their families. Local officials have skimmed about 40 percent of the entrance fees paid by visitors to Kenya's Maasai Mara. Some of the large lodges in the park have routinely underreported occupancy rates in order to avoid remitting taxes to the authorities (Bonner 1993, 134–36). In all countries where poaching has been a serious problem—Tanzania, Kenya, Congo, Zambia—government officials at the highest levels have been involved in the ivory trade (Simmons and Kreuter 1989, 47).[9]

In 1977 such corruption forced Kenya to ban hunting altogether.[10] The hunting ban resulted in the forfeiture of significant revenues for Kenya, because the going price for hunting an elephant in South Africa is $12,000, and a rhino fetches $28,000 (Anderson and Hill 1995, xii). In 1989 a pair of uncarved elephant tusks sold for an average of $2,000 (Simmons and Kreuter 1989, 47).

Few of the benefits of the parks trickle down to rural Africans living in the vicinity. Unless they become poachers themselves, rural Africans typically see little benefit from wildlife conservation. Pervasive corruption siphons off the revenues to officials at all levels of government, and only a small fraction of tourism earnings accrues to local residents. More than 100,000 tourists a year visit Kenya's Maasai Mara, generating millions of dollars in revenue. Yet a 1988 study of the Mara found that less than 10 percent of gross tourism revenues accrued to the locals (Bonner 1993, 220). Although the benefits to adjacent villagers are elusive, those villagers pay the price of living with the wildlife: their personal security and property are constantly threatened by marauding wild animals.

Quite apart from poaching, attempts to contain wildlife in parks and nature reserves have sometimes turned out to be disastrous because of habitat imbalances that are created when numerous species are confined in a given area. Unable to range widely across the countryside in search of food, elephants put enormous pressure on the land in the parks. They destroy trees and other vegetation, leaving the land exposed to the elements. By the late 1960s the elephant population in Kenya's Tsavo National Park had grown to an unsustainable level of 40,000. Some conservationists suggested culling the elephants in the park, but culling is anathema to many in the West.[11] In any event, the culling was not done, and a severe drought hit Tsavo in 1969–70, causing some 6,000 to 10,000 of the park's elephants to die of starvation (Leakey and Lewin 1996, 204; Kreuter and Simmons 1995, 148). Meanwhile, in areas outside protected reserves, the opposite problem occurs—namely, bush encroachment due to diminished cyclic thinning by elephants and overgrazing by domestic livestock. Maintaining elephant-induced ecological processes over wider areas than those encompassed by national parks would help to promote biodiversity on the African savannas; but it would inevitably exacerbate human–elephant conflicts if property rights did not provide positive incentives for rural Africans to conserve elephants (Kreuter and Simmons 1995, 148).

In property-rights terms, the problem of elephant conservation is that legal title to the elephants is typically vested with the state in which they occur (the so-called range state) or its designated agencies, but because of low funding and internal corruption most states' ownership rights are not effectively enforced. When elephants are, in effect, an open-access resource, as they continue to be in much of Africa, there are no owners to insist that potential users (consumptive or nonconsumptive) pay the opportunity costs of their use.[12] Moreover, the public-good nature of elephant existence means that it is impossible to charge a price for existence value. Because individuals cannot be excluded from enjoying the benefits of knowing elephants exist in the wild, they face an incentive to under-reveal the true value they place on the elephants' continued existence and to enjoy a free ride in nonconsumptive use of the elephants (Kreuter and Simmons 1995, 150).

However, the lot of the elephant has varied across Africa. Although Kenya, Tanzania, Congo, and Zambia have seen a diminution of their elephant herds due to poaching, corruption, and poor management, the countries of southern Africa—Zimbabwe, Botswana, Namibia, and South Africa—have practiced successful conservation, and hence southern African elephant populations are not at risk.[13] The fundamental reason for the greater success in southern Africa is that, for the most part, wildlife authorities there have not pursued the rigid separation strategy of East Africa, with all its attendant flaws. Consumptive utilization has been the way of the southern Africans. This approach rests on the recognition that people cannot realistically be separated from wildlife and that, so long as people and wildlife do live together, the future of the wildlife depends crucially on the willingness of the people to tolerate the animals in their midst. If rural people are to support conservation, they must gain tangible benefits by doing so (Bonner 1993, 216, 223). Farmers' antagonism toward the animals can be mitigated only if they are compensated for the damages caused by elephants and other wildlife. Consumptive utilization embeds wildlife in the economic life of local cultures (Kreuter and Simmons 1995, 157–58).

The most successful programs of consumptive utilization are founded on the creation of community-based usufruct rights to wildlife, which mitigate the effects of its open-access status (Kreuter and Simmons 1995, 150).[14] These rights are vested in the local people who live among or adjacent to the animals. At least some of the revenue from commercial utilization of the wildlife, whether it comes from tourism, hunting, or the sale of animal products such as meat, hides, or ivory, accrues to local communities. Because the communities may sell access to "their" wildlife to the operators of hunting or photo safaris, they obtain real benefits from the animals and have a strong incentive to invest resources to protect them from poachers (Kreuter and Simmons 1995, 150).[15]

CONSUMPTIVE UTILIZATION

According to Raymond Bonner (1993, 33, 286), the two most cost-effective and successful conservation programs in Africa are the community guard program in the Kaokoveld in northwest Namibia and CAMPFIRE (Communal Areas Management Program for Indigenous Resources) in Zimbabwe. Both are based on the principle of consumptive utilization.[16]

In the Kaokoveld, poaching started on a large scale in the mid-1970s after a heavy influx of whites into the area. The community guard program, the brainchild of Garth Owen-Smith, a Kaokoveld conservationist, was devised in 1982 as a response to the poaching as well as to the poverty of the Himba and Herero people, pastoralists who had lost 80 percent of their stock in the devastating drought of 1979–82. The Himba and Herero depended increasingly on international aid for survival, and they began to expand their game hunting. Owen-Smith realized that conservation would succeed only if the local people received some tangible benefits from the presence of the wildlife. Using funds provided by the Endangered Wildlife Trust, a private organization based in Johannesburg, Owen-Smith recruited volunteer rangers, who were paid the equivalent of $25 per month plus food and household supplies. The program also encouraged villagers to make crafts for sale to tourists. A local Conservation and Development Committee was established, its revenue derived from a tax of $10 per tourist paid by tour companies, safari operators, and game lodges. The Himba and Herero volunteer auxiliaries turned out to be far more successful at identifying and tracking poachers than were the overextended and underfunded government officials. The task of the auxiliaries was to give information on the whereabouts of poachers to the local headmen, who apprehended and punished the perpetrators. The program succeeded in curbing poaching, and by 1987 so many animals had returned to the area that the villagers experienced something of a tourism boom (Bonner 1993, 21–22, 26, 31, 33).

The CAMPFIRE project in Zimbabwe operates on a much larger scale. During the colonial period, when Zimbabwe was known as Southern Rhodesia, tribal trust lands—referred to as communal lands—were set aside for rural blacks. Those lands include some of the poorest in the country; in many such areas wildlife is the only valuable resource.[17] CAMPFIRE was started in 1982 by the Zimbabwe government, but it languished at first for lack of funds. The project really came to life in 1988 when it was adopted by a local nonprofit development organization, Zimbabwe Trust. The primary goal of the trust's founders was not conservation but to find an approach to promoting development and alleviating poverty more workable than those of international aid agencies, which seemed mainly to produce more dependency. Zimbabwe Trust allocated funds to help rural Africans in the communal lands to establish their own wildlife management programs that would incorporate consumptive utilization practices to increase their wealth and improve their nutrition (Bonner 1993, 253, 262–63). The program's main objective was to establish cooperatives with territorial rights over well-defined communal resource areas (Kreuter and Simmons 1995, 160). Under CAMPFIRE, villagers in communal lands may cull wildlife for meat, sell hunting concessions, and set up tourist joint ventures. Previously impoverished people in areas such as Nyaminyami on the shores of Lake Kariba and Guruve on the northern border with Mozambique have been able to increase their incomes substantially through CAMPFIRE. Wildlife revenues have enabled villagers in those regions to build schools, clinics, and corn-grinding mills.[18] Unlike park entrance fees and other game-reserve revenues throughout Africa, these funds do not accrue to the central government treasury but go directly to the communal wildlife management trusts. In 1989 Nyaminyami used the funds to hire and equip twelve game rangers at a salary of $100 per month, creating one of the best-paid, best-equipped ranger units in Africa (Bonner 1993, 268). Wildlife revenues are also allocated to compensate villagers for damage inflicted by wild animals.[19] Besides yielding hunting revenues, tourist operations offer enormous economic potential to the communal lands of Zimbabwe.[20] Bonner (1993, 263) cites a 1990 estimate that Nyaminyami, originally one of the poorest areas of Zimbabwe, would be generating $500,000 a year from wildlife by the mid-1990s.

Apart from CAMPFIRE, Zimbabwe has attempted to foster custodial and participatory relationships between rural people and protected areas in general (Simmons and Kreuter 1989, 49). In the 1980s the Zimbabwe Department of National Parks and Wildlife Management gave peasant communities the right to hunt specified numbers of elephants and other game. The communities are allowed to exercise the right themselves or sell hunting permits to commercial operators. Under this system, villages that successfully increase their wildlife herds are rewarded with a greater number of hunting permits, so their incentive to preserve herds is further enhanced (McPherson and Nieswiadomy 1998,

3). For example, one subsistence community received hunting permits for elephant and buffalo in exchange for relinquishing some of the community's land and voluntarily refraining from poaching in Gona-re-Zhou National Park. The permits were sold to a safari operator; part of the proceeds was used to develop community facilities, and the rest was distributed directly to community members who had lost crops to animal damage. When animals that destroy property are killed by National Parks personnel, income from the sale of their hides and ivory accrues to neighboring communities (Simmons and Kreuter 1989, 48).

Consumptive utilization approaches to conservation have been tried also in East Africa, although on a much smaller scale than in southern Africa. When Masai pastoralists were evicted from Amboseli Park in Kenya in 1974, they responded with a campaign against the wildlife, spearing the leopards and rhino, which are great tourist favorites. To stop the slaughter, the government promised annual payments to compensate the Masai for the loss of prime grazing land and watering spots and for tolerating wildlife on their property adjacent to the park. When the payments were made, the poaching stopped.[21] In 1992, when Richard Leakey, the head of Kenya's wildlife department, devised a revenue-sharing scheme that paid 25 percent of Amboseli's gate receipts to people living near the park, "the first thing they did with the money was hire fourteen of their own game rangers to protect the wildlife—the very wildlife that is such a menace to them, which suddenly becomes an acceptable menace when there is money to be made from it" (Bonner 1993, 230). Also exemplifying consumptive utilization is a scheme devised in 1985 by safari operator Robin Hurt. Appalled by the decimation of elephants and rhino in Maswa Game Reserve on the southern edge of Tanzania's Serengeti National Park, Hurt developed a plan to pay villagers living on the outskirts of Maswa for turning in poachers, rifles, and snares, to pay them to work as rangers, and to give the village council a share of his safari revenue for community projects. Hurt's program succeeded. In its first six months, sixty-eight poachers were caught and convicted and seven poachers' camps were discovered and destroyed, five inside Serengeti, and the local village of Makao acquired a shining new corn-grinding mill as a clear quid pro quo (Bonner 1993, 228, 235, 249–50).

The difference between these few consumptive utilization experiments in East Africa and their southern African counterparts such as CAMPFIRE and the Kaokoveld project is that in East Africa, the benefits received by rural Africans are partial. Rather than having full usufruct rights to the wildlife as in CAMPFIRE, the villagers in Kenya and Tanzania receive only a small portion of the income generated by their conservation efforts. The pattern throughout much of Africa is that wildlife proceeds go into national treasuries, which allocate expenditures in accordance with the preferences of the politically influential urban elite (Bonner 1993, 274). Successful conservation, however, requires that the benefits of consumptive utilization accrue directly to those interested and affected parties whose decisions are instrumental in controlling poaching and protecting wildlife (Vorhies 1996).

THE INTERNATIONAL PROPAGATION OF SPECIAL-INTEREST POLICIES

As a first step toward understanding the origins of the international ivory-trade ban, we now describe a conceptual framework for dealing with the question of how the beliefs and preferences of small interest groups—even if they do not, at the outset, reflect broad public opinion or scientific knowledge—can spread throughout a political system, and even from one national polity to another. Several rational-choice models explain how the policy position or preference of a relatively small number of individuals can propagate itself and gain the acceptance of a much larger number. This propagation process, variously characterized in terms of reaching a critical mass or threshold (Kuran 1995; Witt 1989) or a tipping point (Margolis 1998), as herd behavior (Devenow and Welch 1996), or as an "informational cascade" (Bikhchandani, Hirshleifer, and Welch 1992), generally implies a sudden and often difficult-to-predict bandwagon phenomenon.

Here we draw on the threshold approach of Timur Kuran (1987a, 1987b, 1989, 1991, 1995) and on our own previous analysis (Kaempfer and Lowenberg 1992). Individuals are assumed to obtain utility from conforming with certain beliefs even if conformity requires a sacrifice of income or other goods. For example, Robert Higgs (1987) has argued that an individual's utility depends not only on a basket of goods consumed but also on "the degree to which one's self-perceived identity corresponds with the standards of one's chosen (or merely accepted) reference group, that is, with the tenets of the ideology one has embraced" (43). Groups, in essence, reward their supporters with selective incentives, such as the right to share in a feeling of group identity or "political presence" (Uhlaner 1989). Individuals obtain "reputational utility" from supporting the policy espoused by a certain group, and that utility increases with the size of the group or the number of its supporters (Kuran 1987b).[22]

Reputational utility increases with group size because the private payoff to a nonactivist from supporting a group's policy goal increases as more people endorse that goal. Individuals "fortify their reputation as supporters of a given cause by rewarding other supporters and by withdrawing favors from opponents" (Kuran 1987a, 645). People wishing to draw attention to their decision to support a group do so partly by complimenting and rewarding other supporters, because such actions carry more weight than a mere verbal declaration. "Given that an individual can win praise from both the members and supporters of a pressure group, society comes to believe that the benefit from supporting a group rises with the size of the group's following" (Kuran 1987b, 61).[23] Therefore, the individual obtains greater utility from joining a larger rather than a smaller group of peers because the larger group creates a greater sense of group identity. Of course, as Kuran points out, the individual's desire for a good reputation, and the material and psychic rewards that go along with it, must be tempered by his integrity: a rational individual will falsify his personal preferences in order to display outward support for a group only to the extent that the disutility he obtains from compromising his personal beliefs does not outweigh the reputational utility thereby attained.

The proportion of the population believed to support a given policy or interest group is referred to by Kuran as the "collective sentiment" (1989, 46). Each individual has a private threshold level of population-wide support that will induce him to join the supporters. One individual, for example, might be willing to join if 40 percent of the population has already done so, whereas another individual might require 60 percent of the population to outwardly favor an outcome before he will contribute to their efforts. Kuran (1991) shows that, depending on the cumulative distribution of private thresholds, it is possible for a critical mass of population-wide support for a group to exist: if the perceived collective sentiment exceeds that critical mass by even one individual, support for the group could quickly spread to embrace a large percentage of the population.

The notions of preference falsification and collective sentiment are important in explaining how a sudden and surprising surge of support for a position, previously thought to have few adherents, might easily occur. In the context of a political revolution, such as those in Iran in 1979 and eastern Europe in 1989, many individuals typically conceal their true preferences for political change until they are convinced that a sufficiently large proportion of their fellow citizens is favorably disposed (Kuran 1991). The concept of preference falsification can also help us understand instances of collective conservatism, when individuals who privately support a given change in policy fail to speak out in favor of it because they believe that most of their contemporaries are opposed (Kuran 1987a).

The availability of reputational utility does not negate the importance of free-riding in any public-good situation. The actual collective outcome sought by a group might not enter into the individual's decision calculus at all, and yet he might choose to support the group in order to obtain reputational rewards or avoid reputational sanctions, such as ostracism. It follows that interest groups or political parties can foster greater support if they can persuade individuals that their platforms are already popular. This relationship helps to explain why the leaders of groups or parties often expend a great deal of effort trying to convince people that their policies command the support of a considerable portion of the public (Kuran 1987b, 72; Uhlaner 1989, 272).

The leadership of an interest group might propagate public support for its policies by three distinct methods. First, the group might succeed in lowering private thresholds for collective action, perhaps by saturating the public with information or publicity designed to alter private preferences in favor of the group's objective. Second, the group might increase the reputational rewards available to individuals contributing to the group's cause, perhaps by convincing potential contributors that the group has a good chance of success in attaining political influence or power to make appointments in government. Third, the group might try to convince individuals that a critical mass of citizens already supports its policies. Any one of the three—a lowering of private thresholds, an increase in reputational utility, or an increase in collective sentiment—can initiate a bandwagon process that propagates support extensively among the population.

Such a bandwagon process may have an important international dimension. Specifically, events abroad or the pressures of foreign interest groups can serve as catalysts for all of the three mechanisms. First, individuals might revise their private beliefs and preferences when they discover that foreigners publicly profess support for some policy objective. An individual's private preferences are shaped by his private beliefs, which in turn depend partly on the beliefs expressed publicly by other people. Kuran (1987a, 655) cites a large literature in psychology that shows that cognitive limitations require the individual to rely on beliefs conveyed by others in formulating his own private beliefs. The greater the number of people who appear to hold an opinion, the greater the extent to which private beliefs and preferences will be altered to match that opinion. Lobbying by foreign special-interest groups or the adoption of certain policies by foreign governments can be a source of information for individuals in a given country; that information might lead to a change in private preferences with regard to the policies of the domestic government, and consequently to a lowering of private thresholds for collective action aimed at changing those policies.

Second, foreign events or lobbying might produce an increase in the reputational utility afforded individuals who support certain domestic interest groups by increasing the effectiveness of those groups in rewarding their contributors with selective incentives. Thus, a signal of foreign support for the policies of a domestic group could be perceived as raising the probability of the group's eventually achieving its goal, which in turn might encourage activists in that group to work harder and devote more effort to organizing collective action.

An increased expectation of success could also mobilize nonactivist individuals, attracted by the reputational benefits of victory. To the extent that an individual's ability to enjoy the fruits of enhanced group identity or political visibility depends on having actively contributed (walked a picket line, say, or joined a demonstration), individual participation is increased (Uhlaner 1989, 265, 274).

Third, foreign interest-group lobbying might produce an increase in collective sentiment for the policies advocated by a domestic interest group. Such foreign pressures create the perception among individual citizens of the domestic nation that some policy change is deemed desirable by many people abroad. If that is the case, it implies that many individuals in the domestic country—a large proportion of one's own population—probably think similarly and are willing to take action to secure the policy. As pointed out by Kuran (1989, 54, 64), one of the roles of an interest-group leader is to create the belief that almost everyone privately supports the group and that, in reality, opponent groups have only the smallest bases of support. One way to achieve that alteration of collective sentiment is to expose the pervasiveness of preference falsification, so as to convince nonactivists that a substantial percentage of the population actually supports the group's goal. Events abroad or foreign pressures can help to persuade individuals of the plausibility of such claims of widespread support.

GENESIS OF THE CITES BAN

We now consider the dynamics of conservationist interest groups' efforts to preserve the African elephant from a perceived threat of extinction. Following Bonner (1993), we document a bandwagon process in which successful pressure by a handful of small interest groups in the West quickly spiraled into increased support for an anti-trade conservation strategy adopted or advocated by like-minded groups throughout the world. Bonner demonstrates how the imperative facing conservation groups to protect their membership bases and sources of funding in the face of pressures from competing groups led to a remarkably rapid policy transition, from nearly universal support for the ivory trade to virtually universal condemnation. That bandwagon process clearly illustrates the types of propagation mechanisms analyzed in the preceding section.

CITES came into existence in 1973. It is the most comprehensive international conservation agreement, the modern successor to the 1900 Convention for the Preservation of Animals, Birds and Fish in Africa. Under CITES, trading is prohibited only in species that are "threatened with extinction," or endangered. These species are listed in appendix 1 of the convention. Appendix

2 lists the species that are not yet endangered but might become so if trading is not controlled. In 1977 the African elephant was placed in appendix 2, which allowed for limited trade in ivory and hides under a system of permits (Simmons and Kreuter 1989, 47). Because the CITES secretariat was given no enforcement power, implementation of the treaty depended on individual signatory countries. With weak enforcement of CITES controls, poaching and trade in ivory increased. Nonsignatory countries became entrepôts for illegal ivory. Moreover, the treaty applied only to raw ivory, not to ivory that had been worked. That loophole meant that a tusk needed only to be cut, or slightly carved, to be exempt from CITES.[24] In 1985, to strengthen CITES controls on ivory, a quota system was established in which each ivory-exporting country would determine how many tusks it would export each year based on "sustainable off-take" (Simmons and Kreuter 1989, 47). The quota system was abused, however: some countries announced preposterously large quotas. Nevertheless, despite the quota system's evident weaknesses, by 1988 most conservationists believed that it needed more time to work before it could be declared a failure (Bonner 1993, 96–97).

Implicit in the ivory quota system was the assumption that African countries should have the right to earn income from the controlled sale of ivory because that income could then be used to fund their conservation efforts. In the 1980s, the overwhelming majority of wildlife experts and conservationists supported consumptive utilization, sometimes referred to by conservationists as "sustainable utilization." In 1980 the principle had been formally incorporated into the World Conservation Strategy, a landmark conservation manifesto endorsed by mainstream conservation organizations such as the World Wildlife Fund (WWF; subsequently renamed the World Wide Fund for Nature), the largest conservation organization in the world, and the International Union for the Conservation of Nature and Natural Resources (IUCN), as well as by the U.N. Environment Program (Bonner 1993, 98). By 1988 "it would have been hard to find a conservationist with any zoological background and experience in Africa or with elephants who believed that a ban on the ivory trade was the way to save the African elephant" (Bonner 1993, 87). In early 1989 the CITES secretariat argued that a transfer to appendix 1 "would not contribute to the conservation of the African elephant, and may in fact be counterproductive" (quoted in Kreuter and Simmons 1995, 153). Even a limited consumer boycott of ivory products was not supported in the scientific community (Bonner 1993, 54). But although endorsement of ivory trading was widespread among conservationists, it certainly was not among animal rights and preservation advocates such as the Humane Society and the more extremist Friends of Animals (FoA), which lobbied vigorously against consumptive utilization.

Bonner (1993) describes how conservationists with impeccable scientific credentials, who were opposed to an ivory-trade ban, were "overcome by the public pressure and emotion and concerns about money" (34). They discovered that calling for a ban brought in more funding than any other cause and that any organization that failed to climb on the ivory-trade ban bandwagon risked losing members to more extremist competitors. The African elephant turned out to be a "flagship species": it could draw donations needed for other conservation activities that by themselves would attract little interest (Kreuter and Simmons 1995, 153; Leakey and Lewin 1996, 210).[25] In terms of the analysis in the previous section, emotive publicity campaigns to "save the elephant," together with competition for funds, led many individual conservationists to falsify their preferences for consumptive utilization in order to outwardly support the ivory-trade ban.

Heightened attention had been drawn to the African elephant by a preliminary report released in 1989 by the Ivory Trade Review Group (ITRG), financed by the WWF and Wildlife Conservation International, which claimed that elephants had decreased from 1,343,340 in 1979 to 631,930 in 1989. The report indicated that elephant populations had fallen during that decade by 77 percent in East Africa and by 44 percent in Central Africa. The accuracy of the estimates was questioned by some biologists, however, on the grounds that most of the data were derived from informed guesses rather than from scientific census figures. In addition, increasing elephant populations in Zimbabwe, Botswana, and Kenya's Amboseli Park were not acknowledged in the report. Nor did the report take into account that elephant populations had been rising in many areas before 1970 and that the subsequent decline in the 1980s might have been a natural adjustment to shrinking elephant-carrying capacity of the land caused by increasing human population and declining forage resources (Kreuter and Simmons 1995, 151). In any case, the ITRG report, noting that ivory prices had been rising as a result of increasing demand for ivory products, especially in Asia, concluded that the ivory trade posed a threat of extinction to the African elephant: "It is the ivory trade and hunting for ivory, and not habitat loss or human population increase, that is responsible for the decline in [African] elephant numbers" (quoted in Simmons and Kreuter 1989, 46).

According to Bonner (1993), the campaign to ban the ivory trade started in earnest in February 1988, when the African Wildlife Foundation (AWF), a small conservation organization based in Washington, D.C., mailed out an "Urgent Memorandum" to its supporters, informing them that the "insatiable greed of the ivory hunters" was responsible for a "slaughter" of African elephants and declaring 1988 the "Year of the Elephant" (53–54). The foundation asked for tax-deductible gifts, which started flowing in almost immediately. Three months later the AWF held a press conference at the National Zoo in Washington, D.C., and urged the public not to buy ivory. At that time the AWF was not in fact calling for a ban on the ivory trade. It merely wanted to draw attention to the plight of the African elephant at the hands of poachers and to increase Americans' awareness that ivory comes from elephants (54, 87). The magnitude of the public response to the press conference surprised even the AWF, although it is not all that surprising in the context of the threshold analysis of the previous section. Recall that one way to mobilize support for a group is to lower private thresholds for collective action through a successful publicity campaign. As we have noted, individual preferences and beliefs are shaped in part by perceptions of others' beliefs. By portraying the consumption of ivory products as morally indefensible, the AWF was able to convince many people to come out in support of an ivory boycott.

The AWF, however, was only a small organization. In 1988 it had a budget of $2 million and a staff of six. By contrast, the WWF, with its well-known Chinese panda logo, had 5 million members internationally and annual donations exceeding $200 million. In 1988 the U.S. chapter of the WWF alone had a budget of $23 million, a staff of two hundred, and 450,000 members. At first the WWF did not support an ivory-trade ban. In April 1988, at a meeting convened by the WWF in Lusaka, Zambia, the delegates criticized the AWF's Urgent Memorandum as "emotional and inaccurate." In a May 1988 internal memo, WWF–U.S. wrote that although elephant populations had declined, the species was not yet endangered (Bonner 1993, 90, 94).

But the WWF faced a difficult conundrum: although most of its own conservation professionals strongly opposed an ivory-trade ban, explaining to the public the basis of their opposition would not be easy. Consumptive utilization is a difficult position to defend, from a public-relations perspective, because it implies that individual elephants might have to die for the sake of the survival of an entire elephant population. The WWF did not tell its members and supporters that it funded programs in Africa in which communities made money from selling wild-animal products and from the right to hunt wild animals. The WWF was afraid that people gave contributions to conservation organizations because they wanted to see animals preserved, not utilized (Bonner 1993, 99–100). Indeed, there is some justification for this viewpoint: for Westerners whose chief nexus with the African elephant is one of existence value, the death of even a single elephant is utility-reducing. The fund-raisers of the WWF argued that the organization would have to support an ivory-trade ban or risk losing large numbers of members and contributors to other organizations such as FoA that had no scruples about opposing consumptive utilization.

As noted previously, individuals who privately support a given policy position sometimes refuse to champion it outwardly because they believe that most of their contemporaries are opposed. Such preference falsification clearly played an important role in the propagation of support for the international ivory-trade ban. According to Bonner's account (1993), when the WWF-International convened its April 1988 meeting in Lusaka to discuss what its elephant conservation strategy should be in light of pressures from other groups to embrace an ivory-trade ban, some delegates evidently made statements that they would not have made publicly—statements in some cases contrary to what they had said publicly—because the meeting was closed (89). Once Kenyan officials had announced that they wanted a ban, however, some local conservationists who had previously argued against such a strategy did not speak out. "It was not that they had changed their views; indeed, they had not. But they were not about to challenge what had become the popular position. . . . None of the other conservationists in Kenya who believed the ban was a bad idea—and most felt that way—had the courage of their convictions" (130).[26]

The debate among conservationists was transmitted from country to country. In the fall of 1988, as the WWF-International came under increasing pressure to formulate a policy toward the African elephant, memoranda and comments were faxed back and forth between the international office in Gland, Switzerland, and the various national organizations. The latter wanted a coherent policy statement that they could use to respond to letters from members and questions from the media. Finally, it was decided that the WWF's position should be that it opposed the killing of elephants "except where absolutely necessary for the conservation of the species" (Bonner 1993, 108). National organizations in the Netherlands, Finland, and New Zealand found this compromise acceptable. But WWF–U.S. balked, arguing that *any* policy endorsing the killing of elephants would be problematic for public-relations reasons, especially in light of the vicious attacks on such a policy that could be expected from animal-rights organizations.

Meanwhile the AWF escalated its "Year of the Elephant" campaign in February 1989 with a full-page advertisement in the *New York Times* intended to deter consumers from purchasing ivory jewelry. "Today, in America, Someone Will Slaughter an Elephant for a Bracelet," read the caption, over a picture of an elephant with its face hacked off (Bonner 1993, 117–18).[27] Four days later a coalition of animal-rights groups, including the Humane Society, FoA, and the Animal Welfare Institute, held a press conference in Washington, D.C., announcing that it was filing a petition with the Interior Department to have the elephant declared an endangered species under U.S. law, thereby halting ivory imports. The Humane Society petitioned the U.S. Fish and Wildlife Service to formally recommend appendix 1 listing for the African elephant, but U.S. officials were reluctant to comply without a proposal from an African country for such a listing (Kreuter and Simmons 1995, 154). Meanwhile, an organization called Defenders of Wildlife, which previously had shown little interest in the African elephant, published a chapter in its magazine, *Defenders*, characterizing the poaching of elephants as "genocide." In April an organization called the International Wildlife Coalition placed an advertisement in the *New York Times* bearing the caption "African Chainsaw Massacre" and claiming that the elephant would be extinct by 1997. Three days later, FoA placed an advertisement in the *New York Times* excoriating Sotheby's for proposing to auction two large pairs of elephant tusks worth $16,000 a pair. Both FoA and AWF claimed responsibility for persuading Sotheby's to cancel the auction (Bonner 1993, 118–19).

The effects of these shrill campaigns can be interpreted in light of the threshold analysis of the previous section. Recall that one of the mechanisms available to an interest group to propagate support for its preferred policy is the raising of the level of collective sentiment, that is, the percentage of the population that is believed to support the group's policy. Clearly in this case the emotive publicity helped to convey an impression of overwhelming public support (who could possibly be in favor of a chainsaw massacre?), so that any remaining opponents of an ivory-trade ban would feel increasingly isolated in their views.

Interest groups also quickly discovered that emotional appeals were phenomenally successful in raising money. The *Defenders* chapter brought in $40,000, a large sum for an organization with 80,000 members. When the AWF launched its elephant campaign, it had only 24,000 members and was struggling financially. Within a year its membership had nearly doubled, and its donations had increased by 66 percent. The AWF's advertisement in the *New York Times* brought in $42,526 from 1,200 people. Later the advertisement appeared in *USA Today* and generated almost $26,000. The International Wildlife Coalition's advertisement raised $25,000. In other countries, too, elephant campaigns turned out to be bonanzas for conservation groups. In Britain a mailing by WWF–U.K. raised more than $500,000 (Bonner 1993, 120).

Again, these fund-raising successes are explicable in terms of the threshold analysis of the previous section. When soliciting donations, interest groups offer selective incentives in return. Donors receive subscriptions to the groups' publications, bumper stickers, and other awards that identify them as participants in the groups' campaigns and foster a sense of identification with the cause. As noted previously, the ability to capture reputational utility normally depends on making some sort of material contribution to a group's effort, and the amount of reputational utility the group can dispense to each supporter rises with the size of the group's following. Therefore, the willingness to make donations rises exponentially as the number of donors expands. Moreover, the emotional nature of the campaigns in this case created an aura of righteousness around contributors, which further raised the value of the reputational utility awarded (or the cost of reputational sanctions imposed on noncontributors).

The bandwagon was gaining momentum, and conservation groups that did not jump aboard risked being left behind in the competition for members and contributions. In frustration, the chief fund-raiser for WWF–U.S. faxed the International in Switzerland, "We are in danger of losing our position with elephants" (Bonner 1993, 121). Bonner's account makes clear that the fear of WWF officials was not that the elephant was threatened, but that the WWF was! Curtis Bohlen, senior vice president of WWF–U.S., came down firmly on the side of the fund-raisers in favor of an ivory-trade ban (124, 126). On June 1, 1989, WWF–U.S. held a press conference in Washington, D.C., at which it "strongly endorsed" the proposals to place the elephant in appendix 1 of the CITES treaty. A few days earlier Bohlen had called the International to indicate what WWF–U.S. intended to do. He threatened to publicly upstage the IUCN if it did not support the ban (Kreuter and Simmons 1995, 154). The message was: "Go along or be embarrassed" (Bonner 1993, 139). The WWF-International went along, holding a press conference in Geneva and faxing a memorandum to all WWF national organizations instructing them to "follow the line we are taking as closely as possible in order to avoid any further stories of splits in the WWF family over the ivory issue" (139). In this way, mobilization of support for the ivory-trade ban was transmitted from one nation to another through the success of like-minded groups abroad and the raising of public awareness of the issue.

On June 5, 1989, President Bush announced that ivory could no longer be imported into the United States. The 1988 African Elephant and Conservation Act had given the president the authority to ban imports of ivory from countries that violated CITES provisions or dealt in illegal ivory. Now he simply extended that authority to ban all ivory from all countries (Bonner 1993, 140).[28] A few days after President Bush's announcement, Prime Minister Margaret Thatcher banned ivory imports into Great Britain. The interest-group campaigning had been just as intense in Britain as in the United States. In fact, WWF–U.K. had four times as many members, as a percentage of the population, as did WWF–U.S. The European Union also banned ivory imports, and Japan and Hong Kong, the destinations of most raw ivory, instituted some controls as well (Simmons and Kreuter 1989, 46).

Western animal-rights and conservation groups pressured African officials to formally propose an ivory-trade ban at the 1989 CITES meeting. Such a formal proposal coming from African governments would not only enhance the likelihood of Western governments' going along with a ban but would also increase the reputational utility awarded to supporters of the interest groups in question by increasing the probability of their ultimate success. Founders of the London-based Environmental Investigation Agency, with funding from the Washington-based Animal Welfare Institute, worked to persuade Tanzanian conservationists to support a ban and even drafted the letter sent by Tanzania's Wildlife Conservation Society to the Tanzanian president asking him to propose an appendix 1 listing for the elephant (Kreuter and Simmons 1995, 154). After the Tanzanian proposal was released, Kenya quickly followed suit.[29] Not coincidentally, these East African countries were the ones that had experienced the greatest declines in elephant populations in the 1980s. The trade ban "became a convenient way to avoid scrutiny of the widespread participation in illegal trade by Kenya's leading politicians" (154). According to the ITRG's population estimates, elephants had decreased by 74 percent between 1979 and 1989 in the predominantly East and West African countries voting in favor of the ivory-trade ban, whereas elephants had increased by 9 percent during the same period in the predominantly southern African countries voting against the ban.[30] In essence, countries with declining elephant populations voted to impose an anti-trade policy on countries with sound wildlife management programs whose elephant populations were stable or increasing (152).

Kenyan wildlife director Richard Leakey justified the ban by arguing that to permit even limited trade would leave "an open door to further catastrophic poaching" (Leakey and Lewin 1996, 210). Leakey claimed that as long as ivory was "available in the marketplace" it would have economic value, and therefore elephants would be "exploited" without regard to the fate of the species (205). However, this argument reveals a fundamental misunderstanding of how markets work. Ivory has economic value because people desire to consume it. Attempts to abolish that value by keeping ivory from the marketplace are futile, because a market, illegal if need be, always arises for a valued good, despite the best efforts of regulators.

The ivory-trade ban effectively passed the full social cost of internationally adopted conservation policies onto the African range states, while virtually all of the benefits accrued to Westerners who, despite benefiting from the existence value of the elephants, have failed to compensate the range states with financial support during the CITES ban (McPherson and Nieswiadomy 1998, 12). By foreclosing all commercial use of elephants, the trade ban accorded rights to elephants but violated the legitimate rights of native people to manage their own resources (Kreuter and Simmons 1995, 155). African range states continue to allocate considerably higher proportions of their territory to wildlife preservation than do most Western countries.[31] But elephants living outside the parks receive no protection. Their survival in the long run depends on the creation of institutions that enable people to use their land and their wildlife jointly[32] and to attain access to markets once property rights are established (158).

By removing the legal market for ivory, the CITES ban imposed significant costs on the Africans who were already successfully practicing consumptive utilization. To estimate the income lost by the Nyaminyami Wildlife Management Trust because of the ivory-trade ban, warden Elliot Nobula calculated the total weight of the tusks of those elephants that had been killed, not by hunters, but because they were damaging farmers' fields, and concluded that Nyaminyami lost $20,000 during the first eight months of 1990. In addition, but for the ban, Nyaminyami would have culled a few elephants (constituting a sustainable offtake of 3 percent) and sold the ivory and skins. That sale alone would have garnered Nyaminyami Z$250,000, enough to increase every family's income by at least 25 percent (Bonner 1993, 271). Overall, the ivory-trade ban deprived Africans throughout the range states of $50 million in annual revenue from ivory sales and another $50 million in earnings for African ivory carvers (Barbier and others 1990, cited by Kreuter and Simmons 1995, 159). That $100 million is equivalent to the estimated amount needed annually for effective protection of elephants in Africa's parks and reserves.

Economic theory predicts that when the sale of a valued commodity is prohibited, its price inevitably increases; individuals with a comparative advantage in avoiding detection (usually criminals and corrupt public officials) take over the formerly legal market; and, in the case of a common-property resource such as elephants, the quantity of the resource shrinks, and eventually it disappears (Simmons and Kreuter 1989, 48). Yet the proponents of the ivory-trade ban argued that it would cause demand and prices of ivory to fall, leading to reduced illegal elephant hunting. Although legal purchases of ivory in signatory states have indeed stopped since the trade ban, there is some evidence that illegal trade and trade among nonsignatory states have increased. Consistent data on ivory prices are difficult to obtain because most of the trade now takes place on the black market.[33] Nor is it clear whether the CITES ban has been successful in reducing poaching. In the Zambezi Valley of Zimbabwe, illegal elephant hunting has escalated significantly since the 1989 ban was implemented. By contrast, Kenya attributed its reported 32 percent increase in elephants between 1989 and 1991 to reduced poaching because of the ban. (The 32 percent increase is implausible, however, given a maximum annual population growth rate of about 4 percent for elephants; more likely the increase reflected immigration or an inaccurate census. See Kreuter and Simmons 1995, 159.) Where noticeable decreases in poaching have occurred, they reflect increased levels of law enforcement rather than the closing of legal markets for ivory.

CONCLUSION

Using a rational-choice approach, we have identified three mechanisms whereby widespread support for an interest group's policy position might be propagated internationally: reductions in private thresholds for collective action; increased reputational utility from group participation; and enhanced collective sentiment in favor of the policy. Clearly, all three were instrumental in bringing about the international ivory-trade ban.

NOTES

1. On free-market approaches to environmental and conservation issues, and on differences of interpretation among practitioners of these approaches, see Cordato 1997 and Hill 1997.
2. In February 1999 the ban was relaxed to permit Botswana, Namibia, and Zimbabwe to ship some ivory to Japan ("U.N. Legalizes Sale").
3. In Kenya's Laikipia region, for example, elephants have not only trampled crops and devastated grazing land but also torn up water pipes and smashed dams. Elephants destroy the simple structures erected by villagers to store their harvested corn, and they typically linger near the water holes where the village women fetch water (Bonner 1993, 28, 213–14).
4. For a critique of the notion of existence value, see Nelson 1997.
5. Fewer than 5 percent of the visitors to Kenya's parks are Kenyans, and the percentage of indigenous use is higher in Kenya than in any other African country except South Africa (Bonner, 1993, 221).
6. An example is the case of the Masai who were evicted from the Serengeti National Park and Ngorongoro Crater in Tanzania (Bonner 1993, 179–93).
7. Expenditure is low in absolute terms but relatively high as a percentage of government budgets. This percentage ranges from 0.2 percent in Botswana to 0.45 percent in Tanzania and 0.6 percent in Zimbabwe. By contrast, the U.S. government spends 0.15 percent of its total budget on management of protected areas (Kreuter and Simmons 1995, 157).
8. In 1989 the newly appointed head of Tanzania's wildlife department launched a serious campaign against elephant poaching, as a result of which wardens were caught with ivory in their possession. It turned out that the wildlife department itself was actively involved in poaching (Bonner 1993, 134).
9. Traders in Kenya arranged with poachers in advance to buy ivory, some of which originated from elephants killed in Tanzania and Sudan and then smuggled into Kenya. When government inspectors asked store owners for proof that they had legally acquired wildlife products, the owners simply bribed the inspectors (Bonner 1993, 216). In the 1970s a Kenyan game department official reported that between 10,000 and 25,000 elephants were killed for their ivory each year, and that two assistant ministers responsible for wildlife management were involved in poaching and smuggling (Bonner 1993, 51; Simmons and Kreuter 1989, 47). Tanzania banned all trade in tusks in 1987, yet a year later a member of parliament was caught with 105 tusks in his official truck (Simmons and Kreuter 1989, 47).
10. Many professional hunters had been shooting more animals than their permits allowed and then bribing rangers not to report them to the game department (Bonner 1993, 242).
11. And not only in the West. Kenyan conservationist Richard Leakey presents an impassioned argument against both culling and trophy hunting. See Leakey and Lewin 1996, 209.
12. The problem with resources that are not controlled by a single agent, as in the open-access or

common-property case, is the "use it or lose it" incentive. For example, the American bison was hunted nearly to extinction because each hunter had an incentive to kill an animal before someone else did (McPherson and Nieswiadomy 1998, 2). Effective governance of a common-property resource by local communities becomes especially problematic when the resource is mobile across community boundaries (Hill 1997, 391–92), as is the African elephant.

13. It is not coincidental that the countries experiencing the most rapid declines in elephant populations are those with the most political instability, wars, coups, repressive governments, corruption, and attenuation of private property rights. McPherson and Nieswiadomy (1998) point out that countries lacking democratic institutions and secure property rights are unlikely to be able to control poaching successfully. Moreover, political strife and war lead directly to elephant slaughter as soldiers seek ivory revenues to finance their campaigns. In an empirical cross-country study of the determinants of elephant populations in Africa, McPherson and Nieswiadomy (1998) demonstrate that political instability and unrepresentative government are associated with decreasing herd sizes, whereas countries that recognize private property rights to wildlife have experienced, on average, a 15 percent higher annual growth rate of elephant populations.
14. Economists have long recognized that private property rights generally favor conservation of wildlife species whereas collective rights encourage poaching. See, for example, Vorhies and Vorhies 1993. Under collective rights there is no residual claimant who can potentially prosper by superior game management (Hill 1997, 393–94). See Kremer and Morcom 1996 for a formal analysis of the conditions under which alternative preservation policies toward common-property resources such as elephants are likely to produce desirable results.
15. On the effectiveness of community-based wildlife management programs, see Gibson and Marks 1995.
16. Our description of both programs relies heavily on Bonner's (1993) account. Subsequent applications of the consumptive utilization approach include Zambia's Administrative Management Design for Game Management Areas (ADMADE) and Luangwa Integrated Rural Development Project (LIRDP), Botswana's Natural Resources Management Project (NRMP) and Tanzania's Selous Conservation Program, which started in the early 1990s (McPherson and Nieswiadomy 1998, 7–8).
17. In 1989, 10,000 elephants lived on Zimbabwe's communal land (Simmons and Kreuter 1989, 48).
18. In 1989 Nyaminyami earned enough from hunting revenues to cover the costs of its conservation program. In that year, the governmentally determined trophy fee for an elephant was $3,750 (Bonner 1993, 268).
19. Nyaminyami's compensation schedule paid Z$20 for the loss of a goat and Z$20 for a 90-kilogram bag of corn or sorghum (in 1989 Z$2 was equivalent to US$1). When elephants or buffalo trample crops in the field, the local wildlife trust pays a settlement based on average yield (Bonner 1993, 273).
20. The Nyaminyami Wildlife Management Trust was offered 10 percent of after-tax profits by a tour company in exchange for the right to lease land on Lake Kariba's Bumi Bay for a game-viewing camp. Another company offered 5 percent of its gross income, amounting to Z$100,000 a year, for the right to develop a luxury camp, also on the shores of Lake Kariba. These examples contrast starkly with the experience of East Africa, where tourism revenues from game parks generally failed to trickle down to neighboring residents to any significant degree (Bonner 1993, 219–23, 273–74; Simmons and Kreuter 1989, 47).
21. Later the payments stopped, too, when the government ran out of funds.
22. The notion of reputational utility is due to Akerlof 1980.
23. Along similar lines, Howard Margolis (1990) demonstrates that "social motivation" occurs when individuals follow a path-dependent rule of behavior, choosing to allocate resources at the margin to the pursuit of social outcomes if the amount of resources they have already spent on such outcomes is small relative to the amount that others are spending.

24. A carving industry for illegal ivory developed in Dubai and the United Arab Emirates (Bonner 1993, 96).
25. Conservation of a flagship species can also help in the conservation of other species by preventing habitat loss. On the notion of flagship species as it pertains to tigers, see 't Sas-Rolfes (1998, 17).
26. According to Bonner (1993), one would have expected Richard Leakey, once he became Kenya's wildlife director, to quickly lift Kenya's ban on hunting, because he had often made assertions that people who live with wildlife must benefit if the wildlife is to be preserved. Bonner writes: "It is not only because Leakey is unwilling to subject himself to international opprobrium that he has cut his convictions to fit popular opinion." It is also because Kenya's powerful tour operators are opposed to hunters (243).
27. The advertisement had been prepared free of charge by the advertising firm Saatchi and Saatchi (Bonner 1993, 118).
28. At the behest of big-game hunters in the United States, however, the Interior Department enacted regulations permitting the importation of trophy tusks from sport hunting in Zimbabwe and South Africa. The justification was that those countries had healthy elephant populations and strong management programs–precisely what South Africa and Zimbabwe had argued in Lausanne in an unsuccessful effort to be exempted from the ivory-trade ban (Bonner 1993, 270).
29. In a famous act of symbolism, on July 18, 1989, Kenyan President Daniel arap Moi set fire to a 12-ton pyre of elephant tusks valued at nearly $3 million that had been confiscated from poachers, to demonstrate Kenya's dedication to ending the trade in ivory (Simmons and Kreuter 1989, 46; Leakey and Lewin 1996, 201).
30. East Africa's parks lost 56 percent of their elephants between 1979 and 1989, and outside the parks 78 percent disappeared. Fourteen non-southern African countries lost more than 60 percent of their elephant populations between 1979 and 1994 (McPherson and Nieswiadomy 1998, 1). By contrast, Botswana's elephant population rose from 20,000 to 51,000 between 1979 and 1989, and Zimbabwe's elephant population increased from 30,000 to 43,000 over the same period (Simmons and Kreuter 1989, 46) and stands at 60,000 to 80,000 today (McPherson and Nieswiadomy 1998, 2). Botswana and Zimbabwe voted against the ban.
31. Botswana devotes almost 18 percent of its land to wildlife, Tanzania more than 13 percent, and Zimbabwe just under 13 percent. The corresponding figure for the United States is 8 percent (Kreuter and Simmons 1995, 157).
32. Land is communally owned in much of Africa.
33. Although Leakey claims that the world price of ivory fell from $120 per pound to $4 after implementation of the ban (Leakey and Lewin 1996, 210), Kreuter and Simmons (1995, 158–59) cite contrary evidence of rising ivory prices.

REFERENCES

Akerlof, George A. 1980. A Theory of Social Custom, of Which Unemployment May Be One Consequence. *Quarterly Journal of Economics* 94 (June): 749–75.

Anderson, Terry L., and Peter J. Hill. 1995. From a Liability to an Asset: Developing Markets for Wildlife. Introduction to *Wildlife in the Marketplace*, edited by Terry L. Anderson and Peter J. Hill, pp. xi–xv. Lanham, Md.: Rowman and Littlefield.

Barbier, Edward B., Joanne C. Burgess, Timothy M. Swanson, and David W. Pierce. 1990. *Elephants, Economics and Ivory*. London: Earthscan Public.

Bikhchandani, Sushil, David Hirshleifer, and Ivo Welch. 1992. A Theory of Fads, Fashion, Custom, and Cultural Change in Informational Cascades. *Journal of Political Economy* 100

(October): 992–1026.
Bonner, Raymond. 1993. *At the Hand of Man: Peril and Hope for Africa's Wildlife*. New York: Knopf.
Cordato, Roy E. 1997. Market-Based Environmentalism and the Free Market: They're Not the Same. *The Independent Review* 1 (Winter): 371–86.
Devenow, Andrea, and Ivo Welch. 1996. Rational Herding in Financial Economics. *European Economic Review* 40 (April): 603–15.
Gibson, Clark C., and Stuart A. Marks. 1995. Transforming Rural Hunters into Conservationists: An Assessment of Community-Based Wildlife Management Programs in Africa. *World Development* 23: 941–57.
Higgs, Robert. 1987. *Crisis and Leviathan: Critical Episodes in the Growth of American Government.* New York: Oxford University Press.
Hill, Peter J. 1997. Market-Based Environmentalism and the Free Market: Substitutes or Complements? *The Independent Review* 1 (Winter): 387–96.
Kaempfer, William H., and Anton D. Lowenberg. 1992. Using Threshold Models to Explain International Relations. *Public Choice* 73 (June): 419–43.
Kremer, Michael, and Charles Morcom. 1996. Elephants. NBER Working paper no. 5674, July.
Kreuter, Urs P., and Randy T. Simmons. 1995. Who Owns the Elephants? The Political Economy of Saving the African Elephant. In *Wildlife in the Marketplace*, edited by Terry L. Anderson and Peter J. Hill, pp. 147–65. Lanham, Md.: Rowman and Littlefield.
Kuran, Timur. 1987a. Preference Falsification, Policy Continuity, and Collective Conservatism. *Economic Journal* 97 (September): 642–65.
———. 1987b. Chameleon Voters and Public Choice. *Public Choice* 53: 53–78.
———. 1989. Sparks and Prairie Fires: A Theory of Unanticipated Political Revolution. *Public Choice* 61 (April): 41–74.
———. 1991. The East European Revolution of 1989: Is It Surprising that We Were Surprised? *American Economic Review* 81 (May): 121–25.
———. 1995. *Private Truths, Public Lies: The Social Consequences of Preference Falsification.* Cambridge, Mass.: Harvard University Press.
Leakey, Richard, and Roger Lewin. 1996. *The Sixth Extinction: Patterns of Life and the Future of Humankind.* New York: Anchor Books.
Margolis, Howard. 1990. A Note on Social Equilibrium. Working paper. University of Chicago, Graduate School of Public Policy Studies, March.
———. 1998. A Formal Model of Ethnic Conflict. Working paper. University of Chicago, Graduate School of Public Policy Studies, March.
McPherson, Michael A., and Michael L. Nieswiadomy. 1998. African Elephants: The Effect of Property Rights and Political Stability. Working paper. Department of Economics, University of North Texas, July.
Nelson, Robert H. 1997. Does "Existence Value" Exist? Environmental Economics Encroaches on Religion. *The Independent Review* 1 (Spring): 499–521.
Simmons, Randy T., and Urs P. Kreuter. 1989. Herd Mentality: Banning Ivory Sales Is No Way to Save the Elephant. *Policy Review* 50 (Fall): 46–49.
't Sas-Rolfes, Michael. 1998. Who Will Save the Wild Tiger? PERC Policy Series No. PS-12. Political Economy Research Center, Bozeman, Mont., February.
Uhlaner, Carole Jean. 1989. "Relational Goods" and Participation: Incorporating Sociability into a Theory of Rational Action. *Public Choice* 62 (September): 253–85.
U.N. Legalizes Sale of African Ivory to Japan. *Seattle Times*, February 10, 1999.
Vorhies, Frank. 1996. Economics, Policy, and Biodiversity. Paper presented at the Economics, Policy, and Natural Resource Management Southern African Regional Workshop, Pretoria, South Africa, September.
Vorhies, Frank, and Deborah Nolte Vorhies. 1993. Managing the African Elephant as a Resource. Working paper. EcoPlus, Johannesburg, South Africa.

Witt, Ulrich. 1989. The Evolution of Economic Institutions as a Propagation Process. *Public Choice* 62 (1989): 155–72.

Acknowledgments: We are grateful for helpful comments received from Charles Becker, Romie Tribble, Jr., and other participants in a session of the Southern Economic Association meetings in Atlanta, Georgia, held in November 1997. Funding for Professor Lo wenberg was provided by California State University, Northridge.

PART IV

Entrepreneurship, Property Rights, and Land Use

11

Free Riders and Collective Action Revisited

RICHARD L. STROUP

The free-rider problem associated with public goods was recognized by David Hume, even before the time of Adam Smith's writings. Each citizen who can enjoy the benefit of a public good has an incentive to try to lay the whole burden of provision on others, whenever the exclusion of nonpayers is very costly or impossible. Hume recommended in 1739 that government provide the goods in question, such as bridges (Musgrave 1985). Two and a half centuries later, economists typically recommend a similar solution (Arrow 1970; Atkinson and Stiglitz 1980; Auerbach and Feldstein 1985; Cornes and Sandler 1986; Nicholson 1989; Samuelson 1954).

The public-provision prescription is seldom questioned, although today's economists and policy analysts, having been exposed to public-choice logic and empirical analysis, do recognize that government is an imperfect institution. Government provision of public goods, it is conceded, will not be free of problems (Shleifer 1998). For example, the rational ignorance of voters is widely recognized, and so too is the disproportionate influence of organized special-interest groups. Lobbyists and their campaign contributions are the facets of the problem that receive the most attention.

Even though problems associated with the imperfection of government are commonly recognized, it is seldom noted explicitly that the root of those problems is precisely the same as that of the free-rider problem associated with private production of public goods. The formation and successful control of a government program in the public interest, for any reasonable definition of that nebulous term, are themselves public goods. Who will pay the price in time, effort, and other lobbying costs to originate a program and to control it in the interest of the general public? Does efficiency in serving the public have a constituency? Adam Smith pointed out long ago that no individual can be expected to seek the public interest. Markets work to exhaust the gains from trade and cooperation because each individual has an interest in finding and capturing any and all such gains. Of course, when free riders can enjoy a public good without payment or trade, production and the potential gains from it may never occur. Efforts to originate government programs and to control them in the public interest are no different. As Gordon Tullock (1971) put it, "The public decision-making process is a procedure for generating a public good; and the persons involved in it, whether they are the voters, judges, legislators, or civil servants, all can be expected to treat it as any other public good" (917).

Tullock recognized the point very clearly.[1] He perceived the likelihood of shirking, in the form of spending too little time and effort in researching the issue subject to public decision and in the form of utilizing the decision maker's own preferences rather than the interest of the public in general. For these reasons, public decisions will not necessarily promote the well-being of the general public. Indeed, a program justified in the name of producing a public good may in fact be utilized by special interests to help only themselves, harming the public in the process. Analyses of programs gone awry are common, but the free-rider problem that surely causes many of these problems is seldom mentioned. Tullock's observations, though published in a prominent economics journal, seem not to have made a large impression on policy-relevant discussions by economists since that time. The problem of free riding seems to this day to be discussed almost entirely in the context of *market* failure.

One constructive use of Tullock's basic insight can be in systematic side-by-side comparisons of the incentive problems built into private provision of public goods, on the one hand, and those built into every case of public provision of any good (and of public regulation), on the other. This is the sort of comparison called for by James Buchanan (1987), by Kenneth Shepsle and Barry Weingast (1984), and by Neil Komesar (1994). A careful and realistic evaluation of the incentives facing participants in the public policy process—that is, of the free-rider problem inherent in all politically directed public activity—would be useful in comparing alternative institutions whether the output of the public policy were a public good or not, and in determining whether a change in policy might provide a superior result.

PUBLIC PROVISION OF PUBLIC GOODS: SOLVING THE FREE-RIDER PROBLEM OR EXPANDING IT?

Economists who discuss public goods and the free-rider problem use many examples. Among the most common are national defense, public health measures such as mosquito abatement, and roads and bridges (Varian 1984, 253; Atkinson and Stiglitz 1980, 486–87; Nicholson 1989, 727). Each is subject to the most common problem cited: the lack of any producer's ability to exclude beneficiaries in a low-cost fashion, a condition that generates the free-rider problem, resulting in an expectation that the good will be underprovided. The standard solutions offered are government provision of the good, through purchase or production, or government subsidization of its private provision. Each, however, introduces many free-rider problems of its own.

The production of goods and services, whether in the private sector or by the government, is a complex undertaking. In meeting the demand for a good, a starting point is to define specifically the quantity and the qualities of the good to be provided.

Decisions on What to Produce and How to Produce It

Consider national defense. What is the proper type and level of national defense? What sort of fleet should the navy build and support, and how large should it be? Where should the ships and their support be based in order to provide the best defense for a given expenditure? Similar questions must be answered about the air force and its airplanes and about the army and its forces.

Each of these decisions has intensely important ramifications for military suppliers and for departments within the military bureaucracy. The interest of each of these groups is likely to be well represented in the decision process, both directly and in lobbying. Members of Congress and relevant members of the current administration will be strongly lobbied on behalf of each supplier group and probably by each bureaucratic department.[2]

Who, on the other hand, will persistently lobby for the diffuse interest of the general public by, for example, identifying and then lobbying for the most cost-effective set of resources, or deployment of those resources, to deter potential foreign aggressors? Citizens who are employed by defense contractors or who live in an area where a defense contract is locally important may become active when their specific issue is being debated. For them, economic benefits for themselves and their localities will loom large, whereas the search for cost-effectiveness will surely be secondary. Certainly the firms, the chambers of commerce, and other organized interests will lobby intensively, often with large budgets to do so, as the narrow issues of specific interest to them are being considered.

Most citizens and most groups, however, are not apt to be involved when a specific defense procurement or deployment issue is settled. Each may well recognize that the many decisions on defense procurement and deployment are important; but for each person, the cost of learning about the issue and becoming involved is borne privately by the citizen, whereas the payoff for making better decisions in the service of cost-effectiveness is spread among the general public. The classic free-rider problem presents itself very strongly, even with respect to decisions about the goals of public provision of national defense.

The same problem appears in complex decisions on government provision of mosquito abatement. Which wetlands should be treated? Should aerial spraying of mosquitos be utilized? Which chemical pesticides are acceptable to reduce mosquito populations? Which is best in each situation? Again, those with income directly at stake or with strong views on these questions will probably be heard. But by whom will the general public's diffuse interests be strongly and persistently represented? Each ordinary citizen is likely to act as a free rider when the level and description of government provision of a public good is being decided. And such decisions are only the beginning of the government provision process.

CONTROL OF THE PRODUCTION PROCESS

A large literature in economics shows that in the private sector, where minimizing the cost of producing anything (of a chosen quantity and description) is necessary to maximize the profits of a firm, organizing for cost-minimizing production nonetheless remains a complicated process. Thrainn Eggertsson, in his review and extension of this literature, lists the following activities that typically must be undertaken in the production of a good in a modern economy; each is applicable to production or regulation by the government as well:

1. The search for information about the distribution of price and quality of commodities and labor inputs, and the search for potential buyers and sellers and for relevant information about their behavior and circumstances [Who might be the least-cost provider of each good and service needed?]
2. The bargaining that is needed to find the true position of buyers and sellers when prices are endogenous [What prices can actually be negotiated on behalf of the public?]
3. The making of contracts [What specifications and stipulations should be included to reduce future performance problems while controlling costs?]
4. The monitoring of contractual partners to see whether they abide by the terms of the contract [Are all parties to the contract complying with its terms?]
5. The enforcement of a contract and the collection of damages when partners fail to observe their contractual obligations [When should contract

problems be renegotiated? When should they be taken to court instead? How should these matters be handled, once decisions are made?]

6. The protection of property rights against third-party encroachment [When should the rights of suppliers and those regulated be respected? How should they be protected, and at whose cost?] (Eggertsson 1990, 15)

The successful performance of each of these activities requires thoughtful and diligent action, good judgment, and intelligent responses to the recognition of mistakes and altered conditions. How will each unit of a bureau be organized to allow each decision maker the latitude to operate intelligently in changing circumstances, while reducing the ability to shirk or to act opportunistically under pressure on behalf of external interests or of the bureaucratic unit's own narrow concerns or even venal interests?[3]

Entrepreneurship is a key to efficiency in a world where technology and relative prices change rapidly. Private firms are constantly adjusting their own organizations to handle changing problems and opportunities. For them the carrot is profit; the stick is failure to survive under competition. Feedback to them from the product market tends to be constant. Each buyer has a personal stake in monitoring the product in order to seek better products and to avoid paying for inferior goods. For managers of publicly traded corporations, there is constant feedback from the stock market as well, concerning both current practices and decisions about the future of the corporation. In the capital market, each investor providing capital to a firm or holding its stock has an incentive to monitor that firm's policies (or to pay a specialized investment advisor to do so) in order to buy a larger stake if the firm's prospects brighten or to bail out by selling ownership rights if adverse events are detected. In the public sector, however, there is no such personal incentive for efficient decision making and typically no such monitoring and feedback. Each voter has a voice, but exit from tax payment is usually difficult or costly, even if the taxpayer studies and disapproves of the government agency's policies or its products.

Who will oversee the efficiency of day-to-day and year-to-year operation of each unit of government in the public interest? Ultimately, voters are in charge. But the individual as a voter, unlike the same individual as a buyer of a product or a share of corporate stock, can seldom benefit *personally* from obtaining better knowledge and applying it more diligently. The individual buyer in a market can easily exercise an option such as selling stock or buying a different product, employing information gathered on the details of how a product will work or how a private corporation is being managed.

The same individual as voter has a more serious problem. Rational voter ignorance is a well-known and well-documented phenomenon. Voters tend to be free riders on the vigilance of others on any issue that is not of unusually concentrated personal interest to them. And there is no capital market that constantly assesses a government program's current performance or the plans for the future. Instead there is constant pressure from special interests, tending to mold each program and its operation to serve their narrow interests.

Some of these problems can be diminished by partially "privatizing" provision, using tax finance and low-bid private sources of supply, or by localizing the provision of goods, so that citizens can more easily choose to "vote with their feet" (Tiebout 1956) in search of better government outcomes. Neither of these tactics eliminates the free-rider problem in the public sector, but together they can reduce it by introducing competition enforced by the possibility that, periodically at least, buyers can exit.

Control of Regulators as a Public Good

We have seen how, when a bureau or a government program is established and given access to the public purse through the budget process, control of the program in the public interest is a public good and subject to the free-rider problem. When the bureau is also given broad regulatory powers, control can be even more difficult. In addition to setting the specific program goals and the general methods of achieving those goals within a budget, regulators are often able to utilize methods that increase costs to others but do not force the spending of the bureau's own funds.

We can expect narrow bureaucratic interests, backed by specific constituencies and afflicted by "tunnel vision,"[4] to attempt to expand their activities well beyond those desired by the broader public. The bureaucrats and their clients see and value the benefits of their program, but do not see, or have no reason to value, the alternative goals given up to gain incremental advantages to their own program. Regulators who can order costly measures to be undertaken, without the check of a budget, will also utilize costly methods, failing to economize except to enhance their own program. No check on cost is present, other than political pressure. Competition of the sort that eliminates waste seldom exists in government. Further, the presence of bureaucratic rule, or "red tape," which is needed because the zeal of an owner or a residual claimant is not present to force cost control, further reduces the ability of bureau leaders to streamline procedures and reduce costs. And as noted previously, because the control and monitoring of government action in the public interest are themselves public goods, the potential controllers and monitors—voters and their elected representatives—tend to free ride rather than to zealously protect the public interest.

As a result, government programs often impose very high costs on society even when the programs imposing the costs produce little in the way of demonstrable, intended benefits.[5] In fact, a bureau can go so far as to cause negative results for the overall goal of its own program. One example is the program to enforce the Endangered Species Act.

The Endangered Species Act

Animal species listed as endangered or threatened are generally protected by the U.S. Fish and Wildlife Service (FWS) under the Endangered Species Act (ESA). The ESA as now written and interpreted gives the FWS the responsibility and authority to preserve each listed species and its habitat without regard to cost. Any habitat, public or private, declared to be important to threatened or endangered species is in effect placed under constraints dictated by FWS project biologists, without regard to cost to the land's private or public owner and without compensation to the owner. FWS habitat decisions may be extremely costly to private resource owners or to other agency missions, but the cost is exempt from the deliberations of the congressional budget process. No compensation is paid. Congress, which took the credit for helping to save species when it approved the ESA many years ago, declines to monitor the program's true costs and benefits. Some members say that they had no idea, when they voted for passage, that the costs for landowners would be so great. But few have even attempted to revise the ESA. Most are, understandably, riding free while taking credit wherever they can.

At least three consequences flow from this arrangement. The first, which was (according to the courts) intended by the Congress, is that there is no limit to the land that may be devoted to the biologist's mission to preserve land from any use other than as habitat for a listed species. The second consequence, presumably unintended, is that land, logically treated by the project biologist as having zero price, is substituted for other, on-budget factors of production, such as professional services, that could be utilized to find less costly conservation plans. The cost of other factors of production would be borne by the FWS, whereas land costs are borne by the landowners. In effect, factor prices for the production of habitat protection are severely distorted. Tunnel vision precludes each program leader from seeing and fully valuing the total social cost of the program's elements. The project managers also ride free. Control of the program in the interest of the general public is, as always, a public good.

The third result, exacerbated by the cost imposed on landowners by the first two, is that prior to being contacted by the FWS, private landowners have an incentive to manage their land in such a way that listed species will not find the land attractive. When no members of the species are present on or near his land, a landowner can avoid bearing the direct costs of the ESA. Landowners can also gain by cooperating with their neighbors to learn and teach techniques for preemptive habitat destruction and even by applying social pressure on those nearby to apply the habitat-modifying techniques. These actions occur even when, as is frequently the case, the species themselves do no harm and would have been welcomed on the land absent the uncompensated financial harm imposed by the ESA.

There is increasing evidence (Stroup 1997; Lueck and Michael 1999) that preemptive habitat destruction is important. Lueck and Michael (1999), for example, find statistically that when a colony of the listed red-cockaded woodpecker is near a forest, costly measures are taken by the landowner to make that forest less attractive to other colonies of red-cockaded woodpeckers. The birds prefer old-growth forests for their nests, so the landowner is more likely to harvest the trees before they reach the normal harvest age.

The unintended negative effect of preemptive habitat destruction by landowners has been noted by close observers of habitat preservation for some time. The effect could be mitigated by putting land-use acquisition on budget for the FWS. That would cause the program biologists and their supervisors to face market prices for land, restoring both the total cost and the true land cost as a factor of production to program decision makers. Less costly and more effective strategies for habitat preservation would be encouraged.

A second unintended negative effect of the ESA program is serious harm to voluntary programs that have been successful over the past several decades. Long before the ESA, wood ducks on the eastern flyway were saved from threatened extinction partly by the voluntary placement of nest boxes by landowners. Eastern bluebirds were saved in a similar way. More recently, the preservation of wetlands habitat in the form of prairie potholes has been effected very economically through the solicitation of landowners' cooperation, again expanding available habitat for important wildlife. These and many other successes would be exceedingly expensive, if not impossible, now in the case of listed species, due to landowners' fear of the uncompensated takings that might easily result under the ESA.

Changing the ESA to avoid uncompensated takings would eliminate both of these unintended negative effects. However, environmental groups oppose such a change, perhaps because the current arrangement is a powerful tool for another purpose: to stop the use of certain tracts of land for timber harvest, development, or even intensive agriculture. The free-rider problem severely reduces the incentive of most voters and elected politicians to carefully monitor the actual results of the ESA program and to demand changes that would make it more effective and minimize its cost.

State Forest Management

When the benefits of the good produced in the public sector are concentrated on a few beneficiaries and thus have less of the character of a public good, the beneficiaries have a strong incentive to monitor the program. Such close monitoring occurs, for example, in state land programs that are intended to make money for the state's school system.

State land programs that sell timber have been shown to operate much more economically than the U.S. Forest Service program, which is supposed to produce benefits of many kinds for the general public in the national forests. Donald Leal (1996) has shown that the several state programs he examined make money, while the Forest Service timber-sale programs on similar sites nearby operate at a substantial loss. Leal also demonstrated that in Montana, which conducts environmental audits on both state and federal timber lands, the state forest management was environmentally sounder as well.

State programs that are ordered simply to make money for the schools are easier to monitor than their federal counterparts. One need only check: did they make money or not? And watching them are specific interest groups that capture the benefits when the state programs operate more efficiently. The gains they make are concentrated on an organized beneficiary, the government school system. The good the state forest programs produce, in other words, has less of the public-good aspect than the federal program, whose proceeds go into the federal treasury. It may seem ironic that government provision of goods that are less collectively enjoyed and more like private goods should logically be better monitored and thus more efficient, but Leal's evidence is consistent with that logic.

PUBLIC GOODS AND PUBLIC BADS: THE POTENTIAL FOR HARMS BEYOND RESOURCE WASTE

The fact that control of a government program in the public interest is a public good implies that groups with far narrower interests may gain control. The actions of such groups not only cause resource waste (from the general public's viewpoint) in the presence of voter shirking because of free riding in the political and bureaucratic processes; but control by those with narrow agendas may indeed cause positive harm. That outcome can easily be produced by regulatory agencies, as illustrated by the case of the FWS and the ESA. Similar outcomes may result from public goods provision.

The classic public good is national defense, as provided by a large standing army. It is intended to protect citizens against threats from hostile foreign nations. But it may be used in ways that increase, rather than reduce, the likelihood of citizens' dying at the hands of hostile foreign nationals. One of the controversial questions surrounding the Vietnam War is whether tens of thousands of Americans would have died at the hands of foreigners if the large standing military force had not been available for quick deployment to serve the aims of the administration sending them. More recently, one wonders whether the likelihood of terrorist attacks in the United States might be increased by certain deployments of the U.S. military abroad. If military forces were smaller, that particular risk would be reduced. Without far more study than we will ever give to the question, most citizen-voters, including this writer, cannot know whether a larger standing army would increase or decrease the risks of harm at the hands of foreigners. In the light of Tullock's basic insight, who would be expected to answer this question on behalf of the general public (if a credible answer is even possible) and then to lobby strongly—perhaps against the interests of military suppliers—for adoption of the policy that best serves the general public?

Production of public goods to reduce natural risks can also result in increased risk, when free riding with regard to control of the program leaves decisions in the hands of narrowly interested parties. Mosquito abatement is a case in point. Is a reduction of mosquito populations worth the risk of extensive aerial spraying of the sort that is so controversial in California? Is it worth disturbing a wetlands ecology? Few citizens have sufficient knowledge to vote intelligently on the issue, even if given a chance in a referendum. Public provision of mosquito abatement, however, unlike private abatement, generally means that the official decision maker is not personally liable if that official's program damages affected citizens. The standard of care is likely to differ as a result of public provision.[6] Again, fighting one danger by establishing pubic provision of a good may worsen other dangers. The free-rider problem among voters, and the poor oversight that results, makes public provision of "protection" from a specific risk potentially dangerous.[7]

IMPLICATIONS OF THE FREE-RIDER PROBLEM FOR PUBLIC POLICY AND VOLUNTARY ACTION

The free-rider problem and its resulting reductions in the efficiency and effectiveness of government actions have important implications. Better understanding of these matters could lead to improvement in institutional choice by society generally as well as by individuals as they choose which institutions to support voluntarily.

Some goods with public-good aspects are provided by a variety of institutions—private individuals, private clubs, and government agencies. For example, wilderness or habitat for endangered species can be provided by individuals, clubs such as the National Audubon Society or the Sierra Club, and government agencies.

In any case, provision requires individuals who are sufficiently well paid or passionate to work hard or make other sacrifices to bring about the production of the public good. Even government provision requires voluntary private sacrifices. For example, establishing public-sector habitat protection in the first place requires strong lobbying by clubs such as the Audubon Society or the Sierra Club. To maintain such programs, lobbying must continue; otherwise the programs will be defeated or distorted by those seeking other goals from the federal budget and federal lands.

The clubs mentioned in fact spend most of their time and money in lobbying for government provision. Might their efforts be more productively spent organizing and campaigning for individual private landowners to help or, instead, on a campaign seeking help from individuals for the private club's efforts to obtain and administer such lands themselves as owners?

The theory of clubs suggests, and observation confirms, that club decisions tend to be made by like-minded individuals who have been drawn together by shared preferences.[8] In clubs, decisions about how the good should be defined and managed are made by like-minded individuals rather than by a median voter in a public election, by politically sensitive (and compromising) politicians, or by bureaucrats under political control. In contrast, government provision must, at least formally in a democracy, be "controlled" by voters, few of whom are likely to be as passionate and motivated and therefore as knowledgeable about the good to be provided as are the self-selected, dues-paying members of a club.

Expanding the responsibility for provision to the level of the general public gains tax support from all taxpayers only by trading off the ability to retain the focus and intensity of like-minded club members. Whether the trade-off is worthwhile can be debated. However, there is some empirical evidence, to be discussed later, that suggests voluntary provision is qualitatively different and can be qualitatively superior.

Each institutional approach requires at least partial solution of the free-rider problem to obtain cooperation among enthusiasts. Each requires work and sacrifice. Government efforts may be the most effective means for some individuals and some groups, who are relatively good at lobbying and other forms of political action. For them, formation of a club that amounts to a special-interest lobbying group may yield the highest return on their resources.

Yet even when they successfully elicit government action and the provision of the good is assigned to a bureaucratic unit, continuing provision is not guaranteed. Defense of the bureau's budget and control of its actions in the service of specific desired goals are never-ending challenges. Public-sector victories are never bought—only rented. Whether majoritarian interests (and the median voter, who typically is uninformed about provision of the good) prevail or minoritarian special interests (such as those of contracting providers) prevail, the battle is never permanently won, for the courts have consistently held that one legislature cannot bind the next. In the words of one environmental leader, whose specialty is preserving land for wild habitat, "You know what I like? A deed in the courthouse."[9]

The preceding analysis suggests that for an enthusiast seeking to expand the provision of a public good, the same amount of organizational and fund-raising effort might be more successful if applied directly to seeking private provision of that good. The analysis does not indicate that such will necessarily be the case. Depending on the circumstances, the shortfall in payments from some who benefit, and thus the shortfall in provision, caused by the free-rider problem in private provision will be a greater or smaller problem than the reductions in the value (reduced quantity and quality delivered by public provision) that result from the multiple free-rider problems in the public sector. Even with its tax-revenue advantage, however, public provision may still deliver a lesser amount of the good, as the following example illustrates.

WHEN PROVISION OF RELIGION BECAME PRIVATE

As a case study, consider the provision of religious services. At least from Adam Smith's time forward, some have maintained that the church provides positive benefits to the community, beyond those provided to churchgoers. In this view, people who participate in religion become better neighbors, more upstanding citizens, and more honest trading partners (Olds 1994, 280). Such public-good aspects of religious observance, it has been claimed, justify taxpayer subsidies. Even today, many nations have an established (that is, state-supported) church.

Two centuries ago, when Connecticut and Massachusetts were considering disestablishment (ending the subsidy to each state's established church), the argument was made that without taxpayer support, religion would wither. Kelly Olds found, however, that when disestablishment occurred in Massachusetts and Connecticut and all subsidies were abolished, indicators of religious activity rose rather than fell. The statistical evidence indicates that both church membership and the demand for preachers rose rapidly (Olds 1994, 277). Many churches prospered, and their methods of finance varied. The services provided undoubtedly changed. Church leaders, after all, now had to appeal more to individual congregations and less to legislators and state voters as a group. Whatever the reason, the provision of church services increased substantially. Church membership in the United States as a whole rose for nearly all of the nineteenth and twentieth centuries (Iannaccone 1998, 1468). Further, the United States, without an established church, has greater church attendance and a greater percentage of citizens proclaiming themselves to be religious than any of the industrialized nations that provide public funding for their religious establishment (Iannaccone, Finke, and Stark 1997).

Evidently, at least in some cases, private provision allows and encourages appeals to more subgroups than does government provision, which is bound by the need for support by the median voter or by the most powerful interest groups. Indeed, public provision may undercut private support for an activity by replacing it, as it seems to do dollar for dollar in the case of support for university education (Becker and Lindsay 1994). In any event, increased public provision of a good is not an unalloyed benefit to the mission supported, and may on balance harm that mission.

CONCLUSION

Gordon Tullock first emphasized the importance of the free-rider problem in the public sector. Recognizing this problem should alter our expectation that when the private sector provides a good "imperfectly," citizens have the option to utilize the government in order to achieve an ideal delivery of the good. The public-sector approach is one candidate among several as an institutional means of providing a particular public good. However, the pervasiveness of the free-rider problem in the public sector has several implications:

- Establishing a government program to provide a public good requires voluntary private effort. Keeping the program funded requires continued voluntary private effort.
- Control of a government program in the general public's interest requires voluntary private effort in competition with the many narrow interests that will seek to control the program.
- When taxpayers fund a program, the product will differ from a privately provided product when the preferences of voters as a whole differ from the preferences of those with the greatest zeal and the strongest interest in

the activity. The latter group could control a private effort but must compromise with those having less interest when public provision is utilized.

- Private clubs and other private providers have more freedom to choose their methods, and they have full financial responsibility for the cost, so we can expect them to be more efficient providers.
- In light of the preceding points, one expects that private provision may produce more or better-quality public goods. The evidence conforms to that expectation in specific cases such as provision of religious services.

No human institution works perfectly. When a decision about provision of a specific good in a specific case is to be made, each available institutional alternative should be realistically compared.[10] To look at only one option and point out its weaknesses, and hence to conclude that another option must be superior, is exactly the sort of mistake a student makes by supposing that products ought to be manufactured according to absolute advantage rather than comparative advantage. The same factors that make one sector weak may simultaneously weaken an alternative sector. When an externality involves many people, then market provision may be problematic, but so may be government provision. The understanding that "public decisions are public goods," first brought to light by Tullock, and the related considerations discussed in this chapter should be borne in mind as institutional arrangements are compared.

NOTES

1. Tullock's chapter spurred my efforts in the present chapter to push the analysis further and to make it somewhat more concrete.
2. The rent-seeking literature points out the costs to society of competition among special interests to control various aspects of public policy. See Rowley, Tollison, and Tullock 1988.
3. Niskanen (1971) has shown that in certain circumstances the bureaucratic unit may, due to a monopoly of cost information, be able to put the legislature, at budget time, on an all-or-none demand curve so as to maximize the unit's budget. Today that case is generally thought to be an extreme one, even by Niskanen (see Blais and Dion 1991); but the tendency, and some ability to indulge it, may still exist.
4. "Tunnel vision" is the term utilized by U.S. Supreme Court Justice Stephen Breyer (1995) to explain why each regulatory agency tends to go well beyond any social optimum in its activities. It does not experience the opportunity costs of its actions, and it avoids many ordinary constraints in the exercise of regulatory powers granted by the legislative branch. For examples of how tunnel vision affects government programs and program costs in the context of potential risks from toxic chemicals, see Stroup and Meiners 2000.
5. For a concise overview of the high-cost, low-benefit nature of much federal environmental regulation, see Crandall 1992.
6. Precisely this problem helped to generate the Love Canal hazardous-waste problem at Niagara Falls, New York, and led to the creation of the Superfund program. Recognition of potential

liability had caused the chemical company to carefully isolate the buried wastes. When the local school board purchased the property, it was explicitly warned of dangers from the waste, but proceeded to utilize "tunnel vision" in managing the land, ignoring the chemicals and, by digging through the containment walls, allowing them to seep out (Zuesse 1981; Stroup 1996, 5–6).

7. Editor's note: Nothing better illustrates this truth than the pretense of protection afforded by the regulatory programs of the U.S. Food and Drug Administration. See, for example, Dale H. Gieringer, "The Safety and Efficacy of New Drug Approval," Cato Journal 5 (Spring–Summer 1985): 177–201; and Robert Higgs, ed., Hazardous to Our Health? FDA Regulation of Health Care Products (Oakland, Calif.: Independent Institute, 1995).
8. The theory of clubs originated with Buchanan 1965; for a discussion, see Mueller 1989.
9. Brent Haglund, executive director of the Sand County Foundation, as quoted by Anderson and Leal (1997, 52).
10. The case for comparative institutional analyses is made in an important book by Neil Komesar (1994). He stresses that analysis should center on comparisons of how concentrated the payoffs are in each case and how many parties are importantly involved.

REFERENCES

Anderson, Terry L., and Donald R. Leal. 1997. *Enviro-Capitalists.* Lanham, Md.: Rowman and Littlefield.

Arrow, Kenneth J. 1970. The Organization of Economic Activity: Issues Pertinent to the Choice of Market versus Nonmarket Allocation. In *Public Expenditures and Policy Analysis,* edited by Robert Haveman and Julius Margolis. Chicago: Markham.

Atkinson, Anthony B., and Joseph Stiglitz. 1980. Public Goods and Publicly Provided Private Goods. *In Lectures in Public Economics.* New York: McGraw-Hill.

Auerbach, Alan J., and Martin Feldstein, eds. 1985. *Handbook of Public Economics.* Vols. 1 and 2. New York: North-Holland.

Bagnoli, Mark, and Barton L. Lipman. 1992. Private Provision of Public Goods Can Be Efficient. *Public Choice* 74: 59–78.

Becker, Elizabeth, and Cotton M. Lindsay. 1994. Does the Government Free Ride? *Journal of Law and Economics* 37 (April): 277–96.

Blais, Andre, and Stephane Dion, eds. 1991. The Budget-Maximizing Bureaucrat: *Appraisals and Evidence.* Pittsburgh, Pa.: University of Pittsburgh Press.

Breyer, Stephen. 1995. *Breaking the Vicious Circle.* Cambridge, Mass.: Harvard University Press.

Buchanan, James M. 1965. An Economic Theory of Clubs. *Economica* 32 (February): 1–14.

———. 1987. Constitutional Economics. In *The New Palgrave: A Dictionary of Economics,* edited by John Eatwell, Murray Milgate, and Peter Newman, vol. 1, pp. 585–88. London: Macmillan.

Cornes, Richard, and Todd Sandler. 1986. *The Theory of Externalities, Public Goods, and Club Goods.* New York: Cambridge University Press.

Crandall, Robert. 1992. *Why Is the Cost of Environmental Regulation So High?* Policy Study 110, Center for the Study of American Business. St. Louis: Center for the Study of American Business.

Davis, Otto A., and Morton I. Kamien. 1970. Externalities, Information, and Alternative Collective Action. In Public Expenditures and Policy Analysis, edited by Robert Haveman and Julius Margolis. Chicago: Markham.

Eggertsson, Thrainn. 1990. *Economic Behavior and Institutions.* Cambridge: Cambridge University Press.

Haveman, Robert H., and Julius Margolis, eds. 1970. *Public Expenditures and Policy Analysis.* Chicago: Markham.

Iannaccone, Laurence R. 1998. Introduction to the Economics of Religion. *Journal of Economic Literature* 86 (September): 1465–96.

Iannaccone, Laurence R., Roger Finke, and Rodney Stark. 1997. Deregulating Religion: The Economics of Church and State. *Economic Inquiry* 35 (April 1997): 350–64.

Komesar, Neil. 1994. *Imperfect Alternatives: Choosing Institutions in Law, Economics, and Public Policy.* Chicago: University of Chicago Press.

Layard, P. R. G., and A. A. Walters. 1978. *Microeconomic Theory.* New York: McGraw-Hill.

Leal, Donald R. 1996. *Turning a Profit on Public Forests.* PERC Policy Series PS-4. Bozeman, Mont.: PERC.

Lueck, Dean, and Jeffrey Michael. 1999. Preemptive Habitat Destruction under the Endangered Species Act. Working paper. Montana State University, Bozeman, Mt.

McMillan, John. 1979. The Free-Rider Problem: A Survey. *Economic Record* 55: 95–107.

Mueller, Dennis. 1989. *Public Choice II.* Cambridge, Eng.: Cambridge University Press.

Musgrave, R. A. 1985. A Brief History of Fiscal Doctrine. In *Handbook of Public Economics,* edited by Alan J. Auerbach and Martin Feldstein, vol. 1, pp. 1–59. New York: North-Holland.

Nicholson, Walter. 1989. *Microeconomic Theory: Basic Principles and Extensions.* 4th ed. Chicago: Dryden Press.

Niskanen, William A., Jr. 1971. *Bureaucracy and Representative Government.* Chicago: Aldine-Atherton.

Oakland, William H. 1985. Theory of Public Goods. In *Handbook of Public Economics,* edited by Alan J. Auerbach and Martin Feldstein, vol. 2, pp. 485–535. New York: North-Holland.

Olds, Kelly. 1994. Privatizing the Church. *Journal of Political Economy* 102: 277–97.

Pommerehne, Werner W., Lars P. Feld, and Albert Hart. 1994. Voluntary Provision of a Public Good: Results from a Real World Experiment. *Kyklos* 47: 505–18.

Rowley, Charles K., Robert D. Tollison, and Gordon Tullock, eds. 1988. *The Political Economy of Rent-Seeking.* Boston: Kluwer.

Samuelson, Paul A. 1954. The Pure Theory of Public Expenditure. *Review of Economics and Statistics* 36 (November): 387–89.

———. 1955. Diagrammatic Exposition of a Theory of Public Expenditure. *Review of Economics and Statistics* 37 (November): 350–56.

Shepsle, Kenneth A., and Barry R. Weingast. 1984. Political Solutions to Market Problems. *American Political Science Review* 78: 417–34.

Shleifer, Andrei. 1998. State versus Private Ownership. *Journal of Economic Perspectives* 12 (Fall): 133–50.

Steiner, Peter O. 1970. The Public Sector and the Public Interest. In *Public Expenditures and Policy Analysis,* edited by Robert Haveman and Julius Margolis. Chicago: Markham.

Stroup, Richard L. 1995. *The Endangered Species Act: Making Innocent Species the Enemy.* PERC Policy Series PS-3. Bozeman, Mont.: PERC.

———. 1996. *Superfund: The Shortcut that Failed.* PERC Policy Series PS-5. Bozeman, Mont.: PERC.

———. 1997. The Economics of Compensating Property Owners. *Contemporary Economic Policy* 15 (October): 55–65.

Stroup, Richard L. and Roger E. Meiners, eds. 2000. *Cutting Green Tape: Toxic Pollutants, Environmental Regulation and the Law.* New Brunswick, N.J.: Transaction Publishers for The Independent Institute.

Tiebout, Charles M. 1956. A Pure Theory of Local Expenditures. *Journal of Political Economy* 64 (October): 416–24.

Tullock, Gordon. 1971. Public Decisions as Public Goods. *Journal of Political Economy* 79 (July–August): 913–18.

Varian, Hal R. 1984. *Microeconomic Analysis.* 2d ed. New York: W. W. Norton.

Zeckhauser, Richard. 1970. Uncertainty and the Need for Collective Action. In *Public Expenditures and Policy Analysis*, edited by Robert Haveman and Julius Margolis. Chicago: Markham.

Zuesse, Eric. 1981. Love Canal: The Truth Seeps Out. *Reason* 12 (February): 17–33.

Acknowledgments: The research assistance of Charles Steele and Matthew Brown is acknowledged, along with research support from PERC. An earlier version of this chapter was presented at A Public Choice Conference in Honor of Gordon Tullock and published in Public Choice Essays in Honor of a Maverick Scholar: Gordon Tullock, edited by Price V. Fishback, Gary D. Libecap, and Edward Zajac. Boston: Kluwer Academic Press, 1999.

12

Entrepreneurship and Coastal Resource Management

JAMES R. RINEHART AND JEFFREY J. POMPE

Many environmental problems along the coast are fundamentally no different from those in the interior parts of the country. Congestion, noise, and air and water pollution concern Dallas and Kansas City as much as they do Myrtle Beach or Miami. But beaches and estuaries are especially sensitive to economic development, and coastal areas of the United States are experiencing more rapid development than inland areas. The number of people living within fifty miles of the U.S. coastline rose from 61 million to 130 million between 1940 and 1988 (Long 1990, 6). More than 50 percent of Americans currently live within fifty miles of the shoreline, but the figure is projected to rise to 75 percent by 2010.[1] The number of nonresident tourists traveling to coastal areas has also grown substantially. Many, if not most, of our coastal environmental concerns stem from population growth pressure.

Poorly planned development in environmentally fragile coastal areas can cause shoreline erosion, polluted water, noisy and crowded surroundings, and extensive loss of trees, wetlands, fish and other wildlife. Population growth reduces land availability, encouraging developers to fill in marshes, destroying fish and animal habitats. The clearing of land destroys vegetation and trees, which increases runoff and ruins the overall natural beauty of the environment. Damage to streams, marshes, and marine life results from the use of pesticides, fertilizers, toxic chemicals, and other pollutants. Storm water runoff and effluent from sewage treatment facilities also cause trouble. The large amounts of nitrogen and phosphorous that pour daily into estuaries result in algae blooms that remove oxygen from the water, sometimes producing fish kills. Development also leads to the withdrawal of large amounts of water from aquifers, which causes salt infiltration and reduces water quality.

Shoreline erosion is a problem along much of the nation's coastline. The U.S. Army Corps of Engineers estimated in 1971 that 40 percent of the total shoreline of the lower forty-eight states was experiencing significant erosion (Morris 1992, 122). Wind and wave action associated with storms and high tides bring about natural erosion, but people effect much erosion, too. Damming rivers restricts the flow of eroded rock, which is the source of much of the sand on beaches. Destruction of sand dunes, sea oats, trees, and grasses removes natural protection, leaving the shore more vulnerable to ocean waves and currents, thus imperiling property and lives. Barrier islands, landforms that protect the mainland from the direct force of waves and storms, are especially susceptible to erosion. These dynamic islands are constantly eroding or accreting because of changing energy conditions (Leatherman 1988).

Erosion control techniques employing hard devices such as seawalls, groins, and jetties may actually accelerate beach loss rather than prevent it (Platt et al. 1992, 8). Another method of dealing with beach loss involves beach restoration via sand replenishment. This process, which trucks sand from inland pits or pumps sand from rivers or the ocean, is expensive and short-lived. Ocean City, Maryland, replaced nine miles of oceanfront sand at a cost of $51.2 million, only to see the bulk of it wash away during heavy storms in the fall of 1992. Most of the renourishment sand from a 1993 project at Folly Beach, South Carolina, that cost $12 million was washed away within two years ("Beach Programs" 1995). Similar stories abound.

MARKET FAILURE AND GOVERNMENT INVOLVEMENT IN THE COASTAL REGION

Many coastal resources exhibit characteristics of a common pool resource (CPR). This term refers to a resource for which exclusion of potential users is difficult and the use of which by one individual diminishes the amount available to others. When sufficient demand exists and access is not controlled, the result is the classic "tragedy of the commons" (Hardin 1968). Ocean fisheries, which tend to be overfished because users have little incentive to conserve the resource, exemplify this problem. Marshes, estuaries, beaches, and barrier islands all have aspects of CPRs and suffer from what some call market failure.

As estuaries and marshes are owned in common, no one prevents runoff from roadways and developed areas from polluting these critical waters. As development has expanded in coastal areas, increased pollution has reduced the productivity of these valuable resources. In Hilton Head, South Carolina, the recent closure of oyster beds signals such problems. In fact, a third of the 600,000 acres of South Carolina coastline has severe restrictions on oyster harvesting due to pollution (Fretwell 1995, A18). Nationally, the percentage of shellfish beds closed due to pollution has increased from 20 percent in 1965 to 37 percent in 1990 (Stipp 1991, B1).

Sand is another example of a CPR. Wider beaches provide recreational and protection benefits for property owners on and near the ocean. Sand constantly moves, however, and property owners build jetties and other constructions to trap the sand for their own uses. These techniques can deny sand to other areas, reducing benefits for their owners. Numerous controversies have developed over who has the right to sand.

Environmental problems in coastal areas certainly cry out for solutions. The question is: What is the most efficient way to solve the problems? Various methods, including government regulation,[2] government ownership, market incentives, and privatization might be utilized in an attempt to correct the negative effects associated with CPRs.

Until 1960, the government did little to regulate development in coastal areas. For instance, Michael Danielson (1995, 238) reports that in the initial phases of the development of Sea Pines on Hilton Head Island, South Carolina, in the late 1950s and early 1960s, public regulation was minimal and government brought few pressures to bear on developers to behave in an ecologically responsible manner. However, public pressure on the government to become involved intensified in the late 1960s. The National Flood Insurance Act (NFIA), passed in 1968, was intended to promote wise development of flood-prone areas by encouraging floodplain ordinances designed to reduce future flood losses.[3] Additional direction came from the Coastal Zone Management Act of 1972, which provided funding for coastal states to plan, evaluate, and control shoreline erosion.[4] The Coastal Barrier Resources Act (CBRA) of 1982 (PL 97-348) provided a means of slowing development in the coastal zone by removing selected areas from government subsidized insurance and other financial assistance. The Coastal Barriers Improvement Act of 1990 (PL 101-591) further strengthened CBRA.

Also, state governments have initiated efforts to protect coastal areas through regulation and public ownership. In South Carolina, state regulation of coastal areas began with the Coastal Management Act of 1977. The state's regulatory powers were broadened by the 1988 Beachfront Management Act (SC 49-39-250) and revised again in 1990. Permits are now required before dredging or filling activities and dock, pier, and bridge construction can begin. Most other states with shorelines have enacted similar legislation.

PROBLEMS WITH THE GOVERNMENT APPROACH

Recognizing market imperfections in the provision of environmental goods is one thing; improving matters through government regulation is another. A recent study of Oregon's coastal management program (Good 1994) finds that state policies designed to protect beaches are often ineffective. Although the state's policies give preference to hazard avoidance and non-structural means of erosion control, the methods generally used are seawalls and revetments, which can damage neighboring properties. The study suggests revising the state's laws as a means of solving the problems.

The government does not necessarily do a better job than the free market in the provision of environmental amenities. First, it is difficult and costly for the government to measure the benefits and costs of its actions. Second, government bureaucrats generally are not liable for their actions. Government decision-makers do not bear the full weight of the costs associated with their actions: costs are diffused, and cost shifting is rampant. Third, government managers do not and cannot know the value of trade-offs; that is, they cannot know the relative values that individuals place on non-market goods such as environmental quality versus other goods and services. Opinion surveys may help, but because respondents do not actually bear the cost associated with their answers, the results are questionable.[5] Fourth, special interest groups, such as real estate developers, environmentalists, and investors, always attempt to influence government decision-makers to see things their way. Often government policy winds up being compatible with the lobbyists' wishes.[6] Finally, no competitive discipline exists in the government sector.

Unfortunately, in environmental matters government bureaucrats often make matters worse. For instance, government-subsidized flood insurance, provided by the 1968 NFIA, has been a significant stimulus to overbuilding in coastal areas (Miller 1975, 2). Government funded infrastructure subsidized or built by agencies such as the Environmental Protection Agency (EPA) and the Army Corps of Engineers entices developers and builders to become involved to a much greater degree than they would otherwise. For example, the EPA funds a large portion of the capital costs of public waste-water treatment works (Siffin 1981). Public roads, ramps, docks, and bridges produce overcrowding and heavy stress on the environment. Bridges financed by taxpayer dollars have been a key element in promoting the development of barrier islands. Such public works often become the catalyst for coastal development.

Government bureaucrats have an incentive to provide their constituents with what they want at little or no cost to them. According to the "State Beachfront Management Plan" produced by the South Carolina State Legislature, two of the Coastal Council's goals are to "improve public access" and "to develop new access sites" (South Carolina Beachfront Management Act 1990, 3). Charges to users are usually zero or nominal.[7] Yet the coastal areas are already highly developed and crowded. For their efforts, bureaucrats get bigger budgets, staffs, and authority. They do not face the market discipline of competition; nor can they assess their programs by using the measuring rod of profitability.

Markets do have their faults. Market imperfections such as externalities distort efficient outcomes. Also, information can be inaccurate, monopoly power may exist, and social and private discount rates may differ. Yet, when property rights are established, the market tends to produce better outcomes than the government.

BARRIER ISLANDS, ENTREPRENEURIAL ACTIVITY, AND ENVIRONMENTAL GOODS

The absence of property rights gives rise to many of the environmental problems in coastal areas.[8] When ownership rights are not defined, users have little or no incentive to take into account the effects of their actions on the welfare of others. Users rush ahead—the greatest returns going to those who get there first—using more and more of the resource as long as private benefits exceed private costs. Fish grow scarce, water becomes polluted, beaches are crowded and denuded, and wetlands and marshes disappear. However, overuse of environmental resources in the absence of government involvement is not inevitable. With a system of well-defined property rights, market incentives can produce gains in efficiency.

For example, in examining property regimes in southeast Asia, Jeffrey S. Walters (1994) finds that when proprietorship rights are placed in the hands of local marine resource users rather than of government officials, coral reefs and mangrove forests are actually better managed. In another example, W. A. Fischel (1994) finds that without government intervention, private developers provide efficient levels of open space. An argument in favor of zoning is that the proper mix of competing land uses will not result when development is left to private parties. Fischel, updating a study by T. O. Crone (1983), shows that in the absence of government controls, Foster City, California, a private development, provided a proper mix of apartments and single-family homes that maximized the value of land.

Coastal barrier islands present a good example of the privatization solution. In this case, private developers are ahead of legislators and government regulators. Developers can own entire islands; for all practical purposes they own and control much of the natural resource base, including beaches and marshes. Consequently, market incentives encourage developers to factor in environmental values. For instance, if developers can capture the "rents" from a wider beach, they would be willing to expend resources to protect beach areas. As Terry L. Anderson and Donald R. Leal (1993) explain, "the subdivider who puts covenants in deeds that preserve open space, improve views, and generally harmonize development with the environment establishes property rights to these values and captures the value in higher asset prices" (21).

Owners benefit by putting resources to their highest and best use, and suffer by putting resources to a less ideal use. Competitively determined prices reflect the wishes of thousands upon thousands of buyers and the circumstances faced by thousands of producers. Prices measure the net effect of offers by producers and consumers regarding property uses. The government resource manager, seldom having access to competitively determined prices, cannot assess the relative values of goods and services and therefore suffers a serious disadvantage in making decisions about the development of coastal areas.

People had few environmental concerns in much of the coastal zone until coastal development began to accelerate after World War II. Developments built on barrier islands such as Galveston, Texas, Atlantic City, New Jersey, and Miami Beach, Florida, were notable exceptions. In recent years, however, development has increased and concerns over deterioration of the coastal environment have mounted. Simultaneously, though, incentives have arisen for entrepreneurs to produce and market clean water and marshes, wide beaches, low population density, wildlife, and trees. As expressed by Joseph C. Bast and others (1994), "Rising efficiency and prosperity clearly have led to greater public concern for things that once were regarded as unimportant or unaffordable: clean air, clear water, and preservation of wilderness areas, among them" (198).

Figure 1 illustrates the response of developers to the changing market conditions in coastal areas. Increasing land values, resulting from growing demand for coastal property,[9] coupled with rising consumer demand for environmental quality, provided the incentive for developers to protect coastal resources. The movement toward environmental protection, which continues today, sprang from improved scientific knowledge, higher standards of living, and entrepreneurial responses to buyers' demands. Evolving government regulation also has had some impact, but more often than not legislation lags behind market initiatives. Growing affluence and movement along an environmental "learning curve" are driving these markets, and improvements would be even more pronounced were it not for government programs that subsidize bad environmental decisions.[10]

Figure 1: Response of Coastal Developers to the Demand for Environmental Quality

Time	Property Buyers	Property Developers	Environmental Standards
Period I Before 1960	Little environmental knowledge or interest Disposable personal per capita income, 1960: $7,264	Piecemeal development Small lots Streets horizontal and perpendicular to ocean Open communities Building close to sea Seawalls & rock revetments	Sand dune destruction Crowded conditions Polluted water Wetland destruction Commercial atmosphere (e.g., Myrtle Beach)
Period II 1960–1980	Growing environmental awareness and appreciation. Disposable personal per capita income, 1970: $9,875	Community development Larger lots Protection of trees and wildlife Buildings set back from sea Improved building standards	Less crowded Natural beauty protected (e.g., Sea Pines, Seabrook)
Period III 1980–Present	Keen awareness of environmental amenities Environmental shopping Disposable personal per capita income, 1980: $12,005 1990: $14,101	Environmental covenants New technologies Compatibility between development and environment	Control water pollution Beach nourishment & protection Wildlife and vegetation protection (e.g., Dewees Island)

Note: Average United States income in 1987 dollars.

A brief tour through South Carolina island developments such as Sea Pines, Seabrook, Dewees, and Kiawah clearly shows that recreational and residential users of resources have succeeded in outbidding alternative users. Charles Fraser, owner and developer of Sea Pines on Hilton Head Island, convinced his father, a logger who bought the land originally for its timber, to leave the tall pines along the coast, to capitalize on their greater value as a resort amenity (Danielson 1995). Environmentally conscious consumers have often outbid loggers because they place a higher value on the forests for recreational and residential use than do homebuilders.[11] In contrast, many public beaches, such as Myrtle Beach, show just the opposite outcome.

When coastal development began to accelerate a few decades ago, with very few exceptions little thought was given to the value of open space, harmony with nature, or the stabilization value of sand dunes and vegetation. There were few crowded areas and as a result few environmental concerns. Developers often built too close to the ocean; used "hard" erosion-control techniques, such as seawalls, rock revetments, and bulkheads; cut trees; and filled in marshes. However, over time, developers gained a better understanding of the coastal system and responded to the growing appreciation of and demand for environmental goods.

Another beneficial aspect of market-driven environmental decision making is that the provision of environmental goods pushes up the cost of coastal development, encouraging a movement away from the coastal zone as potential residents opt for cheaper land farther from the shore. On the coast, higher prices encourage efficiency, conservation, and ultimately long-term protection. Moreover, developers have incentives to discover and employ new technologies in environmental protection.

Developers also compete with one another to offer more and more environmental goods to their customers. According to Fred Foldvary (1994), "consumers reveal their demands for collective goods through a choice of community, and competition among communities ensures that local collective goods are provided at a minimum cost" (71).

PRIVATE DEVELOPMENT ON BARRIER ISLANDS: SPECIFIC CASES

Along the South Carolina coast, many private developments on barrier islands demonstrate the entrepreneurial process at work. Because the developers own most of the natural resources on the islands, they internalize the costs associated with decisions regarding resource use. They promote and market environmental amenities without the requirement of government oversight and usually with protection that exceeds minimum government standards. These developments have security gates that restrict entry to property owners and their guests. Deed covenants regulate and control building and future development to ensure environmental protection. The enforced agreements create security for property owners and, unlike governmental regulations, are not subject to changing political fashions. We consider four representative examples along the South Carolina coast.

The Sea Pines development covers 3,480 acres of beachfront, forests, lagoons, and sea marshes on the southern tip of Hilton Head Island. Parks, woodlands, golf courses, and tennis courts occupy 2,400 acres. Sea Pines illustrates the "learning curve." Although Hilton Head Island had been bought for the Hilton Head Company by Fred Hack and Joseph Fraser in 1949 for its timber, Joseph's son Charles saw the island's resort potential. He began the island's resort development phase in 1957 and became a pioneer in producing and marketing environmental goods.

The developer used natural materials in construction; laid out lots for excellent views; opened houses to the outside; protected trees, wildlife, and natural vegetation; and wound roads through nicely landscaped properties. According to Michael N. Danielson (1995), "Sea Pines became a training ground for developers, architects, landscape designers, and others who later took their lessons to resorts and new communities across the nation" (34). Fraser introduced strict deed covenants regulating both the developer and individual property owners, which have proven durable and enforceable. In 1985 a court agreed that the covenants legally bind the property owners (Danielson 1995, 287).

Seabrook Island is a beach-ridge barrier island twenty-three miles south of Charleston, South Carolina, and directly south of Kiawah, a sister barrier island. It contains approximately 2,200 acres of land, 3.5 miles of private beach, and it is bordered by three river systems: the North Edisto, the Bohicket, and the Kiawah. Until 1970, the island was undeveloped and used primarily by the Episcopal Diocese of South Carolina as a camp and conference center. The Seabrook Development Corporation, a private company, acquired the land in 1970 and commenced development. Seabrook now has about 2,350 privately owned properties, consisting of 495 single-family homes, 1,003 villas, and 852 undeveloped lots.

Most lots on Seabrook are heavily wooded and attractively spaced along winding streets. Houses are constructed with as little disruption as possible to natural vegetation. The island has a variety of freshwater lakes; numerous marshes, lagoons, and creeks; and abundant wildlife. It has a noncommercial atmosphere and forests of live oaks, pines, palms, and magnolias. Traditional commercial establishments such as grocery stores, banks, service stations, and department stores, as well as churches and schools, are located outside the entrance gates.

Kiawah Island, next to Seabrook Island, is a 10,000-acre barrier island with ten miles of beachfront. Until the early 1900s, it was devoted to indigo and cotton production. In 1952, C. C. Royal of Aiken bought the island for its timber. In 1974 it passed to a buyer with plans to develop it as a resort. With the aid of Charles Fraser and the Sea Pines Company, a plan was devised to develop the island while preserving its natural beauty. Deer, fox, bobcat, waterfowl, and loggerhead turtles still inhabit the island.

Kiawah's gatehouse limits public access. The Kiawah developers spent $1.3 million on a sixteen-month environmental survey before development commenced to determine how to minimize the adverse effects of development on the ecosystem. Ecologists, archaeologists, biologists, historians, land planners, architects, and other scientists and professionals joined in drawing up the Master Land Use Plan. After observing Seabrook's problems with beach erosion, the Kiawah developers also increased housing setback requirements.

Dewees Island is a privately owned barrier island twelve miles northeast of Charleston, South Carolina. Presently being developed by Island Preservation Partnership, the island has 1,206 acres well endowed with trees, water, marshes, and wildlife. To hold down population density and preserve natural habitat, the developer has allowed only 137 single-family houses, occupying 35 percent of the island. Homesites are placed at a greater distance from the beach than required by state guidelines. No wetlands will be destroyed in development of the island, and 350 acres will serve as a permanent wildlife preserve. No commercial facilities are planned for the island, and ferry service provides the only access to it. Only golf carts or electric-powered cars may drive on the island. All roads have a natural sand base, and concrete driveways and walkways are prohibited in order to reduce water runoff. Fencing; inorganic fertilizers and pesticides; and manicured lawns, hedges, and shrubs are disallowed.

Although we consider only a few examples here, many similar cases exist along the South Carolina coastline. Most are being developed with environmental concerns in the forefront. Resort developments near the non-barrier-island coast are also following this path. We have identified more than thirty such developments on or near South Carolina barrier islands.[12]

Another advantage of privately owned developments is that they internalize the costs of maintaining beaches. For barrier islands that provide no public access, such as Seabrook and Dewees, federal and state funds are not available for beach restoration, and property owners must pay all costs.[13] In Myrtle Beach, South Carolina, on the other hand, where beaches are public, federal and state taxpayers pay a significant portion of beach protection costs. The federal government is financing 65 percent of a current multimillion-dollar beach nourishment project at Myrtle Beach, with the state and local governments covering the remainder. Public beaches contribute to state tourism income and provide recreational benefits for the public, justifying some governmental financing. However, much of the recreational and protection benefit of beach nourishment accrues to property owners on or near the beach.[14]

FACTORS AFFECTING THE SUCCESS OF PRIVATIZATION SOLUTIONS

Sea Pines, Dewees, Kiawah, and Seabrook Islands illustrate the evolution of barrier island development. Sea Pines, begun nearly forty years ago, was one of the first developments to market environmental amenities; Dewees represents a more recent trend toward higher environmental standards. Charles Fraser set the tone for this type of environmental marketing on Hilton Head Island in the 1960s. Because Fraser and his partner, Fred Hack, disagreed over the development plan for the island, they divided it and each followed his own path. Hack followed the traditional path of strip development, selling one lot at a time without an overall plan. Fraser, in contrast, thought in terms of the community as a whole, including protection of the coastal ecosystem. Fraser's environmental approach was so financially successful that his ex-partner decided to emulate Fraser and his Sea Pines development (Danielson 1995). Most island developments since that time, especially along the southeastern coast, have copied the Fraser model.

Many interior developments in the United States are also following such a course. Eagle Rock Reserve in Montana, Farmview in Pennsylvania, and Preserve at Hunters Lake in Wisconsin are notable examples of developers' recognizing and capitalizing on the demand for environmental quality (Fakis 1995). Across the country, thousands of such developments are built on lakes, rivers, and mountains with standards that run the gamut from minimal environmental to strict Dewees-type.

By not building as close to the ocean as possible and by protecting shoreline vegetation, property values for the community are enhanced. Although developers find protecting such environmental resources costly, they will voluntarily engage in such activity when they expect to receive private net gains. If property buyers value environmental protection and the developer can capture higher rents, a private developer will protect the environment without government coercion. Our research indicates that private developers are providing environmental amenities beyond the requirements of government regulations. Certain conditions must prevail, however, for such results to occur.

The first condition is that movement toward privatization of barrier islands is closely related to changes in land values. As the value of a CPR rises, precision in the definition of property rights becomes more valuable. The enhanced resource value attracts additional claimants, increasing the potential losses from the CPR as well as the potential returns from better-defined property rights (Libecap 1986, 231). For example, the discovery of rich gold and silver deposits in the American West after 1848 led to negotiation of property rights on previously unclaimed land (Umbeck 1977). Similarly, privatization of coastal barrier islands followed increasing land values, which were driven by growing demand for coastal property. Private developments became more viable because poorly defined property rights meant lower rates of return for developers.

Second, successful developments consist of tracts of land large enough to allow effective control over conditions affecting environmental quality. Small-scale developers have less incentive to provide environmental amenities because of transaction costs and the limited ability to gain from appreciation in land values due to enhanced environmental standards. Small developers bear the costs of environmental protection but the benefits spill over onto adjoining properties. Avoiding piecemeal development, large-scale developers will supply environmental amenities, expecting to capture the benefits through higher land values.[15] The private developer has an incentive to use resources to determine how best to develop the area while minimizing the negative effect on the natural resource base. With many small parcels of land, the free rider problem and the higher costs of controlling externalities through negotiation discourage environmental planning.

Third, the private solution will be more successful where the boundaries of the development are clearly defined and access is more easily controlled. Under these conditions, property owners enjoy the benefits of the CPR and can exclude others. Furthermore, property owners can protect the development from adjacent incompatible uses. Geographical characteristics that limit encroachment, such as a mountain range (Dennen 1976), can substantially reduce costs of exclusion. Dewees Island has the advantage of being an island without a bridge; Seabrook has a single access gate. Although a barrier island has an obvious advantage, interior developments can create similar conditions by using natural barriers such as water and mountains, and man-made barriers such as fences and buffer zones. Because developments such as Sea Pines, Dewees, Seabrook, and Kiawah are self-contained and controlled by a single entity, it is more likely that the developer will capture the benefits of increased land values. The more difficult and costly exclusion becomes and the lower the benefits of exclusion, the less successful private solutions are.

Fourth, environmental goods flow more naturally from privatization when the costs of contracting and enforcing collective decisions are low. For the coastal communities examined here, such costs are minimal because new property owners buying into the community have tastes similar to those of residents already there; hence, a general consensus on how to order the development is more easily reached. Contracting costs are also reduced if restrictive covenants are already in place.

Fifth, privatization works better if buyers are well informed about environmental amenities and quality. Some of the attributes of environmental protection may not be immediately visible to potential buyers, and therefore developers must "spell out" the positive aspects of the development. An examination of the sales pamphlets from developments of this type confirms that developers take pains to inform potential buyers of the environmental amenities available.

Sixth, success is inversely related to the number of property owners, as the costs of contracting and enforcing regulations vary directly with the number of property owners. Most barrier-island communities place limits on the number of members. Barrier islands are not large, and their residents prefer low-density housing.

Although no island can be developed without some negative environmental impacts, islands such as those described here stand in sharp contrast to earlier beach developments such as Myrtle Beach, where trees are cut, land rearranged, and plants and animals removed. But if development is to occur, the market approach can produce environmental amenities without the coercive force of government.[16]

CONCLUSION

Private developers are making significant efforts to protect environmental resources that add to the net collective value of the community. These efforts are simply profit-maximizing behavior by developers responding to property owners' growing demands to protect the environment and preserve the natural landscape. Large-scale developers are more likely than small-scale developers to protect common pool resources because the larger developers are better able to capture the benefits of their actions. This difference is especially evident in coastal areas.

But even in coastal areas, private development cannot solve all environmental problems. For some common property resources, such as fisheries, where access is not controlled and property rights are not well defined, environmental problems continue. Moreover, property rights do not extend beyond the island's environment.[17] For example, upstream dams deprive coastal beaches of sand, and pollution runoff from farm and urban developments damages coastal estuaries. Furthermore, developers do not operate without making mistakes, and in some cases they actually ignore environmental concerns—for instance, it seems clear in hindsight that some buildings were constructed too close to the ocean on Seabrook Island. Utopias rarely, if ever, exist. We can only choose the course of action that promises the best feasible outcome. For coastal development, the market demonstrates significant compatibility between entrepreneurial activity and environmental concern.

Studies of cases such as Sea Pines, Seabrook, Kiawah, and Dewees are useful in determining how successful this approach may be in the long term. For a community of property owners with similar tastes and a strong system of protective covenants, our research suggests that the potential for long-term and widespread success is substantial.

Human beings have a desire for environmental goods such as scenic views, clear skies, clean water, space, and a feeling of harmony with nature. Community developers, such as those on barrier islands, package environmental amenities along with other attractions and receive compensation from property prices in exactly the same way that Wal-Mart profits by selling tennis rackets, golf clubs, electronic equipment, and clothing. This process is under way in many other areas of the country. As consumer affluence and environmental knowledge grow, this movement will accelerate.

NOTES

1. Economic activity in the coastal zone is important to the national economy. In 1985 the National Coastal Research Institute estimated that 31.7 percent of GNP originated in the 413 coastal counties (Morris 1992, 39).
2. The most common regulatory approach by government is the command-and-control, or standards, technique. Under the command-and-control approach, the government dictates the process to be followed or sets specific standards to be met. Alternatively, market incentives, such as taxes on undesirable activities, follow a carrot-and-stick approach to arriv e at appropriate solutions.
3. With this act, governmental response also changed from engineering approaches, such as building concrete seawalls and rock revetments, to direct control of coastal development with various policy instruments (Platt 1994).
4. Most states bordering oceans or the Great Lakes participate in this program. Since 1972, twenty-nine of thirty-five possible states have received funds.
5. It is especially difficult to estimate the value of nonmarket goods, such as environmental amenities. In addition to the contingent valuation approach, commonly used methods are the hedonic technique and the travel-cost approach. See Freeman (1993).
6. Terry Anderson and P. J. Hill (1994) explain how entrepreneurs, rather than "a few farsighted, unselfish, and idealistic men and women," were instrumental in the formation of Yellowstone National Park. The owners of the Northern Pacific Railroad, hoping to capture rents from tourists traveling to the Park, provided the major impetus behind the establishment of Yellowstone, the United States's first national park, in 1872.
7. With market-provided goods, anyone willing and able to pay the real cost for such amenities would have access.
8. In his classic 1960 chapter, "The Problem of Social Cost," Ronald Coase discusses how landowners can voluntarily solve externality problems when property rights are well defined and transaction costs are low.
9. For example, on peak days 75,000 people populate Hilton Head Island, tripling the yearround population. Twenty-five years ago, 3,000 people resided on the island.
10. Building in coastal areas increased significantly as a result of the 1968 NFIA, which encouraged development by providing subsidized flood insurance.
11. A recent case in South Carolina shows how private organizations can also protect valuable resources. A logging company applied to the South Carolina Coastal Council for a permit to

construct a bridge to Sandy Island, not far from Myrtle Beach, in order to cut cypress trees (Paulsen 1995, 1A). Opposition arose immediately. Others preferred that the island's unique cultural history, along with wildlife, including the endangered red-cockaded woodpecker, be preserved, and suspected that the timber objective was only a ruse to develop the island. A political decision was made to deny the permit and a compromise solution led to purchase of the land by the Nature Conservancy, a private corporation that buys sensitive and natural areas for conservation purposes. The Nature Conservancy owns more than 1,300 nature preserves (Endicott 1993, 17).

12. These developments include Bray's Island, Spring Island, Jeremy Cay, Sun City Hilton Head, Rose Hill Plantation, New Point, Fripp Island, Dataw Island, Callawassie, Daniels Island, Dunes West, Wild Dunes Resort, Sea Pines, Ocean Side Village, Fairfield Ocean Ridge, Ocean Creek, Kingston Plantation, Debordieu, Wachesaw Plantation, Wexford Plantation, Tide Pointe, Shipyard Plantation, Seabrook of Hilton Head, Palmetto Hall Planation, Palmetto Dunes, Moss Creek Plantation, Melrose Club, Long Cove Club, Hilton Head Plantation, and Haig Point.
13. Seabrook Island has spent over $3 million on seawalls and revetments (mostly by a few individual property owners) and about $2.5 million on beach nourishment. A project to relocate Captain Sam's Inlet, in an attempt to reduce sand loss, was recently completed at a cost of some $600,000.
14. By examining the monetary connection between beach width and property values, with a hedonic model, Pompe and Rinehart (1994) find that for every additional foot of beach width, $525 is added to property values for oceanfront homes and $234 for properties one-third mile from the beach. These calculations use the mean values of housing for location on the oceanfront (a sale price of $170,430 and a square footage of 2,450) and one-third mile from the beach (a sale price of $75,000 and 1,932 square feet). Clearly, there is a monetary payoff for beach protection and restoration.
15. Capital markets that permit large-scale financing are important. A lack of capital or financing might have contributed to piecemeal development in the early years of barrier–island development.
16. Private solutions may become more valuable, as government regulation is incurring more cost for "takings." The David Lucas case, in which a property owner was awarded $1.5 million when government land-use controls deprived him of the economic value of his two oceanfront lots, is representative of increasing challenges to governmental regulation (Rinehart and Pompe 1995).
17. One might expect to see more and more cooperation between developers in protecting common property such as estuaries or rivers that are shared.

REFERENCES

Anderson, Terry L., and P. J. Hill, ed. 1994. Rents from Amenity Resources: A Case Study of Yellowstone National Park. In *The Political Economy of the American West*. Lanham, Md.: Rowman & Littlefield.

Anderson, Terry L., and Donald R. Leal. 1993. Fishing for Property Rights for Fish. In *Taking the Environment Seriously*, edited by R. E. Meiners and B. Yandle. Lanham, Md.: Rowman & Littlefield.

Bast, Joseph L., Peter J. Hill, and Richard C. Rue. 1994. *Eco-Sanity: A Common-Sense Guide to Environmentalism*. Lanham, Md.: Madison Books.

Beach Programs Let Tax Dollars Wash Out to Sea. 1995. *USA Today*, 18 August, 6A.

Coase, R. H. 1960. The Problem of Social Cost. Journal of Law and Economics 3 (October): 1–44.

Crone, T. O. 1983. Elements of an Economic Justification for Municipal Zoning. *Journal of Urban Economics* 14:184–205.

Danielson, Michael N. 1995. *Profits and Politics in Paradise: The Development of Hilton Head Island.* Columbia: University of South Carolina Press.

Dennen, R. T. 1976. Cattlemen's Associations and the Property Rights in Land in the American West. *Explorations in Economic History* 13: 423–36.

Endicott, E. 1993. *Land Conservation through Public/Private Partnerships.* Washington, D.C.: Island Press.

Fakis, Stephen. 1995. New Communities Make It Easy Being Green. *Wall Street Journal,* 10 November, B14.

Fischel, W. A. 1994. Zoning, Nonconvexities, and T. Jack Foster's City. *Journal of Urban Economics* 35: 175–81.

Foldvary, Fred. 1994. *Public Goods & Private Communities.* Brookfield, Vt.: Edward Elgar.

Freeman, A. Myrick, III. 1993. *The Measurement of Environmental and Resource Values: Theory and Methods.* Washington, D.C.: Resources for the Future.

Fretwell, S. 1995. Pearl of the Oyster Beds Dulled by Pollution Woes. *State Newspaper* (Columbia, S.C.), 12 November, A1.

Good, J. W. 1994. Shore Protection Policy and Practices in Oregon: An Evaluation of Implementation Success. *Coastal Management* 22: 325–52.

Hardin, G. 1968. The Tragedy of the Commons. *Science* 162 (December): 1243–48.

Island Preservation Partnership. 1993. Dewees Island. Dewees Island, S.C.: Island Preservation Partnership.

Leatherman, S. 1988. *Barrier Island Handbook.* College Park: University of Maryland.

Libecap, Gary D. 1986. Property Rights in Economic History: Implications for Research. *Explorations in Economic History* 23: 227–52.

Long, L. 1990. Population by the Sea. *Population Today* 10: 6–8.

Miller, H. Crane. 1975. Coastal Flood Plain Management and the National Flood Insurance Program, A Case Study of Three Rhode Island Communities. *Environmental Comment* 3: 1–14.

Morris, Marya. 1992. The Rising Tide: Rapid Development Threatens U.S. Coastal Areas. *EPA Journal* (September/October): 39–41.

Paulsen, Monte. 1995. The Battle for a Bridge. *The Sun News,* 19 February, IA.

Pivnick, Ester, and June Carney. 1992. *An Exploration of the Natural Treasures of Seabrook Island.* Seabrook Island, S. C.: Seabrook Island Natural History Group and the POT.

Platt, R. 1994. Evolution of Coastal Hazards Policies in the United States. *Coastal Management* 22: 265–84.

Platt, R., H. C. Miller, T. Beatley, J. Melville, and B. Mathenia. 1992. *Coastal Erosion: Has Retreat Sounded?* Boulder: University of Colorado, Institute of Behavioral Science.

Pompe, Jeffrey, and James Rinehart. 1994. *Estimating the Effect of Wider Beaches on Coastal Housing Prices.* Ocean and Coastal Management 22: 141–52.

———. 1995. The Lucas Case and the Conflict over Property Rights. In *Land Rights: The 1990s' Property Rights Rebellion,* edited by B. Yandle. Lanham, Md.: Rowman & Littlefield.

Siffin, William J. 1981. Bureaucracy, Entrepreneurship and Natural Resources: Witless Policy and the Barrier Islands. *Cato Journal* 1: 293–311.

Stipp, D. 1991. Toxic Red Tides Seem to Be on the Rise, Increasing the Risks of Eating Shellfish. *Wall Street Journal,* 25 November, B1.

Umbeck, John. 1977. The California Gold Rush: A Study of Emerging Property Rights. *Explorations in Economic History* 14: 197–226.

Walters, Jeffrey S. 1994. Coastal Common Property Regimes in Southeast Asia. In *Ocean Yearbook* 11, edited by E. Mann Bogese, N. Ginsburg, and R. Morgan. Chicago: University of Chicago.

13

To Drill or Not to Drill

Let the Environmentalists Decide

DWIGHT R. LEE

High prices of gasoline and heating oil have made drilling for oil in Alaska's Arctic National Wildlife Refuge (ANWR) an important issue. ANWR is the largest of Alaska's sixteen national wildlife refuges, containing 19.6 million acres. It also contains significant deposits of petroleum. The question is, Should oil companies be allowed to drill for that petroleum?

The case for drilling is straightforward. Alaskan oil would help to reduce U.S. dependence on foreign sources subject to disruptions caused by the volatile politics of the Middle East. Also, most of the infrastructure necessary for transporting the oil from nearby Prudhoe Bay to major U.S. markets is already in place. Furthermore, because of the experience gained at Prudhoe Bay, much has already been learned about how to mitigate the risks of recovering oil in the Arctic environment.

No one denies the environmental risks of drilling for oil in ANWR. No matter how careful the oil companies are, accidents that damage the environment at least temporarily might happen. Environmental groups consider such risks unacceptable; they argue that the value of the wilderness and natural beauty that would be spoiled by drilling in ANWR far exceeds the value of the oil that would be recovered. For example, the National Audubon Society characterizes opening ANWR to oil drilling as a threat "that will destroy the integrity" of the refuge (see statement at http://www.protectthearctic.com).

So, which is more valuable, drilling for oil in ANWR or protecting it as an untouched wilderness and wildlife refuge? Are the benefits of the additional oil really less than the costs of bearing the environmental risks of recovering that oil? Obviously, answering this question with great confidence is difficult because the answer depends on subjective values. Just how do we compare the convenience value of using more petroleum with the almost spiritual value of maintaining the "integrity" of a remote and pristine wilderness area? Although such comparisons are difficult, we should recognize that they can be made. Indeed, we make them all the time.

We constantly make decisions that sacrifice environmental values for what many consider more mundane values, such as comfort, convenience, and material well-being. There is nothing wrong with making such sacrifices because up to some point the additional benefits we realize from sacrificing a little more environmental "integrity" are worth more than the necessary sacrifice. Ideally, we would somehow acquire the information necessary to determine where that point is and then motivate people with different perspectives and preferences to respond appropriately to that information.

Achieving this ideal is not as utopian as it might seem; in fact, such an achievement has been reached in situations very similar to the one at issue in ANWR. In this chapter, I discuss cases in which the appropriate sacrifice of wilderness protection for petroleum production has been responsibly determined and harmoniously implemented. Based on this discussion, I conclude that we should let the Audubon Society decide whether to allow drilling in ANWR. That conclusion may seem to recommend a foregone decision on the issue because the society has already said that drilling for oil in ANWR is unacceptable. But actions speak louder than words, and under certain conditions I am willing to accept the actions of environmental groups such as the Audubon Society as the best evidence of how they truly prefer to answer the question, To drill or not to drill in ANWR?

PRIVATE PROPERTY CHANGES ONE'S PERSPECTIVE

What a difference private property makes when it comes to managing multiuse resources. When people make decisions about the use of property they own, they take into account many more alternatives than they do when advocating decisions about the use of property owned by others. This straightforward principle explains why environmental groups' statements about oil drilling in ANWR (and in other publicly owned areas) and their actions in wildlife areas they own are two very different things.

For example, the Audubon Society owns the Rainey Wildlife Sanctuary, a 26,000-acre preserve in Louisiana that provides a home for fish, shrimp, crab, deer, ducks, and wading birds, and is a resting and feeding stopover for more than 100,000 migrating snow geese each year. By all accounts, it is a beautiful wilderness area and provides exactly the type of wildlife habitat that the Audubon Society seeks to preserve.

But, as elsewhere in our world of scarcity, the use of the Rainey Sanctuary as a wildlife preserve competes with other valuable uses.

Besides being ideally suited for wildlife, the sanctuary contains commercially valuable reserves of natural gas and oil, which attracted the attention of energy companies when they were discovered in the 1940s. Clearly, the interests served by fossil fuels do not have high priority for the Audubon Society. No doubt, the society regards additional petroleum use as a social problem rather than a social benefit. Of course, most people have different priorities: they place a much higher value on keeping down the cost of energy than they do on birdwatching and on protecting what many regard as little more than mosquito-breeding swamps. One might suppose that members of the Audubon Society have no reason to consider such "antienvironmental" values when deciding how to use their own land. Because the society owns the Rainey Sanctuary, it can ignore interests antithetical to its own and refuse to allow drilling. Yet, precisely because the society owns the land, it has been willing to accommodate the interests of those whose priorities are different and has allowed thirty-seven wells to pump gas and oil from the Rainey Sanctuary. In return, it has received royalties of more than $25 million (Baden and Stroup 1981; Snyder and Shaw 1995).

One should not conclude that the Audubon Society has acted hypocritically by putting crass monetary considerations above its stated concerns for protecting wilderness and wildlife. In a wider context, one sees that because of its ownership of the Rainey Sanctuary, the Audubon Society is part of an extensive network of market communication and cooperation that allows it to do a better job of promoting its objectives by helping others promote theirs. Consumers communicate the value they receive from additional gas and oil to petroleum companies through the prices they willingly pay for those products, and this communication is transmitted to owners of oil-producing land through the prices the companies are willing to pay to drill on that land. Money really does "talk" when it takes the form of market prices. The money offered for drilling rights in the Rainey Sanctuary can be viewed as the most effective way for millions of people to tell the Audubon Society how much they value the gas and oil its property can provide.

By responding to the price communication from consumers and by allowing the drilling, the Audubon Society has not sacrificed its environmental values in some debased lust for lucre. Instead, allowing the drilling has served to reaffirm and promote those values in a way that helps others, many of whom have different values, achieve their own purposes. Because of private ownership, the valuations of others for the oil and gas in the Rainey Sanctuary create an opportunity for the Audubon Society to purchase additional sanctuaries to be preserved as habitats for the wildlife it values. So the society has a strong incentive to consider the benefits as well as the costs of drilling on its property. Certainly, environmental risks exist, and the society considers them, but it also responsibly weighs the costs of those risks against the benefits as measured by the income derived from drilling. Obviously, the Audubon Society appraises the benefits from drilling as greater than the costs, and it acts in accordance with that appraisal.

COOPERATION BETWEEN BIRD-WATCHERS AND HOT-RODDERS

The advantage of private ownership is not just that it allows people with different interests to interact in mutually beneficial ways. It also creates harmony between those whose interests would otherwise be antagonistic. For example, most members of the Audubon Society surely see the large sport utility vehicles and high-powered cars encouraged by abundant petroleum supplies as environmentally harmful. That perception, along with the environmental risks associated with oil recovery, helps explain why the Audubon Society vehemently opposes drilling for oil in the ANWR as well as in the continental shelves in the Atlantic, the Pacific, and the Gulf of Mexico. Although oil companies promise to take extraordinary precautions to prevent oil spills when drilling in these areas, the Audubon Society's position is no off-shore drilling, none. One might expect to find Audubon Society members completely unsympathetic with hot-rodding enthusiasts, NASCAR racing fans, and drivers of Chevy Suburbans. Yet, as we have seen, by allowing drilling for gas and oil in the Rainey Sanctuary, the society is accommodating the interests of those with gas-guzzling lifestyles, risking the "integrity" of its prized wildlife sanctuary to make more gasoline available to those whose energy consumption it verbally condemns as excessive.

The incentives provided by private property and market prices not only motivate the Audubon Society to cooperate with NASCAR racing fans, but also motivate those racing enthusiasts to cooperate with the Audubon Society. Imagine the reaction you would get if you went to a stock-car race and tried to convince the spectators to skip the race and go bird-watching instead. Be prepared for some beer bottles tossed your way. Yet by purchasing tickets to their favorite sport, racing fans contribute to the purchase of gasoline that allows the Audubon Society to obtain additional wildlife habitat and to promote birdwatching. Many members of the Audubon Society may feel contempt for racing fans, and most racing fans may laugh at bird-watchers, but because of private property and market prices, they nevertheless act to promote one another's interests.

The Audubon Society is not the only environmental group that, because of the incentives of private ownership, promotes its environmental objectives by serving the interests of those with different objectives. The Nature Conservancy accepts land and monetary contributions for the purpose of maintaining natural areas for wildlife habitat and ecological preservation. It currently owns thousands of acres and has a well-deserved reputation for preventing development in environmentally sensitive areas. Because it owns the land, it has also a strong incentive to use that land wisely to achieve its objectives, which sometimes means recognizing the value of developing the land.

For example, soon after the Wisconsin chapter received title to 40 acres of beachfront land on St. Croix in the Virgin Islands, it was offered a much larger parcel of land in northern Wisconsin in exchange for its beach land. The Wisconsin chapter made this trade (with some covenants on development of the beach land) because owning the Wisconsin land allowed it to protect an entire watershed containing endangered plants that it considered of greater environmental value than what was sacrificed by allowing the beach to be developed (Anderson and Leal 1991, chap. 1).

Thanks to a gift from the Mobil Oil Company, the Nature Conservancy of Texas owns the Galveston Bay Prairie Preserve in Texas City, a 2,263-acre refuge that is home to the Attwater's prairie chicken, a highly endangered species (once numbering almost a million, its population had fallen to fewer than ten by the early 1990s). The conservancy has entered into an agreement with Galveston Bay Resources of Houston and Aspects Resources, LLC, of Denver to drill for oil and natural gas in the preserve. Clearly some risks attend oil drilling in the habitat of a fragile endangered species, and the conservancy has considered them, but it considers the gains sufficient to justify bearing the risks. According to Ray Johnson, East County program manager for the Nature Conservancy of Texas, "We believe this could provide a tremendous opportunity to raise funds to acquire additional habitat for the Attwater's prairie chicken, one of the most threatened birds in North America." Obviously the primary concern is to protect the endangered species, but the demand for gas and oil is helping achieve that objective. Johnson is quick to point out, "We have taken every precaution to minimize the impact of the drilling on the prairie chickens and to ensure their continued health and safety" (see statement at http://web.archive.org/web/20010721065102/http://www.texasnature.org/news/pressr10.htm).

BACK TO ANWR

Without private ownership, the incentive to take a balanced and accommodating view toward competing land-use values disappears. So, it is hardly surprising that the Audubon Society and other major environmental groups categorically oppose drilling in ANWR. Because ANWR is publicly owned, the environmental groups have no incentive to take into account the benefits of drilling. The Audubon Society does not capture any of the benefits if drilling is allowed, as it does at the Rainey Sanctuary; in ANWR, it sacrifices nothing if drilling is prevented. In opposing drilling in ANWR, despite the fact that the precautions to be taken there would be greater than those required of companies operating in the Rainey Sanctuary, the Audubon Society is completely unaccountable for the sacrificed value of the recoverable petroleum.

Obviously, my recommendation to "let the environmentalists decide" whether to allow oil to be recovered from ANWR makes no sense if they are not accountable for any of the costs (sacrificed benefits) of preventing drilling. I am confident, however, that environmentalists would immediately see the advantages of drilling in ANWR if they were responsible for both the costs and the benefits of that drilling. As a thought experiment about how incentives work, imagine that a consortium of environmental organizations is given veto power over drilling, but is also given a portion (say, 10 percent) of what energy companies are willing to pay for the right to recover oil in ANWR. These organizations could capture tens of millions of dollars by giving their permission to drill. Suddenly the opportunity to realize important environmental objectives by favorably considering the benefits others gain from more energy consumption would come into sharp focus. The environmentalists might easily conclude that although ANWR is an "environmental treasure," other environmental treasures in other parts of the country (or the world) are even more valuable; moreover, with just a portion of the petroleum value of the ANWR, efforts might be made to reduce the risks to other natural habitats, more than compensating for the risks to the Arctic wilderness associated with recovering that value.

Some people who are deeply concerned with protecting the environment see the concentration on "saving" ANWR from any development as misguided even without a vested claim on the oil wealth it contains. For example, according to Craig Medred, the outdoor writer for the Anchorage Daily News and a self-described "development-phobic wilderness lover,"

> That people would fight to keep the scar of clearcut logging from the spectacular and productive rain-forests of Southeast Alaska is easily understandable to a shopper in Seattle or a farmer in Nebraska. That

> people would argue against sinking a few holes through the surface of a frozen wasteland, however, can prove more than a little baffling even to development-phobic, wilderness lovers like me. Truth be known, I'd trade the preservation rights to any 100 acres on the [ANWR] slope for similar rights to any acre of central California wetlands. . . . It would seem of far more environmental concern that Alaska's ducks and geese have a place to winter in overcrowded, overdeveloped California than that California's ducks and geese have a place to breed each summer in uncrowded and undeveloped Alaska. (1996, C1)

Even a small share of the petroleum wealth in ANWR would dramatically reverse the trade-off Medred is willing to make because it would allow environmental groups to afford easily a hundred acres of central California wetlands in exchange for what they would receive for each acre of ANWR released to drilling.

We need not agree with Medred's characterization of the ANWR as "a frozen wasteland" to suspect that environmentalists are overstating the environmental amenities that drilling would put at risk. With the incentives provided by private property, environmental groups would quickly reevaluate the costs of drilling in wilderness refuges and soften their rhetoric about how drilling would "destroy the integrity" of these places. Such hyperbolic rhetoric is to be expected when drilling is being considered on public land because environmentalists can go to the bank with it. It is easier to get contributions by depicting decisions about oil drilling on public land as righteous crusades against evil corporations out to destroy our priceless environment for short-run profit than it is to work toward minimizing drilling costs to accommodate better the interests of others. Environmentalists are concerned about protecting wildlife and wilderness areas in which they have ownership interest, but the debate over any threat from drilling and development in those areas is far more productive and less acrimonious than in the case of ANWR and other publicly owned wilderness areas.

The evidence is overwhelming that the risks of oil drilling to the arctic environment are far less than commonly claimed. The experience gained in Prudhoe Bay has both demonstrated and increased the oil companies' ability to recover oil while leaving a "light footprint" on arctic tundra and wildlife. Oil-recovery operations are now sited on gravel pads providing foundations that protect the underlying permafrost. Instead of using pits to contain the residual mud and other waste from drilling, techniques are now available for pumping the waste back into the well in ways that help maintain well pressure and reduce the risks of spills on the tundra. Improvements in arctic road construction have eliminated the need for the gravel access roads used in the development of the

Prudhoe Bay oil fields. Roads are now made from ocean water pumped onto the tundra, where it freezes to form a road surface. Such roads melt without a trace during the short summers. The oversize rubber tires used on the roads further minimize any impact on the land.

Improvements in technology now permit horizontal drilling to recover oil that is far from directly below the wellhead. This technique reduces further the already small amount of land directly affected by drilling operations. Of the more than 19 million acres contained in ANWR, almost 18 million acres have been set aside by Congress—somewhat more than 8 million as wilderness and 9.5 million as wildlife refuge. Oil companies estimate that only 2,000 acres would be needed to develop the coastal plain (Murkowski 2000).

This carefully conducted and closely confined activity hardly sounds like a sufficient threat to justify the rhetoric of a righteous crusade to prevent the destruction of ANWR, so the environmentalists warn of a detrimental effect on arctic wildlife that cannot be gauged by the limited acreage directly affected. Given the experience at Prudhoe Bay, however, such warnings are difficult to take seriously. The oil companies have gone to great lengths and spent tens of millions of dollars to reduce any harm to the fish, fowl, and mammals that live and breed on Alaska's North Slope. The protections they have provided for wildlife at Prudhoe Bay have been every bit as serious and effective as those the Audubon Society and the Nature Conservancy find acceptable in the Rainey Sanctuary and the Galveston Bay Prairie Preserve. As the numbers of various wildlife species show, many have thrived better since the drilling than they did before.

Before drilling began at Prudhoe Bay, a good deal of concern was expressed about its effect on caribou herds. As with many wildlife species, the population of the caribou on Alaska's North Slope fluctuates (often substantially) from year to year for completely natural reasons, so it is difficult to determine with confidence the effect of development on the caribou population. It is noteworthy, however, that the caribou population in the area around Prudhoe Bay has increased greatly since that oil field was developed, from approximately 3,000 to a high of some 23,400 (see "Oil Development on the Coastal Plain of AWRW" at http://www.anwr.org/case.htm). Some argue that the increase has occurred because the caribou's natural predators have avoided the area—some of these predators are shot, whereas the caribou are not. But even if this argument explains some or even all of the increase in the population, the increase still casts doubt on claims that the drilling threatens the caribou. Nor has it been shown that the viability of any other species has been genuinely threatened by oil drilling at Prudhoe Bay.

CARIBOU VERSUS HUMANS

Although consistency in government policy may be too much to hope for, it is interesting to contrast the federal government's refusal to open ANWR with some of its other oil-related policies. While opposing drilling in ANWR, ostensibly because we should not put caribou and other Alaskan wildlife at risk for the sake of getting more petroleum, we are exposing humans to far greater risks because of federal policies motivated by concern over petroleum supplies.

For example, the United States maintains a military presence in the Middle East in large part because of the petroleum reserves there. It is doubtful that the U.S. government would have mounted a large military action and sacrificed American lives to prevent Iraq from taking over the tiny sheikdom of Kuwait except to allay the threat to a major oil supplier. Nor would the United States have lost the nineteen military personnel in the barracks blown up in Saudi Arabia in 1996 or the seventeen killed onboard the USS Cole in a Yemeni harbor in 2000. I am not arguing against maintaining a military presence in the Middle East, but if it is worthwhile to sacrifice Americans' lives to protect oil supplies in the Middle East, is it not worthwhile to take a small (perhaps nonexistent) risk of sacrificing the lives of a few caribou to recover oil in Alaska?

Domestic energy policy also entails the sacrifice of human lives for oil. To save gasoline, the federal government imposes Corporate Average Fuel Economy (CAFE) standards on automobile producers. These standards now require all new cars to average 27.5 miles per gallon and new light trucks to average 20.5 miles per gallon. The one thing that is not controversial about the CAFE standards is that they cost lives by inducing manufacturers to reduce the weight of vehicles. Even Ralph Nader has acknowledged that "larger cars are safer—there is more bulk to protect the occupant" (qtd. in Peters and Burnet 1997). An interesting question is, How many lives might be saved by using more (ANWR) oil and driving heavier cars rather than using less oil and driving lighter, more dangerous cars?

It has been estimated that increasing the average weight of passenger cars by 100 pounds would reduce U.S. highway fatalities by 200 a year (Klein, Hertz, and Borener 1991). By determining how much additional gas would be consumed each year if all passenger cars were 100 pounds heavier, and then estimating how much gas might be recovered from ANWR oil, we can arrive at a rough estimate of how many human lives potentially might be saved by that oil. To make this estimate, I first used data for the technical specifications of fifty-four randomly selected 2001 model passenger cars to obtain a simple regression of car weight on miles per gallon. This regression equation indicates that every additional 100 pounds decreases mileage by 0.85 miles per gallon. So 200 lives a year could be saved by relaxing the CAFE standards to allow a 0.85 miles per gallon reduction in the average mileage of passenger cars. How much gasoline would be required to compensate for this decrease of average mileage? Some 135 million passenger cars are currently in use, being driven roughly 10,000 miles per year on average (1994–95 data from U.S. Bureau of the Census 1997, 843).[1] Assuming these vehicles travel 24 miles per gallon on average, the annual consumption of gasoline by passenger cars is 56.25 billion gallons (= 135 million X 10,000/24). If instead of an average of 24 miles per gallon the average were reduced to 23.15 miles per gallon, the annual consumption of gasoline by passenger cars would be 58.32 billion gallons (= 135 million X 10,000/23.15). So, 200 lives could be saved annually by an extra 2.07 billion gallons of gas. It is estimated that ANWR contains from 3 to 16 billion barrels of recoverable petroleum. Let us take the midpoint in this estimated range, or 9.5 billion barrels. Given that on average each barrel of petroleum is refined into 19.5 gallons of gasoline, the ANWR oil could be turned into 185.25 billion additional gallons of gas, or enough to save 200 lives a year for almost ninety years (185.25/2.07 = 89.5). Hence, in total almost 18,000 lives could be saved by opening up ANWR to drilling and using the fuel made available to compensate for increasing the weight of passenger cars.

I claim no great precision for this estimate. There may be less petroleum in ANWR than the midpoint estimate indicates, and the study I have relied on may have overestimated the number of lives saved by heavier passenger cars. Still, any reasonable estimate will lead to the conclusion that preventing the recovery of ANWR oil and its use in heavier passenger cars entails the loss of thousands of lives on the highways. Are we willing to bear such a cost in order to avoid the risks, if any, to ANWR and its caribou?

CONCLUSION

I am not recommending that ANWR actually be given to some consortium of environmental groups. In thinking about whether to drill for oil in ANWR, however, it is instructive to consider seriously what such a group would do if it owned ANWR and therefore bore the costs as well as enjoyed the benefits of preventing drilling. Those costs are measured by what people are willing to pay for the additional comfort, convenience, and safety that could be derived from the use of ANWR oil. Unfortunately, without the price communication that is possible only by means of private property and voluntary exchange, we cannot be sure what those costs are or how private owners would evaluate either the costs or the benefits of preventing drilling in ANWR. However, the willingness of environmental groups such as the Audubon Society and the Nature Conservancy to allow drilling for oil on environmentally sensitive land they own suggests strongly that their adamant verbal opposition to drilling in ANWR is a poor reflection of what they would do if they owned even a small fraction of the ANWR territory containing oil.

NOTES

1. According to data from the Federal Highway Administration, as shown in the Statistical Abstract for 1997, average annual vehicle miles was 11,372 for passenger cars and for other two-axle, four-tire vehicles; presumably, the "other" category includes many commercial vehicles that raise the average substantially, making 10,000 a reasonable figure for passenger cars alone.

REFERENCES

Anderson, Terry L., and Donald R. Leal. 1991. *Free Market Environmentalism*. Boulder, Colo.: Westview.

Baden, John, and Richard Stroup. 1981. Saving the Wilderness. *Reason* (July): 28–36.

Klein, Terry M., E. Hertz, and S. Borener. 1991. *A Collection of Recent Analyses of Vehicle Weight and Safety*. Washington, D.C.: U.S. Department of Transportation, DOT HS 807 677, May.

Medred, Craig. 1996. Heated Emotions in So Cold a Place. *Anchorage Daily News*, November 5, C1.

Murkowski, Frank H. 2000. Drilling Won't Make It Less of a Refuge. *Washington Post*, December 10.

Peters, Eric, and H. Sterling Burnet. 1997. Will Minivans Become an Endangered Species? *Brief Analysis* No. 232. Dallas, Tex.: National Center for Policy Analysis, June 4.

Snyder, Pamela S., and Jane S. Shaw. 1995. PC Drilling the a Wildlife Refuge. *Wall Street Journal*, September 7.

U.S. Bureau of the Census. 1997. *Statistical Abstract of the United States*. 117th ed. Washington, D.C.: U.S. Government Printing Office.

Acknowledgment: The author began work on this chapter when he was a visiting scholar at the Federal Reserve Bank of Dallas during the summer of 2000.

14

Externalities, Conflict, and Offshore Lands

Resolution Through the Institutions of Private Property

JOHN BRÄTLAND

How should coastal ocean lands be used? The answer to this question has been the source of intense political and legal conflict in recent decades. The ostensible cause of the conflict is the external costs (externalities) associated with offshore petroleum development. An externality occurs when petroleum producers engage in activities for which they do not bear the full opportunity costs of their actions. The legitimate concerns about environmental externalities are focused most directly on the risks of oil spills arising from blowout accidents on offshore petroleum facilities. Oil spills such as the 1969 Santa Barbara accident are matters of historical record, but it is important to note that since that event most spills have occurred in connection with transportation of crude oil rather than with offshore production operations (Anderson and Leal 2001, 82). Moreover, no serious accident has occurred in connection with exploration and production since the use of blowout-prevention technology has become part of standard universal practice.

A second type of externality appears to account for much of the conflict over the use of these lands. Here, too, one segment of the public is engaging in activities for which the actors avoid bearing the full opportunity costs of their actions: *political externalities* occur because political stakeholders bear little of the opportunity cost of the policies they advocate and succeed in implementing with respect to the use of offshore lands. Thus, government ownership and control have fostered institutions that facilitate and aggravate discord. This chapter proceeds from the premise that both categories of externality are a source of discord and that a reasonable resolution to both can be found in the institutions of private-property rights.

GOVERNMENT OWNERSHIP AND THE DISCORDANCE OF CURRENT POLICY

One might reasonably make the case that modern-day conflict over offshore lands has its origins in the Santa Barbara oil spill, an event that is generally recognized to have imposed genuine environmental externalities. Since that spill, public policy with respect to public offshore lands has been directed toward the implementation of stringent sanctions on petroleum leasing designed to prevent the repetition of such an accident. First, the five-year leasing programs implemented by the federal government include cost-benefit analyses ostensibly to ensure that the social costs do not outweigh the social benefits of leasing. Second, stakeholder participation has been designed to deal with the possibility that the rights and preferences of affected constituencies are not ignored in leasing decisions. The leasing procedure that emerged from this process is routinely implemented within Five-Year Plans mandated by the 1978 Amendments to the Outer Continental Shelf Lands Act (OCSLA). These procedures are designed to assure maximum political participation by all possible stakeholders at the federal, state, and local levels. Also, the Five-Year Plan must satisfy the requirements of Coastal Zone Management Consistency as mandated under the Coastal Zone Management Act. Third, although the petroleum industry has a good record on environmental issues since the Santa Barbara oil spill, legislatures and courts have resorted to broad, sweeping moratoria on leasing in several regions of federal offshore lands, including the federal waters off California.

Cost-Benefit Analysis: Issues of Scientific Legitimacy and Rights

The 1969 National Environmental Protection Act mandates that federal agencies must prepare an environmental impact statement (EIS) "for any major federal action significantly affecting the quality of the human environment" (42 USC Sec. 4321). The Five-Year Plan mandated under the OCSLA lays out planned leasing activity for a particular five-year period—an obvious example of a federal action requiring an EIS. A central element in each EIS done for federal offshore leasing is the assessment of benefits and costs. In principle, cost-benefit analyses are intended to provide decision makers with a supposedly scientifically legitimate estimate of the extent to which the present value of benefits exceeds the present value of estimated social costs. Notice, however, that cost-benefit analysis has never been applied with any particular vigor or rigor to the sweeping moratoria on federal offshore leasing.

In actual practice, cost-benefit analysis tends to serve two distinct purposes: first, it is a pro forma political requirement that must be satisfied before the government can proceed on some major effort, such as the leasing of federal offshore lands; second, it provides purportedly scientific evidence to support what the government has already decided to do.

Hardly any part of this process, however, is immune from sharp criticism. Issues bearing on the choice of discount rates and methods of aggregation have been perennial sources of controversy in attempts to apply cost-benefit analyses to public decision making (Formaini 1990, 39–65; Lind et al. 1982). Amid this unresolved controversy and criticism, there can be no reasonable assurance of scientific legitimacy. Such assurance can be provided only if the time streams of benefits and costs are objectively measurable. By definition, however, the projects the government undertakes are those that presumably would not be undertaken in response to market incentives, and they necessarily involve sacrifices and presumed benefits that are inherently subjective in nature. James Buchanan observes: "The cost-benefit expert cannot have it both ways. He cannot claim 'scientific' precision for his estimates unless he restricts himself rigidly to objectively-observable magnitudes. But if he does this, he cannot claim that his estimates reflect reasonable norms upon which 'social' choices should be based" (1969, 60).[1]

The blunt reality identified by Buchanan becomes painfully clear when one honestly considers the application of cost-benefit analysis to the presumed benefits associated with activities designed to protect the environment. Environmental amenities, as may be affected by offshore petroleum operations, cannot be defined with sufficient operational precision to warrant the imposition of sweeping regulatory sanctions. Each individual's reaction to certain features of the environment will define the individual's perception of what constitutes an environmental amenity. These reactions range from subjective responses to sensory experiences to subjective interpretation of quantitative information. Some individuals may view the absence of unpleasant smells as the principal amenity. Others may focus on some minimum standard of coastal water quality and evidence that subsea wildlife in the area is thriving. For others, the major concern may be the absence of visual blight in the form of offshore facilities. At the same time, certain people may take comfort primarily from an assurance that there will be restitution for damage to property. In other cases, the major source of value may be the knowledge that the risk of an environmental accident has somehow been reduced. Some individuals may find ease of mind in an assurance that no offshore operations exist within so many hundred miles of a certain location. For yet other individuals, environmental enjoyment may be impossible as long as the petroleum industry continues to exist. Where individuals stand in this array of concerns determines what the amenity is for them. Obviously, no objective value with any validity in cost-benefit analyses can emerge from these subjective reactions.[2]

Moreover, cost-benefit analysis has been criticized because of the conflict between individual rights and the utilitarian ethic that dominates its application. If cost-benefit analyses yield positive results, the property rights of those directly affected by the governmental decision are given, at best, secondary weight. "The doctrine underlying cost-benefit analysis is ethically flawed . . . for its willingness to 'tradeoff ' values that should be considered absolute. . . . A right is not something that can be assigned on 'efficiency' grounds; a right is precisely an individual's trump against the claims of efficiency, his protection against social utility monsters. . . . [T]he logic of conceiving the regulatory problem as an *ad hoc* 'social decision' is very much refractory to the logic of rights" (Langlois 1982, 280, 283, 289). Whether one accepts or rejects criticism of this sort, it is clear that policies undertaken on the basis of cost-benefit analyses can engender social antagonism. Do these issues arise in the context of governmental management of offshore lands, and do attempts to involve so-called stakeholders in public decisions resolve the issue of ignored individual rights?

Political Self-Selection of Stakeholders and Their Participation

One might argue that efforts to involve stakeholders in the offshore leasing process represent attempts to deal with the possibility that individuals' rights tend to be ignored or overridden by government policies sanctioned on the basis of cost-benefit analyses. For present purposes, however, the important question is: Who is a stakeholder with respect to the use of public lands? Does the category *stakeholder* include all those who feel that they are affected in some way by land-use decisions? Unfortunately, there is no unambiguous answer to these questions. Does the category refer to peoples' mental state of being traumatized by the loss or the prospective loss of property or amenities? Or does it refer to the irritation, petulance, or anger experienced by people who simply harbor negative feelings toward the petroleum industry? Inclusion of the latter category of people in the allocative decisions for offshore lands may and often does embroil an electorate in political conflict over alternatives uses of offshore lands. Once a large number of people define themselves as parties who care about an issue or as affected parties, politicians and policymakers feel assured that they are dealing with a public-good issue that requires sweeping sanctions (Hoppe 1993, 7–8). The more people care about the actual or possible consequences of an event and express that caring through political agitation, then the more clearly policymakers feel justified in viewing the issue as one that warrants government intervention (Lewin 1982, 207).[3]

Federal legislation not only has fostered the creation of institutions that facilitate public participation by stakeholders, but also has created its own legal framework within which political conflict over the uses of government-owned offshore lands is waged. The Coastal Zone Management Act of 1972 assures significant participation by the residents of the respective coastal states. The act provides grants-in-aid to coastal states for development and implementation of coastal zone management programs. But the act also has provided these states with a means to impede or delay federal offshore leasing. For example, lessees must obtain a federal consistency certification to undertake activities that may affect land or water use in the area designated as its "coastal zone." The certification establishes that the activity to be undertaken by the lessee accords with the program established by the respective states. Not surprisingly, the programs established by the states are necessarily tailored to the political demands of constituent stakeholders.

Unfortunately, no criteria other than political self-selection are used to determine who has a legitimate stake in policy decisions. The ranks of stakeholders are populated by voters with diverse and subjective views on what constitutes an environmental amenity and the way in which they are affected by its presence or absence. Does this political process take the focus off actual environmental issues and, instead, motivate allocative decisions on the basis of the political unpopularity of fossil energy sources or ill will directed at the petroleum industry? Is this approach to environmental policy aimed at dealing with external costs or with the mollification of a certain self-selected political constituency? The answers to these questions seem obvious. Although the approach to policy just described would seem to embrace the essence of democratic participation, the discussion here notes that this participatory process has little to do with rational environmental policy or with the commitment of resources to their highest valued use.

Leasing Moratoria as Acts Imposing Externalities

Environmental conflicts involving petroleum development have been dealt with but not necessarily settled by sweeping moratoria involving hundreds of thousands of acres. Unfortunately, political advocates of these policies are unencumbered by the opportunity costs of these sanctions. In this sense, the moratoria impose major political externalities. In other words, the economic experience of choosing and hence forsaking the value of the next most highly valued opportunity never impinges on the actions of nonowning bureaucrats, politicians, or environmentalists seeking to foreclose certain uses of public lands (Anderson and Leal 2001, 79). Political conflict arises because the weighing of opportunity costs plays no role in settling environmental disputes arising between stakeholders who endeavor to foreclose petroleum development and those who would voluntarily bear the opportunity costs of developing offshore petroleum resources. Hence, self-selected stakeholders have incentives to become extremist in that they have an incentive to exaggerate preferences and overstate claims because whatever the benefits of foreclosing exploration and development, these benefits are provided as a "free good" through the process of political control (Epstein 1995, 301). "[I]f a person can gain by blocking socially useful resource moves through governmental means, then his gain is society's loss. Similarly, if potential users can gain access to the resource through government without paying the opportunity costs of the resource, then low-valued uses may dominate at the expense of more highly valued uses" (Stroup and Baden 1983, 9).

Through a political process, prospective voters define themselves as stakeholders whose notions about appropriate federal-land use become the object of pandering behavior by hopeful politicians seeking election. Because the trade-off between petroleum development and so-called environmental amenities is dealt with through a political process, the actual nature or extent of this trade-off is largely ignored. What is being expressed through the political process often has little to do with demonstrable or provable damage to person or property. Rather, what is expressed is an attitude vented as adversarial political pressure brought to bear on the legislative and regulatory organs of government. This attitude has taken the form of general animosity toward the petroleum industry as a whole rather than of an objection to particular activities in the industry. Because leasing moratoria cover large regions with no allowance for even small amounts of development, political free riders can easily impose external opportunity costs on society. In the case of leasing moratoria, "the present process of decision making treats the value of [the] first acre of public land as the same as the value of the last, making it impossible to acquire information as to the *marginal value* of each acre in its alternative uses" (Epstein 1995, 302, emphasis in original).

The political externalities imposed through political mandates are an ethical red flag with respect to the impact on the rest of economic society. Although this chapter focuses on the uses of federal offshore lands, it is certainly no exaggeration to note that all lands under federal control are potential or actual objects of environmental conflict because those who advocate and impose so-called protective measures for the most part escape the burdens associated with prescriptive regulatory policies. Moreover, these conflicts are not unique to federal lands. Any government-owned land inherently becomes the object of conflict with respect to its alternative uses.[4] In other words, the nature of the conflict explored here pertains to government-controlled lands in general and is not limited to environmental issues.

Costly sanctions are imposed without any framework for ranking uses of public land and for assuring that the moratoria represent the highest valued use for the offshore lands. How might the federal government make such choices in a way that actually weighs opportunity costs? The short but complete answer to this question is that it cannot. Because the land is under government ownership and control, the question has no coherent or operational answer. The actual opportunity costs can never become part of the decision to commit resources to a particular use rather than to another. The requisite sacrifice associated with chosen uses of these resources cannot be borne fully by those presuming to impose decisions with respect to alternative uses. One need scarcely note that if political proponents of moratoria were to bear the opportunity costs of these actions, offshore tracts of land would assuredly be allocated to their highest valued use. But because proponents of leasing bans do not have to bear opportunity costs, no means exist by which to determine whether or not the antileasing burdens are less than or in excess of any realistic assessment of damage to persons or property caused by offshore production facilities.

Under current procedures, even where offshore petroleum development is politically welcomed, no clear assurance exists that such development necessarily represents the highest valued use of these lands. Moreover, the types of land-use decisions that emerge from the political process mean that the data necessary to adapt to change are simply never generated.[5] The leasing moratoria reflect political sentiment but cannot reliably reflect anything about the nature of an economic trade-off and how that trade-off might change over time. Shifts in political sentiments are not and cannot be reliable indicators of such changes. Experience has clarified the need for alternative institutional arrangements for the management of these lands and for revealing the environmental trade-offs associated with competing uses. Government ownership and politically motivated regulatory sanctions provide no prospect of any allocative efficiency free of social conflict. Trade-offs become operational information only as the most valued uses are translated into prices or compensation paid by those who accept the responsibility and opportunity costs associated with ownership.

TOWARD COMPLEMENTARY ACTIONS OF PROPERTY OWNERS TO INTERNALIZE OPPORTUNITY COSTS

Both energy services and environmental amenities require the use of scarce resources and hence are scarce themselves by definition. Desired environmental amenities may be scarce also because of what is relinquished to maintain and enjoy them. Society's recognition of a trade-off between these valued things reflects the reality of that scarcity, the fact that obtaining more of one thing requires a marginal sacrifice in the availability of another thing. But how scarce are these valued things, and how are these scarcities to be reflected in information used to make choices? Current land policy has at least highlighted what means are useless to the attainment of this end.

Conflict arising from the use of any resource usually is attributable to unresolved issues pertaining to the absence of private-property rights. With respect to alternative uses of offshore lands, political decisions are always made by persons without secure property rights in the affected areas. The absence of private-property rights is a clear inducement to conflict because individuals with ostensibly incompatible preferences with respect to use are prompted to utilize political and administrative means to secure certain employments of the land or to preclude others. If lands are under private ownership, conflict is mitigated or eliminated because owners can sell their interest in the lands to those who value alternative but undesired uses more highly. For public offshore lands, such sales are impossible because no citizen has any baseline interest that can be relinquished in an exchange. The reality of scarcity combined with public ownership makes conflict inevitable (Epstein 1995, 300–301).

Clearly, different people perceive differently the scarcities implied in the tradeoff between petroleum development and environmental amenities. Whether stakeholders or not, different individuals attach different degrees of significance to this trade-off, which inevitably becomes a matter of subjective valuation. With private-property rights, "external" costs become opportunity costs that must be reckoned in a decision to choose one employment of resources over another because restitution for invasive damage is an implied right of ownership, as is the right to dispose of property through mutually beneficial exchange. In a monetary economy, exchanges between property owners lead to the emergence of prices that facilitate a rational calculation of the respective costs and benefits associated with alternative actions (Bradley 1996, 47; Mahoney 2002, 39).

Viewed more broadly, under institutions defining and enforcing private-property rights, the opportunity costs associated with environmental trade-offs have calculable meaning only through one or all of the following strategies: (1) enforcement of strict liability and payment of restitution by damaging parties to parties incurring damage, (2) contractual easements between property owners, and (3) acquisition of ownership to control alternative uses. As the basis for public policy, these alternatives must not be viewed as mutually exclusive; rather, they are complementary with respect to internalizing externalities from offshore petroleum development. Property owners may employ all three strategies, depending on the situation they face. As individuals pursue their respective objectives within the institutions of private property, no "social efficiency" is necessarily achieved (Rizzo 1979, 84–86), but these institutions do accommodate conflict-free transactions even between property owners who have ostensibly incompatible objectives or preferences.

INTERNALIZING EXTERNAL COSTS THROUGH TORT ACTION FOR DEMONSTRABLE DAMAGE

The Santa Barbara oil spill of 1969 provides a painful example of the damage that can be inflicted on a community when offshore petroleum producers do not take adequate measures to prevent such damage. Even a commonly held understanding of historical events such as oil spills, however, cannot assure a shared perspective of the likelihood of future accidental damage. Environmentalists are likely to regard such risks as very high and therefore to favor strict and costly prohibitions on the development of offshore lands. Developers, although acknowledging environmental risks, assess the probabilities differently and appreciate the need to keep development and production costs as low as possible. In a traditional regulatory setting, no rational resolution of these conflicting perspectives is achievable. A successful policy must accomplish two tasks in the face of these opposing interests. First, means must be devised by which those who may cause damage are induced to take precautionary measures that reduce the likelihood of accidents. Second, policy must establish institutions within which those who experience actual harm are assured of restitution from those responsible for the harm. Private-property rights provide the only framework within which these tasks can be accomplished simultaneously. Application of this principle to offshore lands necessitates the establishment and enforcement of rules of strict liability on petroleum developers. Some observers take the view that opposition to offshore drilling will never be allayed until affected parties have some assurance that strict liability will be enforced and that restitution will be forthcoming from those responsible for damage (Anderson and Leal 2001, 80).

Several economists have noted the inadequacy of the law of liability and indemnification for damages. In *Human Action,* Ludwig von Mises observes that "where a considerable part of the costs incurred are external costs from the point of view of the acting individuals or firms, the economic calculation established by them is manifestly defective and their results deceptive" ([1949] 1998, 650–51). Walter Block acknowledges the deficiencies in the law as noted by Mises and calls attention to a jurisprudential trend away from the awarding of injunctive relief for property owners who sustain damage because of the actions of others. For Block, the upshot of this historical trend has been that government has been called on to impose prescriptive measures such as legal mandates and regulation (1990, 285). Concordant with the views expressed by Mises and Block, Murray Rothbard expands on the requisite features of the laws with regard to environmental liability. For Rothbard, the principles of justice are grounded on the primacy of self-ownership. On this foundation, his theory of strict liability treats pollution as an act of invasion or aggression by one party against the property, and hence against the person, of another party (1997b, 127). Robert Bradley has examined the application of strict liability to air pollution issues as they arise in connection with petroleum refining. He notes: "To close the loopholes responsible for pollution externalities is to apply—or more accurately, to reapply— tort law to pollution nuisances, be they air, noise, smell, radiation, flare, glare or water. This approach recognizes damage to property and person as invasion and provides restitution to victims, while discouraging—and if need be, enjoining—future occurrences" (1996, 1268).

Rothbard insists on strict causal liability in the case of uncertainties surrounding "environmental risk."[6] With rules of strict liability for damage, incidents inevitably will occur in which environmental damage is sustained. These situations can ultimately be resolved through a process of adjudication in which the damage caused is viewed under appropriate tort law as aggression against another party's property or person. However, under strict liability, the ground rules are defined with some precision. Damage must be proved and shown to have been caused by the actions of a particular party (Rothbard 1997b, 141–42). Under this exacting criterion, the "reasonable man" standard as applied in tort evidence would not be valid, nor would notions of presumptive guilt.

How effectively would the standard of provable responsibility function in the case of external damage imposed by offshore petroleum operators? In the case of the Santa Barbara oil spill, discerning the responsible party was a simple, straightforward matter. A single operator caused the coastal damage.

Moreover, under strict liability as outlined here, the courts would not be called upon to assign property rights in the name of "economic efficiency," as is the case with judicial application of the Coase theorem (Coase 1960). Rather, because pollution is an invasion of another's property, the rights of the respective litigants are already settled prior to litigation.

The ethical principles of self-ownership and the original appropriation (homesteading) of unowned resources have dual implications regarding activities that may affect the environment.[7] What individuals legitimately own must be taken into account when we are considering the nature and extent of tort damage that individuals may sustain. The rights to engage in certain polluting activity may be homesteaded as a "pollution easement," which would then have to be considered in the adjudication of alleged tort damage. A party such as a petroleum developer may have homesteaded a right to a certain minimal level of polluting activity and hence would not be held liable for damage sustained from the activity confined to this "homesteaded level." However, liability would be absolute for proved damage to others from polluting activity exceeding that level. Homesteaded rights to engage in some minimal level of environmental damage should not be a significant concern with respect to offshore petroleum operators. Although accidental oil spills sometimes (actually, rarely) occur, offshore petroleum operations involve no perpetual or continual dumping of effluent into the oceans. The only homesteaded environmental damages would be what some may consider "visual blight" and small amounts of air pollution.[8]

The advantages of reliance on tort law over traditional command and control of regulatory sanctions are clear. First, as noted earlier, the regulatory process is ethically compromised by free-riding constituencies that seek strong sanctions on others but are unwilling to bear the opportunity costs of the mandates they advocate. Case-by-case adjudication avoids much of the influence of parties who are advancing special political interests. Second, direct victim-victimizer confrontations can lead to rational determinations of actual damage, apportionment of liability, and restitution to injured parties (Bradley 1996, 1242). Of these considerations, restitution is usually not provided at all in the regulatory process.

REDUCING PERCEIVED RISKS OF PROSPECTIVE EXTERNALITIES THROUGH VOLUNTARY CONTRACTUAL EASEMENTS

The strict causal-liability approach to internalizing external costs provides the assurance that restitution will be forthcoming from parties responsible for demonstrable damage. In addition, the prospect of liability provides damaging parties with an economic incentive to undertake measures that reduce the likelihood of future accidents that may cause damage. Strict liability may not satisfy fully the concerns of those who may be uncomfortable with existing pollution easements or who may simply fear the prospect of property damage as a consequence of future accidents. As a complementary or augmenting strategy, property owners may enter into *contractual easements* with offshore petroleum developers who may accidentally and inadvertently impose damage. The contractual easement illustrates a type of preemptive action to deal with concerns about possible future accidents.[9] It may take the form of reciprocal contractual arrangements involving payments to petroleum developers in a position to reduce the likelihood of an environmental accident. The reciprocity would arise because payments would presumably be conditioned on the developers' performing some agreed actions to reduce the probability of an accident. The installation of certain equipment or the adoption of precautionary procedures, for example, might satisfy the terms of the easement agreement. In rare cases, the contractual agreement might even include a mutually agreed stipulation that the prospective explorer-developer completely desist from exploration and development activities that pose environmental risks.[10]

Such easement contracts represent one way in which the "bargaining version" of the Coase theorem might function in practice. How realistic are such contractual easements in relation to offshore petroleum development? This approach to dealing with environmental externalities has been criticized because of the assumption that such agreements cannot be successfully consummated where many people may be affected. In such circumstances, the transaction costs are assumed to be too high to allow an agreement that accommodates the interests of all affected parties. Because the transaction costs are presumed to be too high for successful bargaining, some argue that a governmentally imposed regulatory sanction is necessary. The "public-good" aspect of the environmental externality, they assert, forecloses practical reliance on contractual easements. Free-rider issues certainly may reduce the number of cases in which such easements can be workable. Also, with a large number of coastal property owners and a significant number of offshore operators, the transaction costs of contractual easements may preclude their practicality in some cases. Conditions inevitably would arise in which the marginal costs of obtaining the participation of more property owners and more offshore operators would exceed the subjectively reckoned benefits perceived by those seeking to consummate an easement agreement. However, the fact that some property owners may choose to be free riders and some offshore operators may decline participation does not mean that in all such cases insufficient net benefits would be generated so that the effort to arrive at an agreement would not be warranted. Partial participation may well make the easement agreement worthwhile in the minds of those who seek an accord. A functional approach to developing property law for offshore lands would accommodate such easements and enforce the terms of such voluntary contracts.

As noted earlier, the political expression of caring about the real or imagined externalities associated with offshore development is essentially free to a voter. Under current federal control, any political expression of caring is viewed as evidence that significant externalities must exist. Unfortunately, the political process is necessarily ineffective in making distinctions between genuine concerns over demonstrable damage and political expression that has no regard for the opportunity cost of sweeping regulatory sanctions. However, contractual actions undertaken by those who fear environmental harm represent legitimate evidence of the psychic dimensions associated with a possible externality. In other words, such actions identify the nature of the environmental externality to which they are responding. Inaction on the part of others reflects the degree to which they are not affected by perceived prospects of environmental externalities. Hence, this contractual process can accomplish important changes. First, the contractual easement provides an empirical distinction between the parties who are sufficiently affected by the externality and other parties who are not affected or who may seek to be free riders with regard to incremental environmental amenities. Those who may be motivated by nothing more than attitudes or ideologies that are adversarial to the interests of offshore petroleum operators would probably refuse participation in a contractual easement. Second, for the parties who voluntarily seek the contractual easement, the actions ultimately performed are obviously worth the cost. If preferences are not actually demonstrated through acts of exchange, then no scientifically legitimate means can be applied to impute those preferences (Herbener 1997, 84–106; Rothbard 1997a, 212). In any event, this contractual strategy should be available to coastal property owners.

INTERNALIZING EXTERNALITIES THROUGH ACQUISITION OF FULL OWNERSHIP

Private-property rights allow scarce resources to be used and exchanged without conflict among members of society; conflict-free interaction is achieved by establishing mutually binding norms of behavior regarding the use and disposition of these scarce resources (Hoppe 1989, 8). What means should be pursued in establishing private-property rights with regard to federal offshore petroleum resources? Property rights should be sufficiently broad to accommodate the trade-offs between petroleum development and preservation of desired amenities. In other words, the property owner must have complete discretion within the limits of tort law to choose a use for the property that promises to yield the highest value. Should a privatization solution be accomplished through an expansion of the lessees' property rights within a significantly amended OCSLA? Or should the privatization be based on an abandonment of the federal government's presumptive property claim to offshore lands and to the resources those lands may contain?

The Experience of Applying Ownership to Environmental Tradeoffs

Before we can consider these questions, the following questions require our attention: If privatization were allowed, would environmental organizations acquire offshore lands to forestall, control, or foreclose petroleum exploration and development? Can property acquisition also serve the needs of those whose principal interest is environmental protection? In answering no to these questions, critics of privatization have advanced three reasons. First, they claim that those interested in preserving environmental amenities cannot compete with petroleum-producing companies in acquiring ownership of lands bearing petroleum resources. Second, they assert that a market price cannot be placed on the amenities that environmental resources yield. Third, they claim that free-rider problems prevent a full expression of market demand for environmental amenities.

One can counter the first argument by observing that as early as 1995 the funds raised annually by major environmental organizations exceeded $500 million (Epstein 1995, 304). Moreover, as of the years 1997 and 1998, the largest environmental organizations had more than a billion dollars in annual revenue that could be used for acquisition of properties.[11] The charge that amenity value cannot be expressed in market prices must be countered with the observation that federal actions such as the sweeping moratoria of offshore leasing implicitly impose genuine opportunity costs reflected in undeveloped petroleum resources. With respect to the free-rider issue, Anderson and Leal note that membership in and financial contributions to environmental organizations refute the claim of free-rider behavior (2001, 85–86).

Moreover, the property acquisitions of environmental organizations strongly suggest the viability of ownership as a means to internalize external costs. For example, Dwight Lee observes that in situations in which environmental groups have acquired full ownership of "environmentally sensitive areas," they have internalized the costs and benefits associated with alternative uses of the land. Lee presents the example of the Audubon Society's ownership of the Rainey Wildlife Sanctuary, a 26,000-acre preserve in Louisiana. Recognizing that the use of the wildlife preserve has valuable competing uses, the society has allowed some petroleum drilling and production without compromising its fundamental commitment to environmental concerns. He notes: "obviously the Audubon Society appraises the benefits from drilling as greater than the costs, and it acts in accordance with that appraisal" (2001, 218–19). Lee goes on to emphasize, however, that the Audubon Society has taken a much less balanced approach to the alternative use of public lands in which it has no direct property interest, such as the Alaskan National Wildlife Reserve. Lee convincingly makes the case that the key to an efficient and balanced use of the reserve lies in private-property rights by which the true opportunity costs of owners can be reflected in market interactions between those who bear the costs. This conclusion applies with equal validity and force to alternative uses of the resources on the outer continental shelf. Without private-property rights, opportunity costs cannot be manifested in a way that accommodates rational choice from among alternative uses of these offshore lands.

The Audubon Society's management of the Rainey Sanctuary is not an isolated example of aberrant behavior. Another example is provided by the Nature Conservancy's ownership and control of a small but productive oil field in Texas that happens to be one of the last-known breeding grounds for the Attwater prairie chicken, a species that is considered highly endangered.

> Rather than shutting off the petroleum spigots, the conservancy drilled new natural gas wells and let cattle continue to graze on the land—and reaped about $5.2 million in royalties over the last seven years. The Nature Conservancy claims that careful management is allowing it to protect the prairie chicken while working the land to raise money for other conservation efforts. The Texas oil field isn't an exception; nearly half of the 7.2 million acres that the conservancy said it is protecting in the United States is now being grazed, logged, farmed, drilled or put to work in some fashion. (Wilson 2002, 1– 6)

A Variation on Mead's Proposal for Privatization

By what institutional means might privatization of public lands be undertaken? As noted earlier, the answer to this question hinges on how one views the ethical legitimacy of federal property claims to the resources over which ownership has been established by political decree. Acknowledgment of the validity of federal ownership implies that an appropriate solution can be achieved by revamping existing institutions and amending the statutes that authorize the leasing of federally held offshore lands. When Walter Mead and his research associates outlined their proposed changes to the OCSLA, they did not consider the possibility of adapting the auction process to deal with the trade-off between environmental amenities and petroleum development.[12] With this environmental objective in mind, one notes that the leasing process might be modified to include bidders with alternative objectives. Under their proposed changes to the OCSLA, the winning bonus bid would be the lessee's sole payment to the government for the right to acquire and hold the lease; royalties and rental payments would be abolished (Mead et al. 1985, 46–47). Also, the OCSLA would be amended to accommodate the issuance of leases in perpetuity with no limit on the length of time that the lease could be active (Mead et al. 1985, 113).

The Mead proposal would be ideally adaptable to the competitive issuance of leases in which individuals and organizations with strictly environmental concerns would compete directly with oil companies for the ownership of the lease. The modification of the lease agreement would be such that the lessee would also have the option of not exploring or developing the tract during the entire duration of the leasehold. In formulating their respective bids, the bidders would take into account their judgment of the opportunity cost of foreclosing alternative uses of the tract.[13] At any time, an oil company might sell a lease to an environmental organization, or vice versa. What would such an adaptation of the Mead proposal accomplish? First, the process by which private property is exchanged would generate price signals that would indicate which alternative uses of these lands are most highly valued. Second, the process would allow owners to choose alternative employments of offshore land in what some may view as mutually exclusive uses. Third, it would provide a conflict-free mechanism for transferring the lands between different uses as environmental perceptions and economic conditions change (Brätland 2000, 17–21).

This variation of the Mead proposal appeals to what is probably a false working premise regarding the behavior of environmental organizations. The assumption implicit in much of the preceding discussion is that if environmental organizations were to acquire offshore tracts in perpetuity, those organizations presumably would not develop the tract for oil and gas production; in other words, environmental amenities and petroleum development must necessarily be mutually exclusive. However, as suggested earlier, there is more compatibility between the two objectives than commonly assumed. The examples of the Audubon Society and the Nature Conservancy suggest that either-or decisions would not necessarily prevail. Both organizations have fostered petroleum production in conjunction with their efforts to attain environmental objectives. Their decisions as owners reflect a careful balancing of benefits and opportunity costs—a balancing of alternative uses that is highly unlikely to occur unless environmental organizations own the natural settings that are the objects of their concerns.

Original Appropriation of Petroleum Reservoirs and Repeal of the OCSLA

Repeal of the OCSLA has a compelling logic that arises from a challenge to the federal government's property claim to offshore petroleum resources. The state's edicts never constitute a legitimate means to establish ownership (Epstein

1985, 10). Moreover, "possession does not come about without an expenditure of resources, and their expenditure makes clear the exclusivity of ownership" (Epstein 1985, 61). A logical inference from these premises is that the federal government does not have proper title to the so-called federal lands, even though its held coercive power to enforce ownership claims is clearly acknowledged. The principle of property that one applies in making this latter inference is that title to previously unowned assets, such as petroleum resources, can be established only by an act of original appropriation. The discovery and delineation of a petroleum reservoir would constitute such an act (Bradley 1996, 69–74).

Application of original appropriation to the petroleum resources of the outer continental shelf would involve the complete scuttling of federal offshore leasing as it has been conducted under the OCSLA. The proposal springs from a criticism of U.S. property law and its adoption of the notion that legitimate ownership of the surface necessarily implies some type of conditional claim to all subsurface resources.[14] No surface owner (in this case the federal government) has performed any act of original appropriation that would establish a presumptive property claim to any petroleum resources ultimately produced from these lands. The lack of a just property claim to subsurface resources necessarily implies the absence of any rights to a royalty share of any subsequent production. In other words, the resource defined by the land surface should be recognized as a separate resource distinct from in situ petroleum; the same legal principles of original appropriation that apply to land surface should apply to in situ petroleum.

Under an application of this proposal, offshore public "lands" would be open to explorers as though the lands had never been under public control. The approach would involve the adoption of a process of original appropriation in which an entire reservoir becomes the exclusive property of its first discoverer, free of royalty obligations or any regulations of production. The discoverer would be obligated to delineate the reservoir fully before unattenuated ownership could be established.

Such a means of securing ownership certainly would meet the requirement that firms seek efficient conservation of petroleum. Choosing to leave the pool unexploited would be a perfectly legitimate decision in connection with economic conservation of the resource. The owner might simply be waiting for the optimal time to develop—an allocative option that is awkwardly foreclosed to lessees under the leasing procedure mandated under the OCSLA.[15] However, the same process of original appropriation would overcome many of the most fundamental policy barriers to establishing a social trade-off between petroleum development and environmental protection because it would help to create an institutional setting in which the discovered oil reservoir might be removed, by the appropriating owner's decision, from the development process through a market mechanism. The critical and central objective is to define a process of property acquisition that permits the owner to reap the benefits of chosen use and to bear the opportunity cost of the most highly valued relinquished use.

Once an ownership claim had been made, the party would have complete control of what ultimately happens with respect to the reservoir. In the case of an owned reservoir, no a priori assumption can be made that petroleum development is necessarily its highest and best use. The owner may strive to achieve objectives that have nothing to do with immediate or even delayed petroleum production. The highest and best use can be revealed only through well-defined, enforced, and transferable rights that allow owners broad latitude in choosing among competing and possibly mutually exclusive uses. Hence, ownership would determine allocation of an appropriated reservoir among the following uses: *(a)* prompt development; *(b)* delayed development in anticipation of higher future petroleum prices or of lower costs of development and production; *(c)* speculative holding for sale at a time that maximizes return; *(d)* holding to preserve permanently what is considered to be the "environmental integrity" of the area in which the reservoir is located; and *(e)* holding for an indefinite period until more information is available on the environmental implications of production. In making such a decision, the owner necessarily bears the opportunity costs of alternative employment (by sacrificing the most highly valued relinquished use).

Through a strategy of original appropriation, the party motivated principally by environmental concerns would need to engage in exploration or contracting for exploration. For the homesteader who chooses uses other than petroleum production, this strategy may be costly because the effort would necessarily involve not only the expense of exploration but also the costs of delineation that, in the case of a large discovery, might involve the drilling of numerous wells. This latter stage in the process of original appropriation may be problematic for the purposes outlined here. The requirement that the discoverer undertake additional development to delineate the reservoir is costly and involves substantial sunk costs. The capital investment in the delineation process must not be so great that it forecloses nondevelopment or nonexploitation as a viable option in choosing alternative uses of the environment containing the reservoir. If delineation wells are required for original appropriation, intended nonexploitation effectively may be foreclosed because capital recovery may become a critical issue in the management of the reservoir. Also, the additional drilling may involve environmental impacts that some prospective appropriators may want to minimize because of a possible intent to claim the reservoir and leave it unexploited.

This problem with the original appropriation (homesteading) of offshore petroleum reservoirs may be more apparent than real when viewed from the perspective of a homesteader motivated principally by environmental concerns. Such a homesteader may choose not to foreclose petroleum development and production totally. In bearing both the opportunity benefits and opportunity costs of ownership, this homesteader may view these two objectives as more complementary than mutually exclusive, as the Audubon Society and the Nature Conservancy have. Were such environmental groups to find themselves as homesteaders of offshore petroleum reservoirs, they might be able simultaneously to satisfy their environmental concerns and to engage in production from their reservoirs. In that case, the capital invested in exploration and delineation would not foreclose intended use but rather complement it.

Outright Purchase of Petroleum Properties

The most obvious and most direct property-acquisition strategy that parties motivated by environmental concerns might pursue is to buy the owned reservoir from a current owner. A party seeking to control the reservoir but not necessarily for the purpose of petroleum production may also exercise this approach. A direct purchase presumably would be made at a price that reflects the capital value of the asset; in this case, the price would be sufficient to cover the opportunity cost of not producing the petroleum. Through free marketability of reservoir ownership, society would be assured that the highest valued use is attained for the asset. Even owners or appropriators motivated by nothing more than contempt for the petroleum industry would be able to foreclose development and production if they were prepared to bear the costs of ownership. Although ownership of the untapped reservoir might change hands several times, and the owner's ultimate intentions might change, secure, well-defined, and enforceable property rights are the only means by which resources can be committed to their highest-valued use. That use in this case might actually be intended nonuse (i.e., a decision not to produce petroleum), but, as emphasized previously, environmental groups that become actual owners of oil-bearing properties are more likely to pursue a more balanced strategy that accommodates some production under carefully managed circumstances. The management behavior of organizations such as the Audubon Society and Nature Conservancy demonstrates that petroleum production and environmental objectives are not mutually exclusive and, indeed, at the margin are quite compatible.

Hence, the process by which owners make choices is critical in yielding the requisite information on the trade-off between competing or ostensibly conflicting uses of offshore lands. Yet "the issue is not simply one of information. *The central issue is the critical interdependence between the market choice itself and the informational content of this process which can only be revealed as the process is allowed to occur.* . . . [The relevant information cannot] be communicated to observers independently of the exchange process within which they emerge" (Buchanan 1979, 86–87, emphasis added). Exchange is predicated on secure private-property rights. In the sorting-out process, voluntary exchange between property owners is critical because it is the most important means by which the requisite information on allocative trade-offs can be revealed (Hoppe 1996, 145; Hülsmann 1997, 42). Other means include voluntary contractual easements and tort actions for proven damages. Through such complementary institutions of private property, those who would foreclose the exploitation of offshore petroleum resources will bear the opportunity costs of such decisions. The only practical way in which to accomplish this task is to limit such decisions to those who are prepared to bear the responsibility of ownership and the consequences of relinquishing particular uses of their property.

CONCLUSIONS

In the foregoing discussion, I have highlighted the long-standing political and legal conflict that has accompanied the uses of public offshore lands. Environmental externalities ostensibly give rise to the dispute, but the fact of government ownership fundamentally accounts for the discord. Resolution of this conflict requires that property rights be defined so that the owner both reaps the benefits of the chosen use and bears the opportunity cost of the most highly valued relinquished use.

Property owners with environmental objectives have demonstrated the desire and the ability to manage assets in a way that achieves both petroleum development and environmental protection.

Given the fact of public ownership, the government has a broad facilitating responsibility to foster private-property rights that avoid conflict over the trade-off between offshore petroleum development and environmental quality. Rules of tort law can be enforced so that when offshore petroleum exploration and development activities cause invasive damage to property owners, the developer responsible for the damage is held strictly liable for making restitution. Coastal property owners who fear possible future damage should have the option of entering into easement contracts with offshore operators that would obligate those operators to implement precautionary measures to reduce the perceived risk of future accidents.

A properly amended OCSLA would internalize prospective externalities by allowing those with environmental concerns to acquire leases competitively; these leases would be issued in perpetuity without royalty obligations or an obligation ever to explore or to develop the leasehold. Complete repeal of the OCSLA would facilitate unencumbered exploration under terms in which first discoverers become the sole owner of an entire petroleum reservoir. Environmental organizations might acquire such property either by becoming explorers and homesteaders of reservoirs or by directly purchasing the reservoirs after discoveries have been made.

NOTES

1. In *Cost and Choice,* Buchanan catalogs the strict and essentially impossible equilibrium conditions that must be satisfied simultaneously before opportunity cost can be objective and measurable. These conditions include the following: *(a)* decisions must be made at the margin; *(b)* decisions must be made on strictly economic or pecuniary grounds (no noneconomic considerations can prompt decisions); *(c)* no unexploited profit or arbitrage opportunities can exist anywhere in the economy; *(d)* future prices, costs, and interest rates are viewed with certainty; and *(e)* decision makers must have no sense of uncertainty regarding the nature of their utility functions (1969, 49–50). The practical implications of Buchanan's observations are that opportunity costs are always a matter of valuation and, hence, are always subjective in any context.
2. One may be tempted to argue that new valuation techniques have overcome this empirical barrier. These techniques have been categorized as *incentive-compatible demand revelation devices* (Mitchell and Carson 1989, 129). Contingent valuation purports to elicit valuations of public goods not traded in markets. The technique employs questionnaires that confront individuals with hypothetical alternatives and ask about willingness to pay or willingness to accept compensation. Murray Rothbard critically observes: "One of the most absurd procedures based on a constancy assumption has been the attempt to arrive at a consumer's preference scale . . . through quizzing him by questionnaires. *In vacuo,* a few consumers are questioned at length on which abstract bundle of commodities they would prefer to another abstract bundle, etc. Not only does this suffer from the constancy error, no assurance can be attached to the mere questioning of people. Not only will a person's valuations differ when talking about them than when he is actually choosing, but there is also no guarantee that he is telling the truth" (1997a, 217). *Demand revelation* is intended to disclose the demand for a public good by imposing on the individual voter the net marginal cost to others of including his preference for the good in the collective decision. The charge to cover this marginal cost has been labeled the Clarke Tax (Clarke 1971). However, because costs are subjective, such demand revelation has no hope of being operational. This verdict is borne out by the fact that, in practice, the Clarke Tax has never been used (Foldvary 1994, 19). These barriers to valuation also invalidate the application of pollution taxes or judicial assignment of property rights; both are devoid of an operational foundation on which to deal with the trade-off at issue.
3. The theory of public goods seems to necessitate an interventionist role for government in dealing with environmental externalities affecting large numbers of people. According to Hülsmann, "The original purpose of public-goods theory was to establish a rational criterion for government intervention. The whole point of the public-private distinction was to delimit the conditions under which it is useful or necessary that government take action" (1999, 17).

4. Issues surrounding environmental conflict certainly extend, for example, to the state ownership of oil and gas lands in offshore state waters.
5. Another way of stating this idea is to say that the uses of these public lands are never truly exposed to the test of economic calculation (Mises 1990, 26–33). Economic calculation requires secure private-property rights and the emergence of prices through a process of monetary exchange of such rights.
6. "Risk is a subjective concept unique to each individual; therefore it cannot be placed in measurable, quantitative form. Hence, no one person's quantitative degree of risk can be compared to another's, and no overall measure of social risk can be obtained. As a quantitative concept, overall or social risk is as meaningless as the economist's concept of 'social costs' or social benefits. . . . Individuals could voluntarily pool risk as in various forms of insurance. . . . Or speculators may voluntarily assume risk . . . as in the case of performance and other forms of bonding. What would not be permissible is one group getting together and deciding that another group should be forced into assuming their risk" (Rothbard 1997b, 136).
7. In explaining the logical foundation of original appropriation, Rothbard observes: "man owns what he uses and transforms. . . . His property in land and capital goods continues down the various stages of production. . . . [A]ll ownership reduces ultimately back to each man's naturally given ownership over himself and the land resources that man transforms and brings into production" ([1982] 1998, 34–40).
8. The principal environmental concern with respect to air pollution relates to ozone deterioration that may arise from the aggregate of facilities operating in the Gulf of Mexico. To date, no significant problem has emerged (personal communication with Dirk C. Herkof, air-quality specialist with the Minerals Management Service, Herndon, Virginia, December 11, 2002).
9. Under private-property rights, coastal property owners also may seek an easement agreement with offshore petroleum developers for another reason: if offshore petroleum developers were to have a homesteaded right to a small level of polluting activity, coastal property owners may seek to contract for a reduction of that level.
10. Such contractual agreements exist in other contexts. For example, in numerous cases, the Nature Conservancy is "paying ranchers, farmers and others to use more environmentally friendly practices" (Wilson 2002, 2).
11. The organizations surveyed include the Nature Conservancy, Ducks Unlimited, the National Wildlife Federation, the National Audubon Society, the Sierra Club and its subsidiaries, the Natural Resources Defense Council, the Environmental Defense Fund, Greenpeace USA, the National Park and Conservation Association, the Wilderness Society, Defenders of Wildlife, Trout Unlimited, and the Izaak Walton League (Anderson and Leal 2001, 86).
12. In *Offshore Lands,* Mead and his colleagues observe: "When the air or water is polluted, it is because no one owns these resources. . . . One way to internalize externalities is to define property rights in such a way that they are enforceable at a relatively low cost" (1985, 38). However, the authors develop no property-based approach to environmental externalities generated by offshore petroleum development.
13. Of course, joint bidding should be permissible in such a system.
14. "Little did early American jurists realize that their acceptance of this conception of mineral-right ownership would lead to a bevy of problems in the unique case of oil and gas. . . . In the case of first title [original appropriation of surface land], it is the surface land that has been transformed, not minerals below. . . . [I]t does not logically follow that the surface homesteader [or legitimate owner through purchase or inheritance] should claim an *a priori* monopoly to exclude . . . [oil and gas] owners beneath him. A tenable theory of first-title rights should have consistent application" (Bradley 1996, 70–71). Under Bradley's proposal, the means of acquiring reservoir ownership is cast as Lockean original appropriation (see Locke [1688] 1948). However, Israel Kirzner's "finders-keepers rule" appears to be an equally legitimate basis for arguing that the party that discovers and delineates the reservoir is the rightful owner (1989, 97–165).

15. Robert Bradley and Walter Mead jointly have proposed a leasing procedure that would accommodate original appropriation of reservoirs under a significantly amended OCSLA. Under their proposal, "large blocks" of land would be auctioned on a bonus-bid basis with no royalty obligation and no diligence constraints on the timing of activity on the lease. Restrictions on qualified bidders would be removed in large part so that lease holding would not be restricted to oil companies. "While title to surface land (and water) would remain in the public domain, the lease rights would remain in perpetuity with the lessee. The grant would preferably be for exploration and production of all minerals, not just oil and gas" (1998, 211). Once a discovery is made, the lessee would become the sole owner of the reservoir through an act of Lockean original appropriation. Clearly, their proposal would accommodate the acquisition of offshore leases by bidders motivated principally by environmental concerns. One notes, of course, that because their proposal is based on significant amendment of the OCSLA, they are implicitly acknowledging the legitimacy of the federal government's ownership claim to the surface.

REFERENCES

Anderson, Terry L., and Donald Leal. 2001. *Free Market Environmentalism.* New York: Palgrave.

Block, Walter. 1990. Environmental Problems, Private Rights Solutions. In *Economics and the Environment: A Reconciliation,* edited by Walter Block, 281–332. Vancouver, B.C.: Fraser Institute.

Bradley, Robert. 1996. *Oil, Gas, and Government: The U.S. Experience.* Lanham, Md.: Rowan and Littlefield for the Cato Institute.

Bradley, Robert, and Walter Mead. 1998. Resolving the Federal Oil and Gas Royalty Valuation Dispute: A Free Market Approach. *Journal of Energy and Development* 23, no. 2: 207–35.

Brätland, John. 2000. Human Action and Socially Optimal Conservation: A Misesian Inquiry into the Hotelling Principle. *Quarterly Journal of Austrian Economics* 3, no. 1: 3–26.

———. 2001. Economic Exchange as the Requisite Basis for Royalty Ownership of Value Added in Natural Gas Sales. *Natural Resources Journal* 41, no. 3: 685–711.

Buchanan, James M. 1969. *Cost and Choice: An Inquiry in Economic Theory.* Chicago: Markham.

———. 1979. General Implications of Subjectivism in Economics. In *What Should Economists Do?* edited by Richard Wagner, 81–91. Indianapolis: Liberty Press.

Clarke, Edward. 1971. *Demand Revelation and the Provision of Public Goods.* Cambridge, Mass.: Ballinger.

Coase, Ronald H. 1960. The Problem of Social Cost. *Journal of Law and Economics* 3 (October): 241–62.

———. [1938] 1981. Business Organization and the Accountant. In *L.S.E. Essays on Cost,* edited by James M. Buchanan and George F. Thirlby, 95–132. New York: New York University Press.

Epstein, Richard. 1985. *Takings: Private Property and the Power of Eminent Domain.* Cambridge, Mass.: Harvard University Press.

———. 1995. *Simple Rules for a Complex World.* Cambridge, Mass.: Harvard University Press.

Foldvary, Fred. 1994. *Public Goods and Private Communities: The Market Provision of Social Services.* Cheltenham, U.K.: Edward Elgar.

Formaini, Robert. 1990. *The Myth of Scientific Public Policy.* London: Transaction.

Herbener, Jeffrey. 1997. The Pareto Rule and Welfare Economics. *Review of Austrian Economics* 10, no. 1: 79–106.

Hoppe, Hans-Hermann.1989. *Theory of Socialism and Capitalism.* Boston: Kluwer Academic.

———. 1993. *The Economics and Ethics of Private Property: Studies in Political Economy and Philosophy.* Boston: Kluwer Academic.

———. 1996. Socialism: A Property or Knowledge Problem. *Review of Austrian Economics* 9, no. 1: 143–49.

Hülsmann, Jörg Guido. 1997. Knowledge, Judgement, and the Use of Property. *Review of Austrian Economics* 10, no. 1: 23–48.

———. 1999. Economic Science and Neoclassicism. *Quarterly Journal of Austrian Economics* 2, no. 4: 3–20.

Kirzner, Israel. 1989. *Discovery,Capitalism, and Distributive Justice.* New York: Basil Blackwell.

Langlois, Richard. 1982. Cost-Benefit Analysis, Environmentalism and Rights. *Cato Journal.* 2: 279–300.

Lee, Dwight R. 2001. To Drill or Not to Drill: Let the Environmentalists Decide. *The Independent Review* 4, no. 2: 217–26.

Lewin, Peter. 1982. Pollution Externalities, Social Cost, and Strict Liability. *Cato Journal* 2, no. 1: 205–29.

Lind, Robert, et al. 1982. *Discounting for Time and Risk in Energy Policy.* Baltimore: Johns Hopkins University Press.

Locke, John. [1688] 1948. An Essay Concerning the True Original Extent and End of Civil Government. In *The Second Treatise of Civil Government and a Letter Concerning Toleration,* edited by J. W. Gough, 1–103. Oxford, U.K.: Basil Blackwell.

Mahoney, Dan. 2002. Ownership, Scarcity, and Economic Decision Making. *Quarterly Journal of Austrian Economics* 5, no. 1: 39–56.

Mead, Walter, Asbjorn Moseidjord, Dennis Muroaka, and Phillip Sorensen. 1985.*Offshore Land: Oil and Gas Leasing and Conservation on the Outer Continental Shelf.* San Francisco: Pacific Institute for Public Policy Research.

Mises, Ludwig von. [1920] 1990. *Economic Calculation in the Socialist Commonwealth.* Auburn, Ala.: Ludwig von Mises Institute.

———. [1949] 1998. *Human Action: A Treatise on Economics.* Auburn Ala.: Ludwig von Mises Institute.

Mitchell, Robert Cameron, and Richard T. Carson. 1989. *Using Surveys to Value Public Goods: The Contingent Valuation Method.* Washington, D.C.: Resources for the Future.

Rizzo, Mario. 1979. Uncertainty, Subjectivity, and Economic Analysis of Law. In *Time Uncertainty and Disequilibrium,* edited by Mario Rizzo, 71–89. Lexington, Mass.: Lexington.

Rothbard, Murray. 1997a. *The Logic of Action One: Method, Money, and the Austrian School.* Cheltenham, U.K.: Edward Elgar.

———. 1997b. *The Logic of Action Two: Applications and Criticism from the Austrian School.* Cheltenham, U.K.: Edward Elgar.

———. [1982] 1998. *The Ethics of Liberty.* New York: New York University Press.

Stroup, Richard L., and John A. Baden. 1983. *Natural Resources: Bureaucratic Myths and Environmental Management.* San Francisco, Calif.: Pacific Institute for Public Policy Research.

Wilson, Janet. 2002. Wildlife Shares Nest with Profit: Nature Conservancy Defends "Working" Landscapes, Says It Can Both Produce Gas and Protect Rare Birds. *Los Angeles Times,* August 20. Page numbers referenced in text of this chapter refer to an Internet-subscription download, September 4, 2002.

Acknowledgments: Responsibility for the views expressed in this chapter is the author's alone. The author gratefully acknowledges thoughtful comments and helpful criticisms from Robert Bradley on an earlier version.

PART V

Urban Environments

15

Is Urban Planning "Creeping Socialism"?

RANDAL O'TOOLE

Socialism is commonly defined as government ownership of the means of production. With the exception of a number of services that are viewed as natural monopolies, such as sewer and water supplies, socialism in the form of government ownership has never achieved prominence in the United States. Instead, governments here have relied on regulation as a way of obtaining the same goals that socialists claim to seek: efficiency, equality, and control of externalities. If this approach is socialism, then urban planning has represented creeping socialism since around 1920. But it has recently accelerated and is now running rather than creeping. Moreover, it has such a head start that lovers of freedom may not be able to halt it, much less turn it around.

Urban planning rests on the ideas that urban residents impose numerous externalities on one another and that planning and regulation can minimize such externalities. Despite their claim of scientific expertise, planners often have little idea what they are doing: cities are simply too complex to understand or control. As a result, the history of urban planning is the story of a series of fads, most of which have turned into disasters. Urban renewal and public housing are two obvious examples.

Ironically, the failure of past planning is the premise for the latest planning fad, variously called new urbanism, neotraditionalism, or smart growth. Smart-growth planners see numerous problems in our urban areas, including congestion, air pollution, sprawl, unaffordable housing, disappearing open space, and costly urban services. These problems they blame on past generations of planners who, say smart-growth planners, got it all wrong. The solution, of course, is to give the current generation of planners more power than ever before because this time they claim to have it right.

SMART-GROWTH PRESCRIPTIONS

Smart-growth prescriptions include variations on the following themes:

- Metropolitan areas should be denser than they are today. In growing regions this objective is achieved by limiting or forbidding new construction on land outside the urban fringe and instead increasing the density of existing developed areas.
- Transportation should emphasize mass transit, walking, and bicycling instead of automobiles. This strategy means few or no new investments in road capacity, combined with considerable investments in transit, preferably rail transit. Investments in roads are often aimed at reducing their capacity, a concept known as traffic calming.
- Land-use planning should focus on making areas more suitable for transit, walking, and bicycling. A major way of achieving this goal is through transit-oriented developments, meaning high-density, mixed-use developments located near rail stations or along transit corridors.
- Developments also should be pedestrian friendly, meaning (among other things) narrow streets, wide sidewalks, and stores fronting on the sidewalk rather than set back behind a parking lot.

Smart growth received a public boost in January 1999, when it was endorsed by Vice President Al Gore. Metropolitan planning agencies across the nation are considering or adopting these or similar smart-growth policies. The Environmental Protection Agency has threatened to deny transportation dollars and other federal funds to many cities that do not adopt such programs.

Smart growth is attempting to reverse two strong trends of the twentieth century. First is the increasing use of personal motorized transportation. As incomes have risen, people who once walked or rode transit have chosen to purchase and drive automobiles instead. Second, and related to the first, is an increasing demand for personal living space, in the form of both house size and lot size. As autos have made transportation less expensive, people have moved beyond central cities and purchased large lots for their homes.

These trends are most obvious in the United States, but they are not uniquely American. All over the world, as incomes rise, people purchase autos and move to low-density suburbs. In the United States, smart-growth advocates blame these trends on government subsidies such as highway funding and mortgage-loan guarantees. But the same trends are observable in western European countries, where the subsidies have been directed to transit and high-density residential development while people desiring autos and low-density housing have been penalized.

As early as 1922, the architect Frank Lloyd Wright saw that new technologies were decentralizing cities. "In the days of electrical transmission, the automobile and the telephone," he said, urban concentration "becomes needless congestion—it is a curse" (Fishman 1988, 286). Today Wright would add jet aircraft and the Internet to his list of decentralizing technologies. In the United States, auto driving per capita has steadily increased by 25 to 35 percent per decade, an average of 2 to 3 percent per year, since at least the 1920s. In parallel, people have increasingly moved to low-density areas. Today nearly half of all Americans live in the suburbs, and half the remainder live in low-density small towns and rural areas.

Since the 1950s, urban critics have complained that suburbs are sterile, lifeless, and placeless. John Keats called the suburbs "conceived in error, nurtured by greed, corroding everything they touch" (1956, 7). "Little Boxes," a 1960s song by the Berkeley writer Malvina Reynolds and popularized by Pete Seeger, labeled the suburbs "ticky-tacky." More recently, James Kunstler described the suburbs as "a trashy and preposterous human habitat with no future" (1993, 105) and "the mindless twitchings of a brain-dead culture" (112). These complaints were largely aesthetic in nature and did not stop people from moving to low-density areas.

Since the 1960s, transportation critics have warned that automobiles are destroying cities. A. Q. Mowbry (1969) warned that highway advocates were planning to "blanket the nation with asphalt" (229). Jane Holtz Kay (1997) claims that the auto has diminished "both the quality of mobility and the quality of life" (19). Yet Americans continue to drive more and more.*

Smart growth represents a merger of the anti-suburb and anti-auto movements. To reverse the driving and suburbanization trends, adherents are willing to impose draconian regulations on urban residents. These include minimum density requirements, strict design codes, and limits on parking and transportation.

Minimum Density Requirements

Density requirements are the next logical step in the zoning regulation that American cities began adopting in the years just prior to 1920. Zoning was originally aimed at protecting property values from externalities (Nelson 1977). No one wants to live next to a dirty, smelly factory. For that matter, people in many

* Editor's note: For an argument that noble motives may underlie the great and growing demand for automobile transportation, see Loren Lomasky, "Autonomy and Automobility," chapter 22 in this volume.

residential areas resisted commercial developments in their midst, and people in neighborhoods of single-family homes opposed the construction of apartments.

Initially, cities adopted four basic zones: industrial, commercial, multi-family housing, and single-family housing. Early zoning was cumulative, so that any use was allowable in industrial areas; any use but industrial was allowed in commercial areas; and so forth. Eventually, refined zoning categories were developed, such as single-family residential on quarter-acre lots, on half-acre lots, and so on, but the cumulative nature was retained. No one objected if someone built on a half-acre lot in an area zoned for quarter-acre lots. After World War II, zoning became increasingly exclusive among the four basic zones. An industrial zone would have only industry; no commercial construction was allowed. But the subcategories remained cumulative: zoning might specify maximum densities, but not minimum.

In contrast, smart-growth zoning is prescriptive. It is completely exclusive, including both maximum and minimum densities. Moreover, it tends to contain many more design requirements, which will be discussed later. The minimum-density requirement can lead to a rapid transformation of a neighborhood, especially when a neighborhood is rezoned from single-family to multi-family housing. Such rezoning is common in the Portland, Oregon, urbanized area, whose regional government adopted a smart-growth plan several years ago.

The west Portland suburb of Orenco was rezoned to very high densities when a light-rail line was built nearby. Many residents owned large lots or second lots adjacent to their homes. Some planned to build a second home on those lots for their children, for their parents, or simply to sell. But the new zoning rule required instead that they build fourplexes or other multi-family housing. Constructing a single home was not allowed.

In Gresham, at the east end of the Portland light-rail line, a neighborhood of single-family homes was rezoned to multi-family housing. If a house burned down, the zoning code required the owner to rebuild in the form of an apartment. Residents who tried to sell their homes soon found that they couldn't find buyers, because banks would not make loans on houses that couldn't be rebuilt after a fire.

Even if a property owner has no plans to build on a vacant lot and does not expect to sell, transformation of a neighborhood from single-family to multi-family dwellings can be very stressful. Increased numbers of people bring increased congestion. The transient nature of the apartment dwellers can lead to crime or a reduction in property values.

Ironically, zoning was originally justified by a 1926 Supreme Court decision that allowed neighborhoods of single-family homes to use the police power of the state to keep out apartments. In neighborhoods of single-family homes, "apartment houses, which in a different environment would be not only entirely unobjectionable but highly desirable, come very near to being nuisances," said the decision in *Euclid v. Ambler Realty* (272 U.S. 365). Now zoning is being used to impose those same nuisances on such neighborhoods.

Smart-growth zones do allow single-family housing, but usually they require that such homes be placed on small lots. Where typical urban lot sizes are about 5,000 square feet (50 by 100 feet) and suburban lot sizes may be much larger, smart-growth lot sizes may be only 2,500 square feet (50 by 50 feet) or even smaller. Smart growth also encourages row houses, although in Portland planners recently indicated that even row houses aren't dense enough: the planners would rather have apartments and condominiums.

Strict Design Codes

In addition to density requirements, smart-growth zoning codes may contain highly prescriptive design codes. Certain designs, it is alleged, promote the use of the automobile and reduce a neighborhood's sense of community. The new design codes are aimed at encouraging alternatives to the auto and encouraging community feeling. A major target of residential design codes is the house with its garage in front—derisively called a "snout house." Requiring recessed garages and front porches is supposed to encourage people to walk instead of drive.

The design code may also specify tiny front and side yards, severely limiting the space people have to park their cars. For larger developments that build the streets as well as the homes, the codes specify narrow streets and limit parking to one side of the street only. Commercial design codes similarly limit parking. Whereas modern supermarkets and shopping malls usually have large parking lots between the street and the store, smart-growth design codes require that stores front directly on the street. Parking, if it is allowed at all, must be hidden in the back. This arrangement is intended to make it easier for pedestrians to reach the store and possibly to discourage auto traffic.

Besides stipulating density and design, smart-growth zoning may require mixed-use developments. For example, one code proposed in a Portland suburb would require four- to five-story buildings in which the bottom story is devoted to retail and commercial uses and the upper stories to residential. Residents could walk to shopping and perhaps even to work. Ideally, such developments would be located near rail stations or along major bus transit corridors.

Parking and Transportation Limits

Smart growth also attempts to discourage auto driving in other ways. Portland is requiring all major shopping and office centers to reduce available parking by 10 percent. Federal law requires major employers in cities with air pollution problems to find ways to reduce their employees' automobile commuting by 10 percent.

One major function that has been socialized in the United States is the provision of highways and streets. Although many nineteenth-century highways were private toll roads, twentieth-century concerns about monopolies led Americans to build and operate virtually all roads through state and municipal governments. Most road funds come from user fees, predominantly fuel taxes. Through such user fees, roads largely pay for themselves, but pricing is inefficient: Users pay the same whether they drive on a dirt road or a practically gold-plated interstate freeway; they also pay the same whether they drive at midnight or at rush hour. Better pricing could reduce congestion, but that fact is not an argument against driving.

During most of the twentieth century, transportation engineers controlled road policy and constructed roads where they were needed. But smart-growth planners say that building more roads only encourages more auto traffic. Their goal instead is to discourage driving by reducing road capacities. They call this strategy "traffic calming." It consists of putting barriers in roads to reduce speeds or flow capacities. Presently, a major suburban arterial road might have two lanes in each direction with a continuous center left-turn lane and auxiliary right-turn lanes near each intersection. The left-turn lanes allow traffic access to side streets and businesses' parking lots. The right-turn lanes allow people to slow and turn without delaying nonturning traffic behind them. Traffic calming might turn such an arterial into a boulevard—a four-lane road with grass and trees in the center. Left turns would be limited to specific intersections, and right-turning traffic would delay cars behind it. The result would be a reduction in speeds and road capacities. The city of Portland is spending $2 million per year on traffic calming.

THE EFFECTS OF SMART GROWTH

Daniel Chirot (1991) claims that eastern Europeans would have accepted enormous restrictions on their freedom if communism had been economically successful. "Almost everything else could have been tolerated if the essential promise was on its way to fulfillment" (21). Similarly, Portlanders and other urbanites tolerate smart-growth regulations because they have been promised that those regulations will improve the livability of their cities. But will the regulations actually do so?

Livability, like sustainability and community (two other smart-growth promises), is a slippery concept. But a review of the smart-growth literature suggests that livability is supposed to comprise less congestion, cleaner air, affordable housing, lower urban-service costs, preservation of open space, and a stronger sense of community. Given the record of central planning in other applications, it is not surprising that smart growth fails almost all of these tests. All except the last are quantifiable, and the available evidence indicates that smart growth produces exactly the opposite of what it promises.

Congestion

The Sierra Club (1998) and other smart-growth advocates claim that urban sprawl—the pejorative term for low-density suburbanization—increases congestion because people have to drive more miles to get to where they are going. In fact, as University of Southern California planning professors Peter Gordon and Harry Richardson (1997) point out, "Suburbanization has become the dominant and successful congestion reduction mechanism" (95).

Smart growth's claim to reduce congestion relies on studies indicating that people living in denser areas that are well served by transit drive less. These studies ignore the self-selection process whereby people who want to drive less tend to live in areas where they can get around without cars. But even if the studies were correct, smart growth would still increase congestion. Doubling an area's density will reduce traffic congestion only if the average person living in that area reduces driving by more than 50 percent. But the smart-growth studies indicate that doubling density reduces driving per capita by only 10 to 30 percent. This outcome will significantly increase congestion.

Portland's smart-growth plan calls for increasing the population density by two-thirds, housing far more people in apartments and transit-oriented developments, and building a total of 120 miles of rail transit but few new roads. In 1990, Portlanders used autos for 92 percent of their urban travel and mass transit for less than 2.5 percent (the remainder is walking and bicycling). Planners optimistically project that their plan will increase transit usage to nearly 5 percent, while walking and bicycling will increase from 5 to 7 percent. This result means that auto's share of the travel market will decline to 88 percent—hardly a significant change. This slight decrease in driving per capita will be overwhelmed by the projected 75 percent increase in population. Planners calculate that the plan will triple traffic congestion, greatly slowing travel times in the region.

Table 1
Average Population Density of Urbanized Areas, by EPA Air Pollution Rating

Pollution Rating	Population Density
Extreme	5,381
Severe	3,027
Serious	2,378
Moderate	2,077
Marginal	1,744
None	1,505

Note: "Urbanized area" is a Census Bureau term that includes the central city of a metropolitan area plus all adjacent land with population density greater than 1,000 people per square mile.

Source: Density (persons per square mile) from U.S. Bureau of the Census 1993; smog ratings from EPA Office of Air Quality and Standards.

Although smart-growth advocates use congestion as a bogeyman to attract supporters, it is hard to see that their goal is anything but a significant increase in congestion. Some privately hope that the increased congestion will lead people to drive less, although the data indicate otherwise. But the planners aren't always so private about the congestion issue. Portland's regional transportation plan says that "congestion signals positive urban development" (Metro 1996, Ch. 1, 20). Earl Blumenauer, formerly a Portland city commissioner and currently its representative in Congress, told National Public Radio that congestion "is exciting. It means business for merchants," apparently because frustrated drivers will stop and shop (Inskeep 1997). The Twin Cities Metropolitan Council has declared a twenty-year moratorium on highway construction in the explicit hope that "as traffic congestion builds alternative travel modes will become more attractive"—including the bus system that happens to be run by the same council (Metropolitan Council 1996, 54).

Air Pollution

It is a chapter of faith among smart-growth advocates that fewer miles of automobile driving will automatically lead to less pollution. But pollution is more complicated than just miles driven. Cars pollute more when they are cold, because catalytic converters don't work until they are warmed up. Engines must work harder and therefore pollute more when they accelerate. Up to about 45 miles per hour for some pollutants and 55 miles per hour for others, cars pollute less at higher speeds. Therefore, a transportation system that results in many short trips at slow speeds in stop-and-go traffic will produce far more pollution than one that results in longer trips in free-flowing traffic averaging 45 miles per hour. Because smart growth is more likely to produce the former conditions, it could significantly degrade air quality. Indeed, Portland planners predict that their plan will lead to a 10 percent increase in smog (Metro 1998).

Table 1 shows that there is a close association between urban densities and air pollution as measured by EPA pollution ratings. The worst pollution is associated with the highest average population densities. The least polluted cities have the lowest densities. The densest urbanized area in the United States, Los Angeles, is also the only city rated as having "extreme" air pollution problems.

Affordable Housing

By building more apartments, condominiums, and homes on tiny lots, smart growth is supposed to result in more affordable housing. But if people don't want to live in those kinds of homes, it doesn't matter how affordable they are. Polls and market data indicate that people prefer homes on relatively large lots (National Association of Home Builders 1999). In most housing markets, the cost of land is a small share of the cost of housing, so the large lots do not make housing unaffordable. But the urban-growth boundaries and other smart-growth density tools create an artificial shortage of land, leading to significant increases in the cost of the type of housing that people want.

The National Association of Home Builders makes quarterly estimates of housing affordability in major urban markets. These estimates are based on the share of households in the market earning enough income to purchase a median-priced home in that market.

When Portland and other Oregon cities drew their urban-growth boundaries in 1979, these areas included an estimated twenty years' supply of vacant land. By 1989, much of that vacant land was still available, and Oregon's urban housing markets were rated among the most affordable in the nation. The vacant land soon became much scarcer, however, and by 1996 Portland-area land prices had sextupled. The Home Builders then rated Portland among the five least affordable housing markets in the nation. By 1998, three of the four Oregon urban areas rated by the Home Builders were among the ten least affordable housing markets, and the fourth was among the twenty least affordable. Oregon cities have grown during the 1990s, to be sure, but so have other cities. Las Vegas, Reno, Boise, and Phoenix are among the many cities that have grown faster than Oregon cities, yet their housing markets are not rated as unaffordable.

To deal with the rising housing costs, the city of Portland has passed an ordinance requiring that any development with more than ten housing units must set aside at least 20 percent of those units for low-income housing. Portland planners estimate that this requirement will result in the construction of about 1,600 low-income units per year. At this rate, it will take more than 65 years to provide housing to all current low-income families. In the meantime, notes the Portland consulting firm Hobson-Johnson, Portland's ordinance will drive up the cost of housing for everyone else—including the low-income people not lucky enough to be immediately housed in a low-income unit.

Urban-Service Costs

Another major smart-growth claim is that low-density suburbanization costs society more than high-density development because of the high cost of extending urban services—sewerage, water supply, roads, and schools—to low-density areas. But the *Costs of Sprawl* studies that claimed to demonstrate this relationship were all based on hypothetical data (Real Estate Research Corporation 1974). Research by Dr. Helen Ladd (1992), of Duke University, compared the *actual* costs of urban services in hundreds of U.S. counties. Ladd found that, at population densities above 200 people per square mile (approximately the density of rural Connecticut), increased density led to higher urban service costs.

The *Costs of Sprawl* studies compared the hypothetical costs of a high-density versus a low-density development on vacant land. Very different, however, is the smart-growth plan of redeveloping existing low-density neighborhoods to higher densities. Such redevelopment can be extremely expensive, because it often requires tearing up existing infrastructure to install higher-capacity services. In 1980, San Diego adopted a plan that encouraged infill development of the inner city and discouraged low-density development in the suburbs. By 1990, the city faced a $1 billion infrastructure shortfall as the existing water, sewer, and other infrastructure could not handle the new, higher densities (Calavita 1997).

Urban-service costs are driven even higher by commercially unrealistic smart-growth zoning codes. Most urban areas have sufficient high-density housing to meet demand. Developers are naturally reluctant to build more such housing for a soft market. To make transit-oriented developments feasible, local governments must provide subsidies and tax breaks.

Portland, Oregon, built light-rail lines in the hope that rail transit would stimulate development, particularly high-density development. Ten years after the first light-rail line was completed, Portland city councillor and smart-growth advocate Charles Hales realized that little development was taking place along the rail line. He therefore persuaded the city council to offer ten years of property-tax breaks to developers. At the east end of Portland's light-rail line, the city of Gresham gave a developer $400,000 worth of tax breaks and outright grants to produce a higher-density apartment structure than the developer had originally planned. West of Portland on the light-rail line, the city of Beaverton gave $9 million of tax breaks and infrastructure subsidies to a transit-oriented development called Beaverton Round. The developer has been unable to find tenants and, near bankruptcy, has asked for and received $3.4 million in additional subsidies (Fentress 1999).

Rail transit itself requires huge subsidies. Portland planner John Fregonese says that light rail "is not worth the cost if you're just looking at transit. It's a way to develop your community at higher densities" (Hall 1995). Although originally promoted as less expensive than highways, rail transit projects being considered by more than sixty U.S. cities typically cost around $50 million per mile—enough to build 2.5 miles of four-lane freeway. Rail transit is often advertised as capable of carrying as many people as an eight-lane freeway, but most U.S. light-rail lines today carry fewer people than one lane of a freeway (Cox 1999).

Preservation of Open Space

For the nature-oriented Sierra Club, the prime value of high-density development is that it protects farms, forests, and open space. But rural open space is not in short supply in this country. According to the Natural Resources Conservation Service (2002), all developed lands, both urban and rural, occupy just 5.5 percent of the lower forty-eight states. According to the 1990 census, urban lands alone occupy just 2 percent (U.S. Bureau of the Census 1993, table 8).

What is in short supply is urban open spaces: the parks, golf courses, urban farms, and even the large backyards regularly used and enjoyed by urban residents. Smart growth targets these urban open spaces for redevelopment. In the Portland urbanized area, for example:

- Clackamas County rezoned a golf course that had been zoned as open space, to promote the development of 1,100 homes and 250,000 square feet of office space—all needed to meet the region's density targets.
- The city of Portland has actually sold city park lands to developers on the condition that they develop them as high-density residential areas.
- More than ten thousand acres of prime farm lands have been targeted for development.
- Some three dozen neighborhoods of mainly single-family homes have been targeted for redevelopment into high-density centers (Metro 1998, map). The large backyards of these homes are all considered developable building sites. Despite these measures, there is no certainty that smart growth will protect large rural open spaces from fragmentation and development. If smart growth leads to congested, polluted cities, many residents may flee to rural areas. A few hundred ten- to forty-acre exurban home sites can occupy as much land as thousands of quarter-acre suburban lots.

Sense of Community

Community is difficult to quantify, but at least one urban sociologist is convinced that high-density cities produce no greater sense of community than low-density suburbs. In the 1950s, Herbert Gans spent two years living in a high-density Boston neighborhood, then two years living in Levittown, New Jersey. Gans (1967) found a great deal of community involvement in Levittown, particularly involvement in zoning and planning decisions. Gans (1982) found no stronger sense of community in Boston's West End. Indeed, he concluded that West Enders felt a loyalty to their ethnic or social group, but "the West End as a neighborhood was not important to West Enders" (392–95; quotation, p. 104).

Gans (1961) also challenged those who claim that suburbs are "lifeless." He observed that in inner-city working-class neighborhoods "the home is reserved for the family, so that much social life takes place outdoors. . . . The street life, the small stores that traditionally serve ethnic groups and other cultural minorities, and the area's exotic flavor then draws visitors and tourists" (172). Meanwhile,

> in middle-class [suburban] neighborhoods, there is no street life, for all social activities take place inside the home. . . . Such neighborhoods look dull, notably to the visitor, and therefore they may seem to be less vital

> than their ethnic and bohemian counterparts. But visibility is not the only measure of vitality, and areas that are uninteresting to the visitor may be quite vital to the people who live in them. (172)

HIDDEN AGENDAS

If smart growth performs so poorly, then why does anyone support it? A close examination of smart growth's supporters reveals that most have hidden agendas. Major advocates include central-city officials eager to restore or maintain the prominence of their cities over the suburbs; downtown interests desiring to reverse the decline of their businesses relative to those in suburban malls and edge cities; transit agencies and employees seeking ever bigger budgets despite transit's falling market share of commuting and other urban trips; "new urban planners" interested in trying their theories out on various cities; urban environmentalists opposed to more freeways and the automobile in general; and engineering and construction firms seeking federal dollars to spend on urban public works such as rail transit.

All of these groups would benefit from suburban congestion. Congestion in the suburbs would make central cities and downtown areas relatively less unattractive. Congestion also is used to justify larger transit budgets, even though transit's market share in most cities is so small—typically under 5 percent of urban trips—that it has little effect on congestion. Increasing congestion leads to demands for planning and new public works to solve the problem. Environmentalists who dislike autos hope that congestion will lead people to choose some other mode of transportation. Congestion is thus a natural goal of smart growth. We should not be surprised if smart-growth proponents make statements such as "congestion signals positive urban development."

THE MOVE TO REGIONAL GOVERNMENT

In most U.S. cities, the smart-growth coalition described in the preceding section has little political power over the suburbs. Most suburbs have a long history of resisting annexation or merger with their central cities. To overcome that resistance, smart-growth advocates support regional government agencies with authority over both the central city and the suburbs.

Some writers are explicit that the purpose of regional government is to prevent local areas from democratically resisting smart-growth proposals. Douglas

Porter (1991) of the Urban Land Institute writes "about the gap between the daily mode of living desired by most Americans and the mode that most city planners and traffic engineers believe is most appropriate" (65), He supports "regional agencies [with] substantial powers to influence local decision making on land use issues" (74) and cites Portland's Metro as an example of such an agency. Metro was created in 1992 by a ballot measure misleadingly titled "limits regional government." The agency has ultimate land-use and transportation planning authority over twenty-four cities and three counties. It has used that authority to give those cities and counties population targets that they must meet by rezoning existing neighborhoods to higher densities.

Though not a smart-growth supporter, the economist Anthony Downs of the Brookings Institution recognizes that a regional government made up of local government representatives "can take controversial stands without making its individual members commit themselves to those stands. Each member can claim that 'the organization' did it or blame all the other members" (Downs 1992, 133). Downs's description fits what has happened in Portland. It is a truism in planning that most members of the public will not get involved until it directly affects their own neighborhoods. Metro was able to write its plan for the Portland area with little public notice or involvement. But cities are encountering fierce opposition from neighborhoods that do not wish to be densified. The cities say that Metro is forcing them to densify. Metro replies that it is not forcing cities to densify any specific neighborhood, only to meet certain goals.

Neighborhood residents are confused and uncertain about how to stop rezoning. Voters in one suburb recalled their mayor and members of the city council, but the new council still must meet Metro's targets. Another suburb voted to ignore Metro's targets, though the vote has no legal effect.

The 1992 Metro ballot measure won partly because its supporters promised that a regional planning agency would prevent Portland from becoming like Los Angeles, which is the most congested city in America (Texas Transportation Institute 1998, table 1). Just two years later, Metro planners compared the nation's fifty largest cities to see which was most like their goals for Portland: high densities with few miles of highway per capita. They discovered that Los Angeles is the highest-density urbanized area in America, with a density 30 percent greater than that of the New York urbanized area (which includes northeastern New Jersey and southwestern Connecticut). Moreover, Los Angeles also has the fewest miles of freeway per capita—about 50 miles per million people compared with an average for U.S. urban areas of about 120. Crowding combined with inadequate highways explains why Los Angeles is so congested.

Metro (1994) admitted that "in public discussions we gather the general impression that Los Angeles represents a future to be avoided." Yet "with respect to density and road per capita mileage it displays an investment pattern we desire to replicate" (7) in Portland. Metro has approved its plan to "replicate" Los Angeles in Portland. But implementation is proving difficult. Portland sub-urbanites are unwilling to accept the restrictions on their freedom that smart growth demands.

Michael McCormick (1997), the author of smart-growth legislation recently passed by the Washington state legislature, lamented such resistance recently at a conference held in Vancouver, B.C. "I like British Columbians because of their willingness to be governed," he admitted. "They accept regulation and I just think, wouldn't it be great if we could have that south of the border?"

The Environmental Protection Agency thinks it has the way to overcome resistance by suburbs that don't want to be under the thumb of a regional government. Harriet Tregoning (1998), the former director of the EPA's urban affairs division, endorsed regional government and supported giving it teeth by withholding federal transportation funds from local governments that refuse to cooperate. At the same time, the open-space funds proposed by Vice President Gore were to be a carrot that was given only to communities that adopt smart-growth policies.

CONCLUSION

Smart growth is a threat to freedom of choice, private property rights, mobility, and local governance. Although smart-growth policies seem drastic, they are really a natural extension of the zoning laws that cities have adopted since the 1920s. Those zoning laws have been made increasingly restrictive over the years, and smart growth will make them even more prescriptive. Smart growth is clearly an example of creeping social regulation, if not creeping socialism.

REFERENCES

Calavita, Nico. 1997. Vale of Tiers: San Diego's Much-Lauded Growth Management System May Not Be as Good as It Looks. *Planning* 63 (3): 18–21.

Chirot, Daniel. 1991. What Happened in Eastern Europe in 1989? In *The Crisis of Leninism and the Decline of the Left: The Revolutions of 1989*, edited by Daniel Chirot. Seattle: University of Washington Press.

Cox, Wendall. 1999. New U.S. Light Rail Volumes Compared to Freeway and Arterial Lanes. http://www.publicpurpose.com.

Downs, Anthony. 1992. *Stuck in Traffic: Coping with Peak-Hour Traffic Congestion*. Washington, D.C.: Brookings Institution Press.

Fentress, Aaron. 1999. Beaverton Pays to Save Round. *The Oregonian*, July 23, C1.
Fishman, Robert. 1988. The Post-War American Suburb: A New Form, A New City. In *Two Centuries of American Planning*, edited by Daniel Schaffer. Baltimore, Md.: Johns Hopkins University Press.
Gans, Herbert J. 1961. *City Planning and Urban Realities: A Review of The Death and Life of Great American Cities. Books in Review*, pp. 170–73.
———. 1967. *The Levittowners: Ways of Life and Politics in a New Suburban Community.* New York: Pantheon.
———. 1982. *The Urban Villagers: Group and Class in the Life of Italian Americans.* Updated edition. New York: Free Press.
Gordon, Peter, and Harry Richardson. 1997. Are Compact Cities a Desirable Planning Goal? *Journal of the American Planning Association*, Vol. 63, No. 1, Winter, 1997: 95–107.
Hall, Dee J. 1995. The Choice: High Density or Urban Sprawl. *Wisconsin State Journal*, July 23.
Inskeep, Steve. 1997. Commuting IV. Washington, D.C.: National Public Radio. Available at http://www.iris.npr.org/templates/story/story.php?storyID=1039432.
Kay, Jane Holtz. 1997. *Asphalt Nation: How the Automobile Took over America and How We Can Take It Back.* New York: Crown.
Keats, John. 1956. *The Crack in the Picture Window.* Boston: Houghton Mifflin.
Kunstler, James Howard. 1993. *The Geography of Nowhere: The Rise and Decline of America's Manmade Landscape.* New York: Simon and Schuster.
Ladd, Helen. 1992. Population Growth, Density and the Costs of Providing Public Services. *Urban Studies* 29 (2): 273–95.
McCormick, Michael. 1997. Speech to the joint convention of the Washington Chapter of the American Planning Association and the Planning Institute of British Columbia. Vancouver, B.C., April 23.
Metro. 1994. *Metro Measured.* Portland, Ore.: Metro.
Metro. 1996. *Regional Transportation Plan Update.* Portland, Ore.: Metro.
Metro. 1998. *Region 2040 Plan.* Portland, Ore.: Metro.
Metropolitan Council. 1996. *Regional Transportation Plan.* St. Paul, Minn.: Metropolitan Council.
Mowbry, A. Q. 1969. *Road to Ruin.* Philadelphia: J. B. Lippincott.
National Association of Home Builders. 1999. *Consumer Survey on Growth Issues.* Washington, D.C.: National Association of Home Builders.
Natural Resources Conservation Service. 2002. *Natural Resources Inventory*, table 1. Available at: http://www.nrcs.usda.gov/technical/land/nri02/nri02lu.html.
Nelson, Robert. 1977. *Zoning and Property Rights: An Analysis of the American System of Land-Use Regulation.* Cambridge, Mass.: MIT Press.
Porter, Douglas. 1991. Regional Governance of Metropolitan Form: The Missing Link in Relating Land Use and Transportation. In *Transportation, Urban Form, and the Environment.* Washington, D.C.: Transportation Research Board.
Real Estate Research Corporation. 1974. *The Costs of Sprawl: Environmental and Economic Costs of Alternative Residential Development Patterns at the Urban Fringe.* Washington, D.C.: Council on Environmental Quality.
Sierra Club. 1998. *The Dark Side of the American Dream.* San Francisco: Sierra Club. Available at http://www.sierraclub.org/sprawl/report98/.
Texas Transportation Institute. 1998. *Urban Roadway Congestion Annual Report 1998.* College Station: Texas A&M University Press.
Tregoning, Harriet. 1999. *Becoming Regional: A Federal Role.* Washington, D.C.: U.S. Environmental Protection Agency. Available at http://www.smartgrowth.org/library/tregoning_ground.html, an EPA web site.
U.S. Bureau of the Census. 1993. *1990 Census of Population and Housing 1990 CPH-2-1.* Washington, D.C.: U.S. Government Printing Office.

16

Eco-Industrial Parks

The Case for Private Planning

PIERRE DESROCHERS

An eco-industrial park (EIP) is a community of companies, located in a single region, that exchange and make use of each other's by-products or energy. Currently, EIPs are being promoted as a means of achieving sustainable development. Proponents argue that such a symbiotic community of businesses produces more environmental benefits than each company could realize on its own. Numerous EIPs have been planned in North and South America, Southeast Asia, Europe, and southern Africa (Ayres 1996; Gertler 1995; Indigo Development 1998; Lowe 1997).[1]

Advocates of EIPs consider the Danish coastal city of Kalundborg a model. There, the main industries and the local government turn by-products into raw materials by trading and making use of their waste streams and energy resources. Although the Kalundborg community and other similar cases developed entirely through market forces,[2] many policy analysts argue that public planners can copy and even improve on Kalundborg. Thus, Paul Hawken (1993) speculates: "Imagine what a team of designers could come up with if they were to start from scratch, locating and specifying industries and factories that had potentially synergistic and symbiotic relationships" (63). Similarly, Sim Van Der Ryn and Stuart Cowan (1996) suggest that the potential of planned industrial ecosystems is even greater than that of the industrial ecosystem that evolved in Kalundborg. Ernest A. Lowe (1997) points out that "while industrial ecosystems must be largely self-organizing, there is a significant role for an organizing team in educating potential participants to the opportunities and in creating the conditions that support the development" (58).

Despite such endorsements, the movement toward public planning of EIPs is misconceived. It rests on a misreading of the Kalundborg experience, and it reflects insufficient knowledge of how market forces historically have promoted resource recovery. In this essay, I show that Kalundborg is simply a contemporary example of how industrial loops have always worked. I compare private and public mechanisms in the development of industrial loops, and I argue that greater reliance on market forces would be a more effective way of replicating the Danish experience.

THE RATIONALE FOR PLANNING ECO-INDUSTRIAL PARKS

Although some people consider all waste a hazard to health and the environment that must be destroyed or prevented, many others consider waste an economic resource (Sinha 1993). Many who view waste as a resource label themselves "industrial ecologists," drawing on an analogy with the natural world, in which living organisms consume each other's wastes. The premise of industrial ecology is that modern industrial economies should mimic the cycling of materials in ecosystems throughout the processes of raw-material extraction, manufacturing, product use, and waste disposal. Industrial ecologists view industries as webs of producers, consumers, and scavengers, and they encourage symbiotic relationships between companies and industries. The ultimate goal of industrial ecology is to reuse, repair, recover, remanufacture, or recycle products and by-products on a very large scale (Allenby and Richards 1994, Ayres and Ayres 1996, Frosch and Gallopolous 1989, Garner and Keoleian 1995, Graedel and Allenby 1995).

Some industrial ecologists and public planners envision EIPs as networks of companies and other organizations that exchange and make use of by-products. By integrating principles of industrial ecology with principles of pollution prevention and sustainable design, such regionally localized firms should, in the view of industrial ecologists, provide one or more of the following benefits, relative to traditional, nonlinked operations: reduction in the use of virgin materials; reduction in pollution; increase in energy efficiency; reduction in the volume of waste products requiring disposal; and increase in the amount and types of process outputs that have market value (Gertler 1995).

THE KALUNDBORG EXPERIENCE

Most industrial ecologists believe that Kalundborg, a small city on the island of

Seeland, seventy-five miles west of Copenhagen, is the first recycling network in history (Garner and Keoleian 1995; Gertler 1995; Lowe, Moran, and Holmes 1996; Schwarz and Steininger 1997). In this city of twenty thousand people, the four main industries—a coal-fired power plant (Asnæs), a refinery (Statoil), a pharmaceuticals and enzymes maker (Novo Nordisk), a plasterboard manufacturer (Gyproc), as well as the municipal government and a few smaller businesses—feed on each other's wastes in transforming them into useful inputs.

This local synergy began to form in the 1970s and now operates as follows. The Asnæs power company supplies residual steam to Statoil refinery and, in exchange, receives refinery gas that formerly was flared as waste. The power plant burns the refinery gas to generate electricity and steam. It sends excess steam to a fish farm that it operates, to a district heating system that serves 3,500 homes, and to the Novo Nordisk plant.[3] Sludge from the fish farm and pharmaceutical processes becomes fertilizer for nearby farms. The power plant sends fly ash to a cement company, and gypsum produced by the power plant's desulfurization process goes to a company that produces gypsum wallboard. The Statoil refinery removes sulfur from its natural gas and sells it to Kemira, a sulfuric acid manufacturer.

A Gradual Development

Consultants did not design, nor did Danish government officials finance, Kalundborg's industrial symbiosis. Rather, it resulted from many separate bilateral deals between companies, each of which sought, on the one hand, to reduce waste treatment and disposal costs and, on the other, to gain access to cheaper materials and energy while generating income from production residue. Even today, no higher level of administration exists to manage the interactions (Lowe 1997, 59). Jorgen Christensen, a spokesperson for Novo Nordisk, is explicit on that point: "I was asked to speak on 'how you designed Kalundborg.' We didn't design the whole thing. It wasn't designed at all. It happened over time" (qtd. in Lowe 1995, 15).

Henning Grann, a Statoil employee, reinforces the point: "The symbiosis project is originally not the result of a careful environmental planning process. It is rather the result of a gradual development of co-operation between four neighboring industries and the Kalundborg municipality" (qtd. in Garner and Keoleian 1995, 28). As Nicholas Gertler (1995) sums it up, the basis of the Kalundborg system is "creative business sense and deep-seated environmental awareness," and "while the participating companies herald the environmental benefits of the symbiosis, it is economics that drives or thwarts its development" (n.p.). Much of Kalundborg's "industrial metabolism" rests on the physical proximity of plants that are compatible in terms of their material flows, plus a few well-established and widespread industrial practices, including the cogeneration of electricity and heat, the hydrodesulfurization of gas, the heating of greenhouses from excess power-plant heat, and the use of standard transportation and purification technologies.

Using Kalundborg as the model, EIP advocates often argue that public planners, following a hierarchy of consciously chosen objectives, can outperform private agents, whose priority is to maximize profit rather than to promote sustainable development (Hawken 1993, Van Der Ryn and Cowan 1996). The idea of planning EIPs has garnered increasing support in academic, business, and political circles. Numerous EIP projects are now under way in North and South America, South Africa, Asia, and Europe. In the United States, the concept has won the support of the President's Environmental Technology Initiative (PETI), the President's Council on Sustainable Development (PCSD), and the Environmental Protection Agency (EPA), which created an Eco-Industrial Park Project in 1994. Several areas have been designated "demonstration sites" for EIPs in the United States.[4]

Reading Too Much into Kalundborg

EIP advocates fail to recognize that Kalundborg's industrial symbiosis is not self-sufficient and is not limited to the industrial park. For example, Statoil (sulfur) and Asnæs (fly ash and clinker), both located in Kalundborg, sell some of their by-products to Kemira and the Aalborg Portland cement company, whose plants are located on the Jutland Peninsula. Gyproc imports its supply of virgin gypsum, still a significant input, from 2,500 miles away in Germany and even from Spain, and in the early 1990s Asnæs fish farms exported to the French market most of the two hundred tons of trout and turbot they produced annually.

Furthermore, many of Kalundborg's plants are subsidiaries of foreign-owned corporations (for example, Statoil is a Norwegian firm, and Gyproc is owned by a Dutch company). In short, Kalundborg is a typical industrial city, a nexus of trade whose firms import and export numerous components and products on a worldwide scale (Gertler 1995, Lowe 1995, Lowe, Moran, and Holmes 1996, Tibbs 1992).

No Kalundborg company ever acted on its own to exploit opportunities that did not fit within its core business, no matter how environmentally attractive those opportunities were. And when government intervention forced a linkage, the venture lost money.[5] Moreover, even though each company considers the others when making decisions, it still evaluates its own agreements independently. There is no "Kalundborgwide" assessment of performance because participating companies believe that such performance would be a complex and unrewarding standard (Lowe, Moran, and Holmes 1996, C7). A final point that EIP planners should note is that the development of Kalundborg's "industrial ecosystem" required environmental regulatory flexibility.[6]

As interest in EIPs has increased, so has research, which has brought other examples of industrial symbiosis to light. In a research project tracking industrial loops,[7] Erich J. Schwarz and Karl W. Steininger (1997) documented the same phenomenon in the Austrian province of Styria. The authors concluded that, as in Kalundborg, cost calculations triggered the development of the Styrian structure. Those findings spurred further research in the Ruhr region of Germany, which resulted in qualitatively similar findings (Schwarz and Steininger 1997, 50). Scholars who study EIPs have also discovered that the same processes have been going on for a long time in major petrochemical complexes such as the Houston Ship Channel (Lowe, Moran, and Holmes 1996, A4). Is Kalundborg unique? No, the Danish city's experience is but a contemporary example of processes that are as old as cities.

INDUSTRIAL SYMBIOSIS IN HISTORICAL PERSPECTIVE

Industrial symbiosis—that is, exchange between firms in which by-products of one industry become the valuable inputs of another—is probably as old as civilization. Certainly long before the advent of modern environmental consciousness and regulation, documents and texts were published containing detailed illustrations of by-product reuse in different industries. Examples include *Waste Products and Undeveloped Substances: Or, Hints for Enterprise in Neglected Fields* (Simmonds 1862) and *The Recovery and Use of Industrial and Other Waste* (Kershaw 1928). Other examples can be found in patent records, graduate theses, and the serial *Waste Trade Directory*, which was published by the Atlas Publishing Company beginning in 1905. The same publisher's *Waste Trade Journal* covered "every aspect of the giant Secondary Materials Industry . . . of the Free World," and its monthly affiliate *Industrial Wastes and Salvage Journal* provided a "complete roundup of national and world markets in new and secondary materials" (407). These periodicals were "revised every year to include vital, complete, up-to-the-minute listings of Dealers, Consumers, Associations, Consultants, Processes and Processors, Equipment . . . covering the Continent and all 50 states" (Lipsett 1963, 407).

The importance of industrial loops was obvious to many commentators of the past. In his classic work *On the Economy of Machinery and Manufactures*, the English polymath Charles Babbage ([1835] 1986) wrote that preventing waste in industrial production often caused "the union of two trades in one factory, which otherwise might have been separated" (217). As Peter Simmonds (1862) observed a few decades later, "in every manufacturing process there is more or less waste of the raw material, which it is the province of others following after the original manufacturer to collect and utilize. This is done now, more or less, in almost every manufacture, but especially in the principal ones of the [United Kingdom]—cotton, wool, silk, leather, and iron" (2). Some years later, the authors of the *Descriptive Catalogue of the Collection Illustrating the Utilization of Waste Products of the Bethnal Green Branch of the South Kensington Museum* also noted that many ingenious persons were busily devising "means by which rubbish may be worked up into a useful product" and that there were "few . . . great manufactures now which have not one or more of these dependent industries attached to them. These secondary products are all examples of one form of the utilization of waste" (Bethnal Green Branch Museum 1875, 4).

Following World War I, some English commentators marveled at the Germans' ability to turn waste products into resources (Spooner [1918] 1974; Talbot 1920). Frederick Talbot (1920) wrote that "the German, when he encounters a waste, does not throw it away or allow it to remain an incubus. Saturated with the principle that the residue from one process merely represents so much raw material for another line of endeavor, he at once sets to work to attempt to discover some use for refuse" (19). Clearly, what is now termed *industrial symbiosis* was prevalent in advanced economies more than a century ago.

Cities as Industrial Loops

Many authors have noted the important role that cities historically have played in resource recovery (Rathje and Murphy 1992; Sicular 1981; Sinha 1993). For example, nineteenth-century Parisian writers such as Baudelaire, Hugo, and Zola wrote moving tributes to the agricultural uses of the urban waste of the French capital (De Silguy 1989). The same processes were going on in all major European and North American cities (Bertolini 1990; De Silguy 1989; Sicular 1981). Catherine De Silguy (1989) attributes the success of nineteenth-century Flemish agriculture, whose yields were three to four times those of French agriculture, to a richer supply of urban products used as compost (78). Countless other examples of resource recovery can also be found in old texts.[8]

Jane Jacobs ([1969] 1970) gives many similar examples from the second half of the twentieth century (107–17). She writes about one producer of book paper who referred to New York City as its "concrete forest," but she argues that viewing cities as "waste-yielding mines" might be a more appropriate metaphor. Jacobs notes, however, that unlike typical mines, whose resources may be depleted over time, cities will become richer the more actively and longer they are exploited because new veins, formerly overlooked, will be continually opened. She comments that "the largest, most prosperous cities will be the richest, the most easily worked, and the most inexhaustible mines" (111).

Viewing cities as mines has a history. For example, Bernhard Ostrolenk (1941) used it in a more literal sense in his classic economic geography textbook: "Even the sources of important raw commodities are changing. . . . The time is not far distant when New York, with its growing production of scrap iron and scrap copper from junked buildings, machinery, automobiles, etc., will be as important a source of raw material for metal industries as is the Mesabi Range or Anaconda" (21).[9]

In the 1920s, Rudolf Clemen noted the conditions necessary for commercially successful waste utilization. Historically, cities have fulfilled most of these stipulations: (1) a practical commercial process of manufacture; (2) actual or potential market outlets for the new proposed by-products; (3) adequate supplies of the waste used as raw material, gathered in one place or capable of being collected at a sufficiently low cost; (4) cheap and satisfactory storage; and (5) technically trained operatives (1927, 1).

Frederick Talbot's comments in the conclusion of his book *Millions from Waste* provide evidence that cities fulfill Clemen's third condition. Talbot (1920) notes that "co-operative and individual methods [of resource recovery] . . . can only be conducted upon the requisite scale in the very largest cities where the volume of material to be handled is relatively heavy. Waste must be forthcoming in a steady stream of uniform volume to justify its exploitation, and the fashioning and maintenance of these streams is the supreme difficulty" (303). Cities typically facilitate cooperation among individuals by facilitating the communication of tacit knowledge, whether technical or commercial, and the development of trust relationships (Desrochers 1998).

The recovery of perishable industrial waste illustrates the historical function of the city in resource recovery. Some of the oldest urban archeological evidence of resource recovery comes from the late Stone Age city of Çatal Hüyük, in central Turkey. It appears that workers who specialized in recovering bones made awls, punches, knives, scrapers, ladles, spoons, bows, scoops, spatulas, bodkins, belt hooks, antler toggles, pins, and cosmetic sticks (Mellaart 1967, 214–15). Some texts written in the Roman era describe shops located near slaughterhouses that turned bones and ivory into items such as pins, tokens, buttons, components of hinges, and wall fittings (Chevalier 1993).

In the early history of eastern U.S. cities, swine were frequently raised near liquor distilleries, where they were fed on the mash (Bogart 1936, 300). One could observe the same phenomenon in Belgium, where beginning in the early nineteenth century most distilleries relocated from the countryside to cities in order to secure markets for their by-products (Dechesne 1945, 51). In 260 New York City stables, cows living on the swill of local distilleries produced most of the milk for the city (Miller 1998, 78).[10]

In his classic study of "the economic basis of urban concentration," conducted in the early twentieth century, Haig (1926) explained why, despite the advent of artificial refrigeration, perishability remained an important factor in determining the location of certain fabricating functions: "Thus if articles which spoil quickly are to be preserved by drying or canning, these processes are usually best performed near the point of extraction. New York City's canneries prove, upon analysis, to be, for the most part, salvage plants designed to preserve the surplus supplies of temporary glutted markets, supplies which would otherwise decay and be wasted. Perishability during some intermediate process of fabrication tends to bind processes together at one place" (191).

Trade and the Open Nature of Urban Industrial Loops

Although urban history supports the idea of industrial ecology, proponents of EIPs often fail to consider that those industrial loops existed because of trade. That is, cities have never been closed or self-sufficient systems, but rather open systems where various inputs and by-products are imported and exported. For example, most of the raw materials used in Çatal Hüyük—with the exception of clay, reeds, and wood—were not locally available (Mellaart 1967, 212).

Other examples of urban recovery in the last two centuries support the linkage of trade and industrial loops. The United Kingdom imported more than six thousand tons of horns and hoofs and approximately ninety-two thousand tons of bones from giraffe, elephant, horse, ox, buffalo, and whale each year in the 1870s for remanufacture (Simmonds 1875, 133–47). Those imports added to a domestic production of bones thought to be between seventy thousand and eighty thousand tons (Bethnal Green Branch Museum 1875, 49).[11] Most of the imports went to London, Birmingham, and Sheffield (Simmonds 1875, 133–47). For example, every year in Sheffield about two million shank bones of oxen were turned into knife handles, spoons, nail brushes, combs, fans, bone flats for button molds, and various other miscellaneous articles. This production was, however, only part of their use.[12]

Much the same process was going on in major cities around the world. In the middle of the nineteenth century, some 375,000 animals a year were slaughtered in New York's "animal district," located a few hundred feet from Times Square. Although the area was probably extremely unsanitary by today's standards, it was yet another prototype of EIP, in which no potential resource was wasted. Bones became handles, buttons, and inputs in textile coloration. Entrepreneurs converted marrow into tallow that chandlers, soap makers, and the rapidly expanding chemical industry found valuable. Sugar refiners and fertilizer producers made use of residual blood. Hooves became gelatin and "Prussian Blue"; hides and hair were valuable commodities; and whatever remained was hog food (Miller 1998, 82). Not just locally produced bones became valuable commodities; many railway cars loaded with buffalo bones arrived in the metropolis for transformation into button molds, knife handles, and other objects (Simmonds 1875, 98).[13]

With the advent of the Chicago stockyards, the supply of animal by-products became so regular and important that chemists set about creating new and different products (Clemen 1927). The earliest innovative work of chemists focused on food products such as oleomargarine and beef extract, but eventually the chemists turned to more distant fields such as pharmaceuticals, explosives, lubrication oils, and cosmetics (Clemen 1923). Before long, "Packingtown" became a model EIP as a number of separate satellite industries that bought unfinished by-products grew around the mammoth slaughtering plants. Large refineries took the nonuniform, steam-rendered lard of packers, refined and bleached it, and sold it on the open market. Soap factories bought various grades of tallow. Glue works made glue from bones, sinews, and various other packing-plant materials. Butterine manufacturers used neutral lard and oleo oil from packing plants to manufacture oleomargarine. Fertilizer plants carted off the pressed tankage and raw or pressed blood, dried it, and sold it as such or manufactured mixed fertilizer (Talbot 1920, Clemen 1927).

These by-product manufacturers, whether or not they were formally affiliated with a giant meat-packing firm, bought their materials in part from packers and in part from outside concerns. For example, in the manufacture of compound lard, the purchase of vegetable oils was necessary. Similarly, the soap industry needed many materials that were not necessarily by-products of the meat-packing industry (Weld, Kearney, and Sidney 1925, 141). The livestock supply that went into the Chicago packing plants came from twenty-seven states (137), and the output was marketed throughout the United States, either directly by the packers or through various brokers (169).

PRIVATE PLANNING IN THE CREATION OF ECO-INDUSTRIAL PARKS

Even though all successful localized industrial parks developed through the normal course of business, many scholars and planners contend that nowadays public intervention is necessary to create EIPs (Hawken 1993; Indigo Development 1998; Lowe 1997; Van Der Ryn and Cowan 1996). The American response largely has been to look to the Danish example as illustrative and to seek more proactive ways to model and imagine eco-industrial possibilities. However, public planning efforts are not likely to outperform market forces. To understand why, it is useful to compare the characteristic features of private and public planning.

Markets are spontaneous orders sustained by an institutional framework consisting of private property rights and the pricing process. Private planning through a market relies heavily on decentralized decision making and a trial-and-error process of discovery and improvement. Unlike public planning, the market itself has no specific purpose and is essentially an arena of voluntary exchanges in which the private goals of individual actors tend to be coordinated (Ikeda 1995). The creation of localized industrial symbiosis is a fairly common outcome of those processes. The reasons can be summarized by examining the relationship of prices and technical innovation to industrial symbiosis.

Prices and Industrial Loops

To understand how private planning leads spontaneously to the creation of industrial symbiosis, one must keep in mind that the ultimate goal of all market actions is to produce marketable goods and services using the least-cost combination of inputs. Firms cannot survive if they waste too many potentially valuable inputs. The history of resource recovery indicates that the goal of cheap production has driven much of it. Babbage ([1835] 1986) observed that "the care which is taken to prevent the absolute waste of any part of the raw material" (217) helped to reduce production costs and encouraged additional investment. A few decades later, the British authors of the *Descriptive Catalogue of the Collection Illustrating the Utilization of Waste Products* also noted that industries were paying attention to the "utilization of waste materials." They wrote, "As competition becomes sharper, manufacturers have to look more closely to those items which may make the slight difference between profit and loss, and convert useless products into those possessed of commercial value, which is the most apt illustration of Franklin's motto that 'a penny saved is twopence earned'" (Bethnal Green Branch Museum 1875, 4).

More than a generation later, the German engineer Theodor Koller ([1902] 1918) observed that "competition is so keen that even with the most economical—and therefore the most rational—labour it is difficult to make manufacturing operations profitable, and it is therefore only by utilizing to the full every product which is handled that prosperity for all may be assured" (vi).

Of course, the tendency of a company to reduce its manufacturing expense by creating new credits for products previously unmarketable is observed today in countless industries (Ayres, Ferrer, and Van Leynseele 1997; Florida 1996; Saunders and McGovern 1993). Market forces promote resource recovery because reused, remanufactured, and recycled materials are generally cheaper than virgin materials for at least three reasons: (1) the value of some residuals can be close to nothing for their producers but of much greater value to somebody else; (2) much processing has already been done in the production of residuals, thereby lowering further processing costs; and (3) residuals are often produced much closer to their potential buyers than virgin materials, thus lowering transportation costs.

So the pricing process, in a context of well-enforced private property rights, serves as a powerful feedback mechanism to prevent the waste of valuable inputs or to find productive uses for by-products. Faced with competition from other producers, all entrepreneurs and managers have no choice but to reduce waste and create industrial symbiosis either within their firms or with other businesses. As Max Muspratt, a past president of the Federation of British Industries, put it in 1928:

> In the days of my childhood, "waste not, want not" was a lesson inculcated upon all young people. Whether there was at once a suitable response in the nursery I am now too old to remember, but the same wise saying has had the constant consideration of every progressive manufacturer for at least a century. . . . Every up-to-date factory has its experts who understand the problems of their particular processes and the character of the waste produced, but it may readily happen in the future, as in the past, that the waste of one industry has no interest for that particular industry and is neglected, but it may be capable of utilisation in some entirely different industry. (qtd. in Kershaw 1928, vii)

Private Property Rights and Industrial Loops

Owners must have reasonably secure expectations of continued ownership if they are to improve or conserve resources. Private property rights are therefore essential to a free-market regime. Less understood is the notion that a private property–rights regime, backed by the common law, historically has been an efficient way of protecting the environment through legal actions for trespass and nuisance (Devlin and Grafton 1998; Meiners and Yandle 1999; Smith 1995). Thus, private owners can often achieve the environmental goals that EIP planners seek through public guidance, and they can do so without creating the drawbacks that accompany regulation.

Liability makes a firm accountable for damages caused to others, including damages caused by pollution. This property-rights framework creates clear incentives for firms to find the cheapest way of reducing their level of discharge into the environment. For example, the early Chicago meat packers initially dumped a significant portion of the nonedible parts of animals into the South Branch of the Chicago River, but the current proved too weak to carry away the by-products. The packers were sued and eventually forbidden to dispose of their refuse that way. They therefore had to transport wastes to a location sufficiently far from the city to be buried, an operation entailing considerable expense. In time, however, new uses were found for these by-products (Clemen 1923).

Of course, the common law is not flawless as an instrument of environmental protection. Multiple polluters, each inflicting small amounts of damage, are unlikely to be held liable, especially when many parties share the damage. Injuries and harms that come after long gestation periods present another challenge. Although parties who can show evidence of injury or imminent harm may have a common-law cause of action, efforts to obtain injunctions for speculative harms such as future cancer are not generally successful. However, where the common law does foster environmental protection, there is no arbitrary distinction between a useful material and a waste, as there often is in regulatory schemes.

As Supreme Court Justice George Sutherland wrote in a famous case in 1926, "Nuisance may be merely a right thing in a wrong place like a pig in the parlor instead of the barnyard" (qtd. in Meiners and Yandle 1999, 926). Similarly, in some instances, "waste is merely raw material in the wrong place" (Talbot 1920, 11), and countless toxic wastes have become useful inputs within the institutional framework of prices and private property (Bethnal Green Branch Museum 1872, 1875; Koller [1902] 1918; Simmonds 1862, 1875; Talbot 1920).

Technical Innovation and Resource Recovery

"Resources are not, they become" (De Gregori 1987, 1241); that is, they are created partly through technical innovation. The same is true of resource recovery. As Charles Lipsett (1963) has stated, "Yesterday's waste has become today's new product or chemical or food, with its own waste which through research and development will become tomorrow's new economic resource" (355). Because of technical innovation (among other things), the market process is in continual flux. Old products and markets disappear, while new ones emerge and make creative use of what were formerly waste products.

Writing about the manufacturing applications of horns, Simmonds (1875) noted that "While many of the former uses of horns for glazing purposes, for drinking cups, for horn-books, and for the bugle of the bold forester, have passed away, other and more elegant and varied applications have been found for this plastic and durable substance" (138). In his 1939 address to the American branch of the Newcomen Society, Oscar G. Mayer, an industry executive and past president of the Institute of American Meat Packers, pointed out that in addition to familiar animal by-products, such as leather, wool, soap, and oleomargarine, new discoveries utilizing animal by-products in the pharmaceutical field were being made almost every year: "What is the value to humanity of such products as pepsin, adrenaline, pituitrin, ovarian extract, pineal extract, insulin and liver extract? . . . Many other compounds of incalculable value will be discovered, for the packer's raw material is an inexhaustible biological well" (Mayer 1939, 18). New uses for waste products are more likely to be found in advanced economies because the sheer diversity of the technical, managerial, and professional capacities of their inhabitants fosters many different ways of turning residuals into resources, while also providing many different potential markets.

PUBLIC PLANNING AND INDUSTRIAL SYMBIOSIS

Public agencies, nongovernmental organizations, and university teams, rather than businesses, lead most formal EIP development efforts. These organizations have taken several approaches. One is to develop an EIP from scratch or build it around a few existing industries by providing a physical site where companies can be located near one another. Another is to create "virtual EIPs"—that is, networks of firms within a region that can exchange by-products without having to relocate (Business Council for Sustainable Development 1997, Kincaid 1999).[14] In either case, such governmental or bureaucratic managers, unlike private agents, must ultimately rely on some form of command and control because even though public planners operate in mixed economies, they do not use the profit-or-loss test to evaluate their performance.

Some authors contend that an EIP development team could outperform market processes. Analysts and consultants have therefore identified various tasks to help create EIPs. These tasks include, for example, recruiting companies to fill a void that may occur when key suppliers or customers move or go out of business, modeling the network of exchanges to reveal new opportunities, and researching technologies and markets for currently unmarketable by-products (Martin et al. 1996, 6–32).

To assess whether EIP developers can outperform private company employees, one must consider the outlook and incentives facing both public planners and private employees. The first difference between the two groups is the manner in which they view the activities of a firm. EIP planning-team members typically view private firms as producers of particular wastes or users of established by-products (Martin et al. 1996). Private company employees, on the other hand, are paid to create the most value from given inputs, not merely to produce a regular supply of waste products. As Henry Ford put it in 1926, "You must get the most out of the power, out of the material, and out of the time" (qtd. in McDonough and Braungart 1998, 83). Therefore, we can expect firms, absent regulatory constraints, typically to reduce their waste flows or to find more productive uses for their waste. Attracting a power plant to a specific location does not ensure that its by-products will be used in the way that public planners imagine. The managers and designers of a power plant may find a more efficient way to extract energy and in the process reduce or eliminate by-products that an older plant might have found profitable to sell. Innovative substitute inputs might become available for a given production process, emitting fewer by-products. Entrepreneurs might develop new and more financially rewarding uses for by-products. The rise of an input price or the lowering of the production costs of alternative producers may make the power unprofitable. Inevitably, the public planners' choices will reflect past experience, not future possibilities. To plan localized waste flows as if they are intrinsically part of the internal structure of firms or not subject to change simply does not accord with historical evidence or with the logic of market processes.

Knowledge of by-products and of production processes—that is, expertise—and how this knowledge affects resource recovery would also differ between a development team and private company employees. As noted previously, at least one author believes that a team of designers could come up with better symbiotic relationships if it started from scratch, locating and specifying industries and factories according to a grand scheme (Hawken 1993). Others argue that an EIP development team can gain a better overview of the waste currently produced than private agents (Schwarz and Steininger 1997, 55). F. A. Hayek (1980) called such faith in the superiority of central planning over decentralized decision making the pretense of knowledge. According to Hayek, the most important knowledge in a market economy is that which people acquire under the particular circumstances of time and place. In a market economy, individuals confronted with specific problems can better tap into such decentralized knowledge because they always know more about the particularities of a given situation in which they are directly involved than does a distant planner.

Even though the EIP experiments are still in their infancy, there is good reason to believe that they cannot avoid certain pitfalls of central planning, especially considering that a fundamental reason for the success of the Kalundborg industrial symbiosis is the "knowledge of the local situation that was not available to staff at corporate headquarters" (Lowe, Moran, and Holmes 1996, C12).

The only kind of knowledge that EIP developers can acquire is a synthesis of what they learn about various by-products from individuals working within firms. The planners classify this information according to broad Standard Industrial Classification (SIC)[15] specifications and then look for possible localized matches by pondering the best-known uses of these by-products. In contrast, in a private firm, employees who have to deal with by-products will typically look at a much smaller set of waste products. In so doing, they can explore more reuse possibilities and contact a larger number of potential customers. Consider, for example, some observations Kincaid (1999) made in a recent by-products survey of producers located in and around North Carolina's Research Triangle area: "Another means of increasing creative thinking about by-products was to foster interaction with people from outside individual facilities. When the interviewers sat down to review the survey booklet with facility representatives, the discussion usually resulted in the identification of promising items to add to survey responses. When the interviewer was able to take a tour of the plant, yet more reusables were usually identified. The creative process was further boosted by discussions between two or more potential partners" (93). Kincaid went on to identify two examples of excited brainstorming that resulted from such meetings:

> Two representatives from a tool manufacturing company visited an amino acids manufacturing plant to discuss a potential acids partnership. After they determined that an acids exchange might be feasible, the tool manufacturing company representatives asked, "What also do you have that we might be able to use?" This query resulted in a walk to where waste fiberboard drums were stored. These drums were lined with plastic bags, and they were originally packed with pouches of desiccant inside to keep the contents dry. The tool manufacturing representative thought his company might be able to use some of the drums, and the two men started enthusiastically brainstorming about who else might be able to use the plastic bags and desiccant pouches. The tool manufacturer sug-

> gested the Adopt-A-Highway program for the plastic bags and marinas for the desiccant.
>
> At a sawdust partnership meeting, a cabinetmaker and a hazardous waste management company representative determined that the latter's company could use the cabinetmaker's sawdust to pack hazardous waste bound for an incinerator. The two men went on to discuss how the sawdust might be used as a spill absorbent as well. This led to animated brainstorming about ways to make socks filled with sawdust for this purpose. (1999, 93)

In short, the SIC, or formal and documented, type of information acquired by conducting by-products surveys in one area is no substitute for direct interaction between technical experts.

An EIP development team typically tries to recruit companies to fill by-product niches (Martin et al. 1996). Private company employees, on the other hand, must factor in many other variables, such as finding adequate labor, materials, and energy supplies, proximity to markets, quality of life and amenities, business climate, capital availability, the need for frequent face-to-face interaction between producers and suppliers, and so forth. Central or EIP planners can never know all these factors in their totality. By-products will be crucial in location decisions only if they are the most important inputs of a firm. In that case, no public planners would be needed to lure a waste-producing or receiving firm because economic incentives would provide sufficient stimulus, as much historical evidence demonstrates.

Another consideration when comparing an EIP development team to private company employees is performance evaluation. The effectiveness of EIP developers will have to be measured, one way or another, by their capacity to create *localized* industrial loops. A private firm evaluates its employees by their capacity to transform inputs into the most valuable output, and attention is not restricted to local markets. One can imagine a situation in which two potential industrial loops are available, one that is local but less financially rewarding, another that is more lucrative but involves shipping by-products to a more distant location. The private employees would send the by-products to the location that pays more, the place where it would be used more productively, thereby saving on the use of other resources. The course of action an EIP development team selects would depend on how the team was to be evaluated. If the team members' performance evaluation were based on a demonstrated capacity to create localized loops, they would have little incentive to use the by-products in the most efficient way by sending them far away.

Ultimately, the justification for a public EIP development team rests on the tenet that private employees will not gather relevant information to create industrial loops. In some situations, of course, they might not. Breaking with an established daily routine is difficult; most company employees have an inward-looking focus, and they may not know what information is available or where to find it or may simply not have the time to get it (Côté and Smolenaars 1997, 72). However, if competition is allowed to reign, the most dynamic firms will get ahead, perhaps driving their less-innovative competitors out of business. Besides, it is a mistake to think that only employees of private manufacturing firms will seek to create industrial loops because numerous brokers historically have been and today still are involved in resource recovery. Both the insiders and outsiders of a firm would, in all likelihood, cover the tasks performed by an EIP development team.

An EIP development team can certainly spot a few business opportunities that have so far escaped the attention of market participants, but it is unlikely that such events would occur often in an industrial setting if employees currently paid to ensure regulatory compliance were instead working on finding creative uses for by-products. If "by-products" and "wastes" could be regarded as any other manufacturing outputs—that is, for their chemical composition and potential uses, not for their place in the industrial pecking order—the private sector would spend much more energy on the creation of industrial loops.

Some company employees and private brokers would almost certainly be charged with finding markets for by-products, much as they find markets for other outputs. They would gather data on possible customers, arrange meetings, hire outside consultants, attend trade shows, subscribe to industry publications, and so forth. They would negotiate deals and sign contracts to transfer by-products, covering in the process standard issues such as quality of supplies, mode and timing of delivery, and legal recourse for failure to comply with previous agreements. Ultimately, market participants not only have more experience and knowledge about particular production processes and by-products than EIP planners, they also have more financial incentives to seek relevant information and to make correct decisions.

REGULATORY OBSTACLES

One of the major reasons that all the actions just described might not occur in the private sector is not the unwillingness of private parties but the regulations that the government imposes in the name of environmental protection. Current environmental regulations are fragmented and inconsistent. Firms collect copious amounts of technical information and data for regulators who enforce compliance with government-mandated standards. Such a framework typically leads to a rigid rule orientation inherently hostile to making allowances for the differences in particular circumstances that unforeseen conditions or changes might occasion (Ikeda 1995; Wallace 1995). The bureaucrats involved in the process typically have no vision of the whole. They concern themselves almost exclusively with their specialty, creating a situation in which compliance with rules and requirements often precludes the realization of economic benefits that companies might generate by trading by-products.

Many analysts now argue that some of these regulations are unenforced or are unenforceable, inefficient, contradictory, and counterproductive; that others have instituted pervasive structural biases against new technology; and that ultimately most have proven extraordinarily costly with little benefit to show for them (Ayres and Ayres 1996; Crandall 1992; Frosch 1995; Heaton and Banks 1997; Landy and Cass 1997; Wallace 1995). Donald Geffen and Alfred Marcus (1994) note that compliance problems tend to take up a great deal of U.S. managers' time because regulations keep changing; moreover, so much effort goes into staying abreast of regulation that little time is left for pollution prevention. George Heaton and Darryl Banks (1997) write that "environmental innovators uniformly lament their treatment in the regulatory process, in which delay, uncertainty, red tape, and skepticism about new compliance techniques are widely acknowledged problems" and that "the most creative segments of the industrial community—new companies, small firms, entrepreneurs—are uniquely disadvantaged by an overall regulatory framework that erects entry barriers against new ideas" (24).

Regulatory Definitions as Barriers to Resource Recovery

By far the most glaring regulatory problems in the creation of industrial loops are the definitions of *solid waste* and *hazardous waste*[16] under the Resource Conservation and Recovery Act (RCRA), which contains more than six hundred pages of complex and intricate regulations.[17] Under the RCRA, a by-product can be a solid waste or a hazardous (solid) waste, or it can avoid a solid waste label. As Gertler (1995) has pointed out, this legal approach has led to a circular definition: "Solid waste is a discarded material, a discarded material is anything inherently waste-like, such that a solid waste is anything inherently waste-like. (Since the three criteria for defining something as a 'discarded material' are connected by 'or' and not 'and,' meeting any one of them results in the substance in question being 'discarded.') Perhaps more importantly, it is somewhat perplexing that recycled materials are defined as discarded, since to 'discard' has the common meaning 'to throw away'" (n.p.).[18] RCRA considers a solid waste "hazardous" if it is ignitable (that is, burns readily), corrosive, or reactive (for example, explosive), or if it contains certain amounts of toxic chemicals (Gertler 1995). In addition to these characteristics, the EPA has developed a list of more than five hundred specific hazardous wastes. So the "hazardous" label refers not only to the inherent properties of a substance, but also to its history relative to its "discarded" status (Frosch 1995, Gertler 1995, Volokh 1995). The ambiguity surrounding solid wastes, hazardous wastes, and secondary materials in the language of the RCRA means that all waste or by-products falling under the statute's definition of solid or hazardous waste are subject to RCRA requirements. Classification of a by-product as solid waste sets in motion a costly permitting process, and the label "hazardous waste" virtually prevents the reuse of the targeted substance, even though it might be chemically identical to or even less hazardous than a "virgin" product. For example, strict environmental laws will likely control a manufacturer that produces waste containing cyanide, a toxic hydrocarbon, or a heavy metal. Unless the firm's managers are willing to invest a great deal of resources in overcoming extremely long and complex bureaucratic barriers (getting permits, collecting data, writing timely reports, being subjected to increased liability, and so forth), the RCRA will most likely prevent the firm from processing that material into a salable product or even transporting it, except to a disposal site.

The automotive industry's anticorrosion measures to protect cars illustrate the negative effects of the RCRA. The anticorrosion process typically creates wastewater rich in zinc. In the past, producers of the sludge from this wastewater sent it to a smelter that recovered the zinc for reuse. But once government regulations in the mid-1980s designated this residual as "hazardous," the regulatory requirements became so stiff that the smelter could not accept it anymore. The zinc now ends up in a landfill, and its required handling gives rise to both an additional cost for its producers and a waste disposal problem for the rest of society (Frosch 1995, 181).

One can also look at a residual from aluminum production, the "potliner" (steel shells lined with insulation and carbon) used in the electrolytic process that converts alumina into aluminum. Such potliners are typically replaced after three to seven years and contain a mixture of carbon, aluminum, sodium, fluoride, silicon, calcium, and trace amounts of cyanide and iron. Historically, various industries reused potliner materials. Mineral wool plants used them as a source of fluoride and as a fuel substitute for coke. Cement kilns used them as a fuel supplement to replace 2 to 5 percent of their coal. Steel plants used the carbon as a fuel source and the fluoride as a substitute for the fluxing agent fluorspar. All this reprocessing ended in 1988, however, when spent potliner was first classified as a solid waste—and eventually as a hazardous waste—despite lack of scientific evidence proving that the previous industrial practices posed any threat to human health (Volokh 1995, 19–20). As an EPA assistant administrator for the Office of Solid Waste and Emergency Response has put it, the RCRA is "a regulatory cuckoo land of definition" where a substance that "wasn't hazardous yesterday . . . is hazardous tomorrow, because we've changed the rule" (qtd. in Volokh 1995, 3).

In short, the regulations governing hazardous waste management impose onerous burdens and responsibilities on those who generate, handle, treat, and dispose of such materials. It is therefore not surprising that very little hazardous waste is recovered. As Alexander Volokh (1995) has pointed out, although about one hundred thousand to one million tons of hazardous waste are recycled in the United States each year, another thirteen million tons are dumped into hazardous-waste landfills. Many analysts therefore view environmental regulation as the single greatest deterrent to the innovative use of by-products.

CERCLA (SUPERFUND)

Whereas the RCRA deals with the regulation of waste-control practices at current manufacturing, transport, and disposal facilities, the Comprehensive Environmental Response, Compensation, and Liability Act (CERCLA)—more commonly known as Superfund—was enacted in 1980 to deal with the cleanup of abandoned sites.[19] This law has a critically detrimental impact on the development of EIPs in old industrial areas.

The enforcement of CERCLA often imposes the legal doctrine of "joint and several strict liability" on everyone having anything to do with the siting and storage of hazardous wastes on so-called "brownfields" (formerly used industrial sites). In practice, this doctrine means that the government can hold anyone who is even peripherally responsible for any portion of the material at a Superfund site financially responsible for the entire cleanup. Restoring a Superfund site usually costs many millions of dollars and may involve decades of litigation. As Rose Devlin and Quentin Grafton (1998) put it, "Technically speaking, all of these firms/individuals are liable to pay up to the full costs of clean-up or any costs stemming from damages to natural resources incurred by the federal or state governments from a hazardous waste site . . . irrespective of whether their actions actually caused the accident" (115).

As a number of difficulties arose from this practice, the 1986 Superfund Amendments Reauthorization Act gave the party that paid for the cleanup costs the right to collect from any others who might have contributed to the damages. Superfund therefore has a major impact on firms' decisions regarding location and handling of by-products.

General Motors is a case in point. The firm is often reluctant to transfer regulated waste to brokers, waste exchanges, and potential users because it cannot get rid of the legal responsibility for the material, and it is not sure it can trust downstream users (Gertler 1995). Besides, the cleanup standards associated with Superfund are generally acknowledged to be unrealistically strict and therefore unnecessarily costly (Stroup 1996). Consequently, developers often fear that if they take on a cleanup effort, not only will they have to spend money to eradicate the last stray molecule of a given substance, but they will also be sued by a future user of the property to clean up contamination that they did not cause. Because the liability scheme of CERCLA is so menacing, land near hazardous-waste sites that has been or might be designated a Superfund site becomes virtually unsalable, despite potential advantages such as location and existing amenities. Vast stretches of urban real estate have been written off for development because of Superfund, in the process starting a cycle of urban blight and promoting "green field" development in distant suburbs (Landy and Cass 1997, 207). Thus, many opportunities for the natural evolution of eco-industrial parks never see the light of day.

In sum, expensive and obtrusive government regulations that have minimal positive effects on public health make the reuse of materials so difficult that in practice a larger waste flow is encouraged. Unlike the common-law approach, environmental and other regulations have resulted in the erection of numerous barriers to resource recovery.

Had Superfund-like rules been in effect in Denmark, the Kalundborg industrial symbiosis would have been a very difficult, if not impossible, task. For example, the flue gas that Statoil pipes to Gyproc and the liquid sulfur that Statoil sells to Kemira probably would not have been approved in the United States because both substances would be classified as "hazardous waste," and the new resources created from these by-products also would have been treated as hazardous under the "mixture and derived from" rule (Gertler 1995, n.p.). Furthermore, the movement of sulfur from Statoil to Kemira and the movement of scrubber-ash gypsum from Asnæs to Gyproc would violate another rule, the ninety-days-storage rule, which again in all likelihood would prevent the profitable reuse of these by-products (Gertler 1995).

Of course, local, national, and international environmental regulations are only one type of public policy that affects the propensity of managers, engineers, and technicians to find creative new uses for their by-products. Antitrust statutes, too, can effectively bar the agglomeration of enterprises necessary to create EIPs; consumer-protection statutes relating to government procurement practices and safety regulations can prevent the use of by-products; and laws governing international trade may prevent their transportation. Zoning ordinances, subdivision regulations, permits for various construction phases, and growth management may also prohibit certain activities. Subsidies targeted at the extraction and transportation of virgin materials discourage the use of by-products (Crandall 1992; Devlin and Grafton 1998; Graedel and Allenby 1995; Lowe, Moran, and Holmes 1996; Roodman 1998; Wernick and Ausubel 1997).

CONCLUSION

Planning a regional community of companies that exchange and reuse by-products or energy has been advocated in academic, business, and political circles. The examples used to justify such an approach, however—such as Kalundborg, Denmark—resulted entirely from market forces. The claim that a more proactive approach would achieve better results than market processes rests on the claim that public agents are more likely to identify and achieve a goal such as the creation of an EIP than private agents whose priority is to maximize wealth. This claim presupposes that public planners are capable of centrally planning and coordinating the activities of numerous public and private agents to conform with their specific objectives. That they can actually do so successfully is doubtful.

Private agents working in competitive businesses tend to minimize the amount of waste they produce, either by using their inputs more productively or by finding new uses for their by-products. The market process, unlike regulation, is a mechanism that inherently and systematically corrects its own errors and encourages innovation. Within a market framework, trading by-products is not a good in itself if a more effective waste solution exists upstream. Where such a solution does not exist, trade is likely to occur.

Ultimately, support for an EIP development team stems from a perception of market failure in developing industrial loops. This perception, however, presupposes that in a market economy, firms have more incentive to cover the costs of by-product disposal than to eliminate them at the source or to find new markets for them. This belief also requires that firms' employees cannot gather useful information as well as an EIP development team would. Both postulates are manifestly false, logically and empirically. Furthermore, current attempts to create EIPs are too narrowly focused and fail to consider that firms are in business to create products that consumers want, using the least-cost input combination, and therefore that firms making locational choices will evaluate many factors besides the proximate reuse of by-products.

Promoting resource recovery where it is economically feasible and environmentally sound is a commonsense action, but the current movement to plan and create EIPs rests on many errors of reasoning and of fact. Greater priority should be placed on removing barriers to reuse of materials. Otherwise, proposals to create EIPs will remain mere academic or public relations exercises.

NOTES

1. Although these communities of companies sometimes bear different labels—such as industrial symbiosis, industrial clusters, environmentally balanced industrial complexes, and localized industrial ecosystems—they all refer to the concept described in this chapter as *eco-industrial parks* (Lowe 1997, 64).
2. Other examples are located in the Austrian province of Styria, the Ruhr region of Germany, and the Houston Ship Channel (Garner and Keoleian 1995; Gertler 1995; Lowe, Moran, and Holmes 1996; and Schwartz and Steininger 1997).
3. Some residual steam was also sent to greenhouses that belonged to Asnæs, but this practice was abandoned after growers elsewhere in Denmark complained of unfair competition because the greenhouses in Kalundborg enjoyed especially low heating costs (Lowe, Moran, and Holmes 1996, C12).
4. Designated sites include the Cabazon Resource Recovery Park (California), Cape Charles (Virginia), Chattanooga (Tennessee), Civano (Tucson, Arizona), East San Francisco Bay (California), Fairfield Park (Baltimore, Maryland), the Green Gold Initiative (Buffalo, New York), the Green Institute (Minneapolis, Minnesota), Londonderry (New Hampshire), Mesa del Sol (Albuquerque, New Mexico), Plattsburgh (New York), Riverside Eco-Park (Burlington, Vermont), Trenton (New Jersey) and Triangle J Council of Governments (North Carolina). See Indigo Development 1998.
5. The Danish government required Asnæs to initiate a fish-farming operation as a way to consume excess sludge. The operation lost money until the government allowed sale of the fish farm to an independent operator, who converted it into a profitable venture. As some observers noted, fish farming "just didn't fit" into Asnæs's line of business (qtd. in Lowe, Moran, and Holmes 1996, C12).
6. Danish authorities approach environmental protection by requiring firms to submit plans detailing their continual efforts to reduce their environmental impact. According to Gertler (1995), the flexibility of this approach, coupled with the fact that the Danish Environment Ministry encourages attempts to find uses for all waste streams on a case-by-case basis, allows firms "to focus their energies on finding creative ways to become more environmentally benign instead of fighting the regulators" (n.p.). Also, the stricter environmental regulations that have been the driving force for some linkages have been performance standards rather than technology standards. Therefore, firms could choose technologies that rendered their waste streams usable as feedstock elsewhere. For a more detailed description of Danish environmental policy-making and especially of its flexible and personal approach as compared to the formulaic and procedure-driven U.S. approach, see Wallace 1995.
7. "The initial procedure was to choose a basic goods company since this could be expected to

represent a supplier and recipient of various kinds of waste. Starting here, the waste streams coming into the plant site as well as originating from it were then followed. For each new supplier and recipient thus identified the procedure was repeated until the geographic system boundary was reached" (Schwarz and Steininger 1997, 50).

8. "The worn-out saucepans and tin ware of our kitchens, when beyond the reach of the tinker's art, are not utterly worthless. We sometimes meet carts loaded with old tin kettles and worn-out iron coal-skuttles traversing our streets. These have not yet completed their useful course; the less corroded parts are cut into strips, punched with small holes, and varnished with a coarse black varnish for the use of the trunk-maker, who protects the edges and angles of his boxes with them; the remainder are conveyed to the manufacturing chemists in the outskirts of the town, who employ them in combination with pyroligneous acid, in making a black die for the use of calico printer" (Babbage [1835] 1986, 11–12).
9. According to a personal communication from an industry specialist, June 1995, New York City also became the premier copper mine in the world, as the advent of fiber optics made the old copper lines useless.
10. Of course, the same phenomenon may be seen today: "Brewing is an example of recycling. Grains are malted and fermented and the extract made into beer. The spent grains from the breweries in our large cities are transported to central plants where they are dewatered, dried, and prepared for animal feed, particularly for milk farms not too far distant, serving the same central population. The grains after drying are high-protein feed" (Mantell 1975, 753).
11. Actually, the demand for bones was so important in the nineteenth century that some merchants had the bones on the battlefields of Waterloo and Crimea picked up and sent to England to be turned into fertilizers (De Silguy 1989, 78).
12. See also Simmonds 1862, 352–60.
13. It must be pointed out, however, that St. Louis was probably the major hub in the buffalo-bones trade (exhibit at the Museum of the Rockies, Bozeman, Montana, July 1999).
14. These arrangements are said to differ from traditional waste exchanges such as Recycler's World, the National Materials Exchange Network, and the Global Recycling Network or from commodity-specific financial exchanges such as the Chicago Board of Trade market for scrap materials in that they are much more proactive in identifying by-products and creating connections (Kincaid 1999).
15. The traditional SIC is currently being replaced by the new North American Industrial Classification System (NAICS).
16. The RCRA defines a solid waste as "any garbage, refuse, sludge from a waste treatment plant, or air pollution control facility and other discarded material, including solid, liquid, semisolid, or contained gaseous material resulting from industrial, commercial, mining, and agricultural operations, and from community activities." EPA officials have encountered difficulty in establishing when a material is "discarded" and have consequently redefined solid waste as "any discarded material not otherwise excluded by regulation or variance"; a discarded material is then any material that is "abandoned," "recycled," or "inherently wastelike." RCRA defines hazardous waste as a "solid waste, or combination of solid wastes which because of its quantity, concentration, or physical, chemical, or infectious characteristics may [a] cause, or significantly contribute to an increase in mortality or an increase in serious irreversible, or incapacitating reversible illness; or [b] pose a substantial present or potential hazard to human health or the environment when improperly treated, stored, transported, or disposed of, or otherwise managed" (42 U.S.C., Sec. 1004[27]).
17. For a much more detailed treatment of this issue, see Dower [1990] 1995; Frosch 1995; Gertler 1995; Landy and Cass 1997; Lowe, Moran, and Holmes 1996; Martin et al. 1996; Powers and Chertow 1995; and Volokh 1995. It must also be noted that in addition to the RCRA, statutes such as the Toxic Substances Control Act (TSCA), the Clean Air Act (CAA), the Asbestos Hazard Emergency Response Act (AHERA), and the Clean Water Act (CWA) also affect the management of solid and hazardous waste (Martin et al. 1996, 5-5).

18. Gertler (1995) also adds that the "reason for all this intrigue is apparently that recycling status exempts a process from the bulk of RCRA regulation, making attaining such a status very attractive to those generating waste. EPA has thus endeavored to separate 'sham recycling,' which is essentially a show at recycling made by generators of hazardous waste in order to avoid the costs of mandated disposal, from 'true recycling'" (n.p.). Needless to say, such an approach ultimately works against resource recovery.
19. For a much more detailed treatment of this issue, see Dower [1990] 1995; Gertler 1995; Landy and Cass 1997; Lowe, Moran, and Holmes 1996, appendix D; Segerson 1995; Stroup 1996; Viscusi 1996; Viscusi and Hamilton 1996; and Volokh 1995.

Allenby, Braden R., and Deanna J. Richards. 1994. *The Greening of Industrial Ecosystems.* Washington, D.C.: National Academy of Engineering.

Ayres, Robert U. 1996. Creating Industrial Ecosystems: A Viable Management Strategy. *International Journal of Technology Management,* Special Issue on Changing Technological Determinants 12 (5–6): 608–24.

Ayres, Robert U., and Lesli W. Ayres 1996. *Industrial Ecology: Towards Closing the Materials Cycle.* Cheltenham, UK: Edward Elgar.

Ayres, Robert U., Geraldo Ferrer, and Tania Van Leynseele. 1997. *Eco-Efficiency, Asset Recovery, and Remanufacturing.* INSEAD Working Paper 97/35/EPS/TM. Fontenavleau, France: INSEAD and the Center for the Management of Environmental Resources.

Babbage, Charles. [1835] 1986. *On the Economy of Machinery and Manufactures.* 4th ed., enlarged. London: Charles Knight. Reprint. Clifton, N.J.: Augustus M. Kelley.

Bertolini, Gérard. 1990. *The Market for Waste: Economics and Management of Trash.* Paris: Éditions L'Harmattan.

Bethnal Green Branch Museum. 1872. *A Brief Guide to the Animal Products Collection.* London: George E. Eyre and William Spottiswoode for Her Majesty's Stationery Office.

———. 1875. *Descriptive Catalogue of the Collection Illustrating the Utilization of Waste Products.* London: George E. Eyre and William Spottiswoode for Her Majesty's Stationery Office.

Bogart, Ernest Ludlow. 1936. *Economic History of the American People.* 2d ed. New York: Longmans, Green.

Business Council for Sustainable Development–Gulf of Mexico. 1997. *By-Product Synergy: A Strategy for Sustainable Development.* Austin, Tex., April. Available: http://www.hatch.ca/sustainable_development/articles/bps_strategy_for_sd_primer.htm

Chevalier, Raymond. 1993. *Science and Technology in Rome.* Paris: Presses Universitaires de France.

Clemen, Rudolf A. 1923. *The American Livestock and Meat Industry.* New York: Ronald.

———. 1927. *By-Products in the Packing Industry.* Chicago: University of Chicago Press.

Côté, Raymond P., and Theresa Smolenaars. 1997. Supporting Pillars for Industrial Ecosystems. *Journal of Cleaner Production* 5 (1-2): 67–74.

Crandall, Robert. 1992. *Why Is the Cost of Environmental Regulation So High?* St. Louis: Center for the Study of American Business, Washington University.

Dechesne, Laurent. 1945. *The Location of Various Manufacturing Activities.* Bruxelles: Les Éditions Comptables, Commerciales et Financières.

De Gregori, Thomas R. 1987. Resources Are Not; They Become: An Institutional Theory. *Journal of Economic Issues* 21 (3): 1241–63.

De Silguy, Catherine. 1989. *The Saga of Trash from the Middle Ages to Today.* Paris: L'instant.

Desrochers, Pierre. 1998. A Geographical Perspective on Austrian Economics. *Quarterly Journal of*

Austrian Economics 1 (2): 63–83.
Devlin, Rose Anne, and R. Quentin Grafton. 1998. *Economic Rights and Environmental Wrongs: Property Rights for the Common Goods.* Northampton, England: Edward Elgar.
Dower, Roger W. [1990] 1995. Hazardous Waste. Reprinted in *Public Policies for Environmental Protection,* edited by Paul R. Portney. Washington, D.C.: Resources for the Future.
Florida, Richard. 1996. Lean and Green: The Move to Environmentally Conscious Manufacturing. *California Management Review* 39 (1): 80–105.
Frosch, Robert A. 1995. The Industrial Ecology of the 21st Century. *Scientific American* 273 (3): 178–81.
Frosch, Robert A., and Nicholas E. Gallopolous. 1989. Strategies for Manufacturing. *Scientific American* 261 (3): 144–52.
Garner, Andy, and Gregory A. Keoleian. 1995. *Industrial Ecology: An Introduction.* Ann Arbor: University of Michigan, National Pollution Prevention Center for Higher Education. Available: http://www.umich.edu/~nppcpub/resources/compendia/INDEpdfs/INDEintro.pdf.
Geffen, Donald A., and Alfred A. Marcus. 1994. *Pollution Prevention—Overcoming Barriers to Further Progress.* Advanced Management Practices Paper, no. 12. Minneapolis: University of Minnesota, Strategic Management Research Center.
Gertler, Nicholas. 1995. Industrial Ecosystems: Developing Sustainable Industrial Structures. M.S. thesis (technology and policy), Massachusetts Institute of Technology. Available: http://www.sustainable.doe.gov/business/gertler2.shtml.
Graedel, Thomas E., and Braden R. Allenby. 1995. *Industrial Ecology.* Englewood Cliffs, N.J.: Prentice Hall.
Haig, Robert M. 1926. Toward an Understanding of the Metropolis. 1. Some Speculation Regarding the Economic Basis of Urban Concentration. *Quarterly Journal of Economics* 40 (1): 179–208.
Hawken, Paul. 1993. *The Ecology of Commerce.* New York: Harper Business.
Hayek, Friedrich A. [1948] 1980. *Individualism and Economic Order.* Reprint. Chicago: University of Chicago Press.
Heaton, George R., and R. Darryl Banks. 1997. Toward a New Generation of Environmental Policy. *Journal of Industrial Ecology* 1 (2): 23–32.
Ikeda, Sanford. 1995. The Use of Knowledge in Government and Market. *Advances in Austrian Economics* 2: 211–40.
Indigo Development. 1998. *Strategic Reviews of Eco-Industrial Park Projects.* Oakland, Calif.: Indigo Development. Available: http://www.indigodev.com/Streview.html.
Jacobs, Jane. [1969] 1970. *The Economy of Cities.* Reprint. New York: Random House.
Johnsen, Bjarne. 1919. *Utilization of Waste Sulphate Liquor: A Review of the Literature.* Ottawa, Ontario: J. de L. Taché.
Kershaw, John B. C. 1928. *The Recovery and Use of Industrial and Other Waste.* London: Ernest Benn Limited.
Kincaid, Judy. 1999. *Industrial Ecosystem Development Project Report.* Research Triangle Park, N.C.: Triangle J Council of Government.
Koller, Theodor. [1902] 1918. *The Utilization of Waste Products: A Treatise on the Rational Utilization, Recovery, and Treatment of Waste Products of all Kinds.* 3d rev. ed., translated from the 2d rev. German ed. New York: D. Van Nostrand.
Landy, Marc, and Loren Cass. 1997. U.S. Environmental Regulation in a More Competitive World. In *Comparative Disadvantages? Social Regulations and the Global Economy,* edited by Pietro S. Nivola. Washington, D.C.: Brookings Institution Press.
Lipsett, Charles H. [1951] 1963. *Industrial Wastes and Salvage: Conservation and Utilization.* 2d ed. New York: Atlas.
Lowe, Ernest A. 1995. The Eco-Industrial Park: A Business Environment for a Sustainable Future. Paper presented at the conference "Designing, Financing, and Building the Industrial Park of the Future," sponsored by the U.S. Environmental Protection Agency, Research Triangle

Institute, and the University of California at San Diego, San Diego, May 4–5.

———. 1997. Creating By-Product Resource Exchanges: Strategies for Eco-Industrial Parks. *Journal of Cleaner Production* 5 (1-2): 57–65.

Lowe, Ernest A., Stephen R. Moran, and Douglas B. Holmes. 1996. *Fieldbook for the Development of Eco-Industrial Parks: Final Report.* Research Triangle Park, N.C.: Research Triangle Institute.

Mantell, C. L., ed. 1975. *Solid Wastes: Origin, Collection, Processing, and Disposal.* New York: John Wiley and Sons.

Martin, Sheila A., Keith A. Weitz, Robert A. Cushman, Aarti Sharma, Richard C. Lindrooth, and Stephen R. Moran. 1996. *Eco-Industrial Parks: A Case Study and Analysis of Economic, Environmental, Technical, and Regulatory Issues: Final Report.* Research Triangle Park, N.C.: Research Triangle Institute.

Mayer, Oscar G. 1939. *America's Meat Packing Industry.* Princeton: Princeton University Press for the Newcomen Society.

McDonough, William, and Michael Braungart. 1998. The Next Industrial Revolution. *Atlantic Monthly* 282 (4): 82–92.

Meiners, Roger E., and Bruce Yandle. 1999. Common Law and the Conceit of Modern Environmental Policy. *George Mason Law Review* 7 (4): 923–63.

Mellaart, James. 1967. *Çatal Hüyük. A Neolithic Town in Anatolia.* New York: McGraw-Hill.

Miller, Benjamin. 1998. Fat of the Land: New York's Waste. *Social Research* 65 (1): 75–100.

Ostrolenk, Bernhard. 1941. *Economic Geography.* Chicago: Richard D. Irwin.

Powers, Charles S., and Marian R. Chertow. 1995. Industrial Ecology: Overcoming Policy Fragmentation. In *Thinking Ecologically: The Next Generation of Environmental Policy,* edited by Marian R. Chertow and Daniel C. Esty. New Haven: Yale University Press.

Rathje, William, and Cullen Murphy. 1992. *Rubbish! The Archeology of Garbage.* New York: Harper Collins.

Roodman, David Malin. 1998. *The Natural Wealth of Nations.* New York: W. W. Norton.

Saunders, Tedd, and Loretta McGovern. 1993. *The Bottom Line of Green Is Black.* New York: Harper Collins.

Schwarz, Erich J., and Karl W. Steininger. 1997. Implementing Nature's Lesson: The Industrial Recycling Network Enhancing Regional Development. *Journal of Cleaner Production* 5(1-2): 47–56.

Segerson, Kathleen. 1995. *Redesigning CERCLA Liability: An Analysis of the Issues.* Policy Study, no. 187. Los Angeles: Reason Foundation.

Sicular, Daniel D. 1981. Currents in the Waste Stream: A History of Refuse Management and Resources Recovery in America. Master's thesis (geography), University of California, Berkeley.

Simmonds, Peter Lund. 1862. *Waste Products and Undeveloped Substances: Or, Hints for Enterprise in Neglected Fields.* London: Robert Hardwicke.

———. 1875. *Animal Products: Their Preparation, Commercial Uses, and Value.* New York: Scribner, Welford, and Armstrong.

Sinha, Abu Hasnat Md. Maqsood. 1993. The Formal and Informal Sector Linkages in Waste Recycling: A Case Study of Solid Waste Management in Dhaka. Master's thesis, Asian Institute of Technology, Bangkok, Thailand.

Smith, Fred L. 1995. Markets and the Environment: A Critical Reappraisal. *Contemporary Economic Policy* 13 (1): 62–73.

Spooner, Henry J. [1918] 1974. *Wealth from Waste.* London: G. Routledge. Reprint. Easton, Pa.: Hive.

Strachan, James. 1918. *The Recovery and Re-Manufacture of Waste-Paper: A Practical Treatise Printed on Paper Made Entirely from Regenerated Waste-Paper.* Alberdeen, Scotland: Albany.

Stroup, Richard L. 1996. *Superfund: The Shortcut That Failed.* PERC Policy Series, PS-5. Bozeman, Mont.: Political Economy Research Center. Available: http://www.perc.org/

publications/policyseries/superfund.php?s=2.

Talbot, Frederick A. 1920. *Millions from Waste.* Philadelphia: J. B. Lippincott.

Tibbs, Hardin C. 1992. *Industrial Ecology: An Environmental Agenda for Industry.* Washington, D.C.: U.S. Department of Energy, Center of Excellence for Sustainable Development. Available: http://www.sustainable.doe.gov/articles/indecol.shtml.

Van Der Ryn, Sim, and Stuart Cowan. 1996. *Ecological Design.* Washington, D.C.: Island.

Viscusi, W. Kip. 1996. Economic Foundations of the Current Regulatory Reform Efforts. *Journal of Economic Perspectives* 10 (3): 119–34.

Viscusi, W. Kip, and James T. Hamilton. 1996. Cleaning Up Superfund. *Public Interest* 124: 52–60.

Volokh, Alexander. 1995. *Recycling Hazardous Waste: How RCRA Has Recyclers Running Around in CERCLAs.* Policy Study, no. 197. Los Angeles: Reason Foundation.

Wallace, David. 1995. *Environmental Policy and Industrial Innovation: Strategies in Europe, the USA, and Japan.* London: Earthscan.

Weld, L. D. H., A. T. Kearney, and F. H. Sidney. 1925. *Economics of the Packing Industry. Part I. History and Organization.* Chicago: University of Chicago and the Institute of American Meat Packers.

Wernick, Iddo K., and Jesse Ausubel. 1997. Industrial Ecology: Some Directions for Research. Paper prepared for the Office of Energy and Environmental Systems, Lawrence Livermore National Laboratory. Prepublication draft, May. Available: http://phe.rockefeller.edu/ie_agenda.

Acknowledgments: A previous version of this chapter was completed while the author was a Research Fellow at the Political Economy Research Center (PERC) in Bozeman, Montana. The author would like to thank PERC's research associates and staff, especially Richard Stroup, Jane Shaw, Roger Meiners, and Daniel Benjamin, for help in completing the project. The usual disclaimer applies.

PART VI

The By-products of Environmental Bureaucracy

17

Regulation by Litigation

Diesel Engine Emission Control

BY BRUCE YANDLE AND ANDREW P. MORRISS

In the past few years, a form of federal regulation has emerged that makes the established rulemaking process look almost benign by comparison. Besides producing a steady flow of new rules to regulate the behavior of firms and industries through a process that follows due-process procedures and adds new pages to the *Federal Register*, the U.S. Environmental Protection Agency (EPA) has been filing lawsuits for the purpose of regulation. Thus, the EPA's enforcement division has added a second engine to the regulatory train. As Robert Reich, secretary of labor in the Clinton administration, has observed, "The era of big government may be over but the era of regulation through litigation has just begun" (1999). This new form of regulatory activity is called *regulation by litigation* (Viscuzi 2002; Yandle, Morriss, and Kosnik 2002).

When agency-sponsored suits are piled on top of agency rules, regulated firms and their customers get caught in a maze so filled with unexpected costs and outcomes that the motivating public-interest goals, if present at the outset, can get lost in a dizzying hurricane of briefs and penalties. After a while, for example, no one seems to be checking on air quality. Along the way, the incentive for regulated firms to settle and thus escape the costly maze becomes all the more attractive.

Reich's comment came after the attorneys general of forty-six states had signed a $246 billion settlement with the tobacco industry that specified, among other things, how the firms would market their products. In addition to providing much-welcomed revenues to hard-pressed states, the settlement leaves the industry regulated by litigation (Rhodes 2003).

As the tobacco suits were going forward, the EPA was suing and settling with the major producers of heavy-duty diesel engines. Under attack by the

EPA's enforcement division, the engine producers were charged with using EPA-approved electronic controls—developed in large part in response to earlier regulation—to "defeat" EPA's mandated engine emissions tests. In a good example of the regulatory doublespeak common at the EPA, the engine controllers were said to have "defeated" the emissions standards by ensuring that the engines met precisely the EPA standards using EPA's tests. Because the EPA's engine test focused only on simulating urban driving conditions, however, meeting the test standard allowed the engine controllers to focus on mileage rather than on emissions under highway driving conditions. In effect, the EPA sued the engine manufacturers because the engine makers had not designed their engines to meet a test procedure EPA had not created.

Despite the legal absurdity of the EPA's position, in 1998 the firms and the EPA signed a $1 billion settlement that tightens the previous regulatory standards and specifies how the industry will regulate emissions of nitrogen oxides (NO_x). Other industries caught in the EPA's regulation-by-litigation maze include major refinery operators, electric utilities, and wood-product firms.

In this chapter, we examine regulation by litigation by focusing on the EPA's suit against diesel engine manufacturers, but we seek to do more than just examine and organize the details of a complex regulatory story. Our purpose is to explain why regulation by litigation emerged when it did and to identify the circumstances in which, given the choice, regulators will select this regulatory instrument when seeking to chasten an industry or to alter industry behavior. We begin with some theoretical background on regulation, showing how theories that claim to predict regulator behavior can be applied to regulation by litigation. Then we turn to the centerpiece of our report, an analysis of the diesel engine episode of regulation by litigation.

LITIGATION AS REGULATION

Theories of regulation may help us to understand the choices regulators make when they decide to take action. For example, public-interest theory (Bernstein 1955) holds that politicians and their appointees systematically seek to serve a broad public interest, always searching for lower-cost ways to provide public benefits. Dissatisfaction with this theory's inability to predict outcomes and a recognition of what in fact seemed to be going on led to the development of the capture theory (Kolko 1963), wherein the regulated end up "capturing" the regulators. Capture theory seemed to gain traction in predicting regulatory outcomes, but it failed to predict which of several competing special-interest groups would win the contest to capture the regulator. George Stigler (1974) and Sam Peltzman (1976) developed positive theories that now form the basis of a rich body of prediction-based empirical work explaining the details of regulation (see Yandle 2001). Fred McChesney (1991) added a refinement to these theories with his "extraction" theory. Instead of explaining the features of regulations added to the legal landscape, extraction theory explains political actions taken to postpone or end regulatory ventures. In these episodes, politicians in effect extract payments from threatened firms and industries in exchange for calling off the regulatory hounds. Finally, the "bootlegger and Baptist" theory (Yandle and Buck 2002) explains how regulation can occur as one supporting group (labeled Baptists) takes the moral high ground while another (labeled bootleggers, who are reinforced by the Baptists) simply seeks to gain competitive advantage.

No one theory of regulation can be viewed as sufficient to explain all regulatory actions. However, out of these theories comes a common notion that concentrated gains and diffuse costs can help explain the regulatory process. All else being equal, regulators will attempt to spread the cost of their action across a large number of relatively powerless consumers while facilitating a concentrated political gain. Firms that have the most to lose or gain will struggle longest and hardest to alter or deflect regulatory costs. Regulation by litigation fits this pattern.

Consider the tobacco episode. The gains from the tobacco settlement are concentrated in a relatively few hands—the attorneys who bring the suits, the tobacco companies that are cartelized by the action, and state politicians who have more money to allocate to favored projects. Meanwhile, the cost of the settlement is imposed on smokers worldwide. Outside the Western world, most smokers are unaware of the settlement and its costs, and in the United States smokers may not know that the higher prices of cigarettes incorporate in part the costs of the settlement.

In a similar way, the EPA's suit against diesel engine producers involved a small number of heavy-duty engine producers, firms such as Cummins, Caterpillar, Detroit Diesel, Navistar-International, and Volvo, as well as a massive number of users who purchase diesel engine–equipped trucks and other machinery. Together, these parties bore the costs. Who were the winners? We argue later that the EPA and the Clinton White House reaped political gains from a high-profile settlement accompanied by multi-million-dollar fines levied against these few large, well-known companies. In this sense, the executive branch of the federal government was the bootlegger who operated under the cover of environmental "Baptist." As a result, the costs of the EPA's action now are spread across a huge number of truck and equipment purchasers.

Were there other winners, in a relative sense? Perhaps. We can also examine the diesel episode in another light. Regulation by litigation also may be used to divide and conquer. Traditional regulation often requires all firms in an industry to adopt similar technologies or to meet similar standards. This coincidence of interest spurs them to create a cartel, enabling them to wrest some benefits from regulation, such as higher prices. Successful suits attacking one or two major firms in the industry will disrupt this regulation-based cartel, raising costs for the firms that are sued and keeping them from opposing regulatory actions that might be taken against other firms in the industry. Indeed, the bruised firms may become quiet supporters of action against their competitors, arguing for a level playing field. Competition in the struggle to develop a "winning" regulatory technology becomes more productive than competition to build more desirable consumer products.

During important political seasons, when symbols can be more important than substance, if not themselves the substance, regulation by litigation can be used to capture valuable political gains that old-fashioned regulation cannot. Regulation by litigation is quicker, requires no extended announcements and related hearings, is not subject to administrative safeguards, and, once in the courts, is more costly to alter through political action. Unlike regulations that may be in the pipeline and not yet final, which can be cancelled by mere notice in the *Federal Register*, suits in progress tend to continue. Then, when settlements occur, the announcement of large civil penalties telegraphs to important constituents the message that enforcement actions are being taken. The size of the settlements can be esteemed as a trophy by those who favor regulation. Regulation-by-litigation trophies can then serve as glittering bookends for the boring pages of traditional regulation that rarely make headlines once the rules are in place and are operating.

THE REGULATORY OPTIONS

Like most regulatory agencies, the EPA has three options when initiating a new regulation.

First, the agency can engage in "regulation by rulemaking." This involves a notice of the proposed rule, a comment period for any and all parties to express their reactions to the agency, and a final notice of rulemaking, which responds to the comments received from interested parties. Once the regulation is final, affected parties bring suit against the agency if they have a basis for doing so.

Second, the EPA may choose a modified and somewhat less-contentious approach: "regulation by negotiation" or "negotiated rulemaking" (Harter 1982). Here, a regulatory agency decides that it will use a formal consensus-building process before coming forward in the *Federal Register* with a proposed rule. The agency determines the composition of a working group that will consider the rule to be developed; names a time, place, and moderator for the negotiation; and announces the details in the *Federal Register* along with a request for additions to the list of parties to be represented and other suggestions for improving the process. After receiving comments, the agency proceeds, seeking to accommodate objections and concerns, with the goal of gaining consensus among the negotiating parties. Following the negotiations, the agency undertakes a traditional rulemaking procedure, using the result of the negotiations as the basis for the proposed rule. Although promoted heavily as a means of reducing conflict in rulemaking, negotiated rulemaking has been disappointing (Susskind and McMahon 1985, 136). The EPA has used the procedure more than any other agency, suggesting that it sees some value to the agency in the process.

The third option, regulation by litigation, is all that negotiated rulemaking is not. Here, the regulator abandons the traditional regulation-by-rulemaking process and heads to the courts. The large hammer of suit, penalties, or settlement is used to achieve what might have been accomplished by other means.

The movement from traditional regulation to litigation in the courts raises a basic constitutional question as to where and how public policy will be made. By constitutional design, elected representatives are designated to make the collective decisions that affect everybody. Decisions to regulate guns, diesel engines, emissions from electric utilities, and drugs are political decisions made by legislative bodies and generally delegated to administrative agencies, not to courts. The courts are available to adjudicate matters of legislative intent and interpretation. They also play an important role in the enforcement of rules that have evolved through the regulatory process. When litigation is used to create substantive rules, however, rather than simply to enforce existing rules, political accountability and oversight are short-circuited.

REGULATION BY LITIGATION: EXAMPLES

Before considering the diesel engine litigation, let us review a prominent episode in which the EPA used regulatory litigation against an industry subject to the agency's traditional regulatory powers: the electricity generator industry. Filed at about the same time as the suit against diesel engine manufacturers, this case also involves compliance with technical regulations that had been in place for years. It also was motivated by efforts to reduce emissions of NO_X, the only criteria pollutant regulated by the EPA that continues to grow in spite of extraordinary efforts since 1970 to control it (Parker and Blodgett 2001).

As the nation's protector of environmental quality, the EPA faced a serious challenge in the early 1990s. Emissions of NO_X, a precursor to ozone, were so large that a major region of the northeastern United States was about to be declared in "nonattainment" with respect to the ozone standard. After such a declaration, the EPA would have to impose burdensome and politically contentious constraints on growth and transportation, so it searched for ways to gain significant reductions of NO_X emissions. One possibility was to take a dramatically new position on the EPA's New Source Review process. The definition of what actions might cause an old plant to become legally "new" and therefore subject to more stringent emission standards plagued both industry and the regulators, creating uncertainty throughout the industry.

While wrestling with the definitions, the EPA began to adopt the regulation-by-litigation tactic in connection with New Source Review requirements, starting with the wood-products industry. The EPA charged wood-products firms with having crossed the new-source line in the past while making modifications and repairs. The firms claimed that they had received EPA approval and permits for the past work, all part of the New Source Review process, but the EPA prevailed, and the firms settled.

Then, in November 1999, the EPA took aim at major electrical utility targets (Domike and Zacaroli 2000). The enforcement division issued notices of violation to seven electric utilities and administrative compliance orders to the Tennessee Valley Authority. The agency charged that since 1977 the firms had modified plants in ways that should have triggered New Source Reviews and therefore possible installation of new-source emission controls. At a press conference announcing the investigation, EPA administrator Carol Browner indicated that the enforcement action was "one of the largest investigations in the history of EPA." Attorney General Janet Reno described the effort as "one of the most significant enforcement actions in our nation's history" (Browner and Reno qtd. in Stagg 2001). The investigation and notice of violations quickly expanded to include thirty-two power plants operating in ten states.

The EPA also investigated the twenty-year history of New Source Review permits issued to firms in the petroleum-refining industry. Suits were filed against thirty-one refineries. By July 2001, the EPA had settled with BP Amoco, Koch Petroleum Group, Motiva Enterprises, Equilon Industries, Deer Park Refining, Marathon Ashland, and Premcor Refining. The record settlements covered one-third of domestic refining capacity. The EPA was moving at full throttle.

Regulation by litigation, at least in theory, can be seen as an end run around safeguards that Congress has put in place to give the legislative branch the final say on regulation. These safeguards now include the Unfunded Mandates Reform Act of 1995, the Regulatory Flexibility Act of 1980, and the Congressional Review Act of 1996. Together, these statutes provide a political mechanism that can be used to rein in overly zealous regulators, but Congress has no explicit mechanism for reining in overly zealous litigators.

Regulating Diesel Emissions by Litigation

While engaged in the New Source Review suits, the EPA also went after manufacturers of diesel engines, again in pursuit of NO_x reductions. Understanding the EPA's choice of regulation by litigation in this case requires some background on the regulation of diesel engines. Starting in 1970, the Clean Air Act required the EPA to establish "national ambient air quality standards" (NAAQS) for a number of pollutants believed to endanger public health or welfare. Unlike emissions standards, which measure the quality of direct outputs, NAAQS measure overall air quality. States must meet these standards by keeping the total emissions from all sources (including natural sources and reflecting population growth) sufficiently low. If air quality falls below one of these standards, states must have a plan to reduce emissions so that air quality will meet the standard. Failure to meet NAAQS triggers costly nonattainment sanctions, such as the withholding of funds for mass transit and transportation planning.

The Clean Air Act has different regulatory regimes for emissions from stationary and mobile sources. Stationary sources, such as coal-fired electricity generators, are regulated through state-issued permits for emissions. Mobile sources are regulated through EPA mandates requiring vehicles or engines (depending on type) to meet emissions controls. The EPA and the states routinely search for ways to reduce emissions further. The states are more likely to champion mobile-source reductions rather than to impose heavy costs on economically important stationary sources, such as plants and power stations.

The EPA and the states depend primarily on environmental modeling rather than on direct measurement to determine the impact of controls on ambient air. The predictions of models are used to establish command-and-control permits and regulations, but in this case the EPA model accuracy is notoriously faulty (Tierney 2002). As a result, faulty estimates, rather than reality, can drive regulatory measures that turn out to be unrelated to improvements in environmental quality.

By statute, states must construct EPA-mandated implementation plans that satisfy the EPA's model of air quality, even though this model may diverge from actual environmental conditions. When the relationship between model results and reality breaks down, then the EPA and the states are placed in a position of needing to "catch up."

In choosing how to do so, the EPA faces different costs and benefits from the three different modes of regulation. For example, various federal laws place some restrictions on the EPA's traditional regulation. These restrictions are especially important for industries with a long product-design cycle, such as trucks and autos. Regulations can be very disruptive and costly if imposed in midcycle. By federal statute, the EPA cannot issue regulations tightening mobile-source emissions standards without providing a four-year lead time for manufacturers. Regulations cannot change for three years after each change and must be issued four model years ahead of their effective date. So if the EPA issues one change to those regulations, its ability to issue additional changes is limited for a time.

Another factor affecting the EPA's choice of regulatory approach is that the Clean Air Act relies heavily on "technology-forcing" regulations for mobile sources; in other words, the law requires implementation of technology that does not exist when the regulations are adopted. By its nature, technology forcing may lead regulators to underestimate the time necessary to produce the needed innovations. Technology forcing creates an incentive to design vehicles to meet standards, rather than opening the door to all possibilities for reducing actual emissions (Bielaczyc and Merkisz 1999), and it requires manufacturers to invest in developing features that customers have not demanded and may even reject.

HEAVY-DUTY DIESEL ENGINES: THE REGULATORY HISTORY

The history of regulation of heavy-duty diesel engines helps us to understand why the EPA moved to regulation by litigation in 1998.[1] Over the past four decades, regulation of heavy-duty diesel engine emissions has gone from almost nothing to a relatively easy opacity test for "smoke" and then through a series of increasingly stringent standards for NO_x, hydrocarbons, and particulates. The amount of NO_x emissions allowed has fallen, for example, from a combined 16 grams per brake horsepower hour (g/bhp-hr) for NO_x and hydrocarbons in 1988 to separate limits of 0.20 g/bhp-hr for NO_x and 0.14g/bhp-hr for non-methane hydrocarbons.

TRADITIONAL REGULATION

Heavy-duty diesel engines are the engines that large trucks use. In the past, when air pollution first became a national issue, large trucks represented a relatively small part of the total truck fleet and made only a small contribution to U.S. air pollution: only 1.75 percent of total particulates, 0.02 percent of carbon monoxide, 1.9 percent of hydrocarbons, 4.8 percent of NO_x, and 0.4 percent of sulfur oxides (SO_x) in the early 1970s.

The first federal regulatory efforts were a series of traditional regulatory restrictions on heavy-duty diesel engine emissions. The Air Quality Act of 1967 established a complex approach to controlling air pollution, based on national ambient-air-quality criteria and state ambient standards. Of most importance for our purposes, the act dealt with a major concern of mobile-source manufacturers: the threat of inconsistent state standards that might force them to outfit vehicles differently for sale in different states. Spurred by smog problems in Los Angeles, California had adopted auto tailpipe emissions standards for hydrocarbons and carbon monoxide in 1966, having already begun regulating mobile-source emissions in 1961. The 1967 act preempted all future state regulation except for California's. Mobile-source emissions standards would be created thereafter only at the national level—and the regulation would include diesel engine emissions.

The initial concern with diesels was "smoke"—the heavy black fumes visibly emitted by many diesel exhausts. The first smoke standards for diesel engines applied to model year 1970. An initial standard was set for model years 1970 to 1973 and a stricter standard for model year 1974 and later. These standards remained the same through model year 1973, even after the passage of the 1970 Clean Air Act Amendments. The 1967 Air Quality Act created the format that diesel engine regulation follows to this day: specific standards for specific pollutants, standard laboratory tests to measure the emissions, and standard conditions under which the tests for emissions are to be conducted.

The year 1970 brought major changes to air pollution regulation. The Nixon administration created the EPA, and air pollution control was transferred to it from the Department of Health, Education, and Welfare. The Clean Air Act Amendments of 1970 established the basic approach to mobile sources that continues today. The statute mandated reductions in mobile-source emissions by 90 percent for hydrocarbon, carbon monoxide, and nitrogen oxides, with an initial target date of 1975. It also set the framework: mobile-source air pollution was to be controlled primarily through federally mandated technology standards on new vehicles. States were left with the regulation of in-use vehicles, a politically difficult issue for them and an authority they were not eager to exercise.

Heavy-duty diesel engines continued to receive less-stringent treatment through the early 1970s, with only smoke being regulated until the 1974 model year. At that time, engine emissions of hydrocarbon, NO_x, and carbon monoxide would begin to be regulated.

Even then, however, the EPA was concerned with potential inconsistencies between test-cycle performance and off-cycle or actual on-the-road performance of engines, although not with respect to heavy-duty engines. The agency issued an advisory circular warning light-duty vehicle and light-duty truck manufacturers that the sophisticated emission control systems then being adopted might be considered illegal defeat devices in certain circumstances. The EPA apparently had an idea that electronic controls might allow trucks to pass the tests but defeat the purpose of the controls in actual practice on the highways.

The 1977 Amendments

The 1977 Clean Air Act Amendments delayed until the 1980s the mobile-source reductions that had been mandated in 1970 but not yet met; they required states to establish inspection and maintenance programs, and they added an explicit requirement that heavy-duty diesel engines should achieve the greatest degree of emission reduction achievable consistent with cost, technical feasibility, noise, energy, and safety factors. That the EPA had done little in the heavy-duty diesel sector may have prompted Congress to specify reductions directly. The amendments called for significant reductions of hydrocarbons and carbon monoxide (by at least 90 percent during and after model year 1983), oxides of nitrogen (by at least 75 percent during and after model year 1985), and particulate matter (during and after model year 1981 or earlier, if practicable). Despite the tightening of the standards, the EPA's proposals in 1980 did not require new technology to meet them.

According to the statute, the new standards could be revised starting in 1979 and again every three years thereafter. Although the statute imposed more stringent standards, "escape valves" also were included, allowing the EPA to revise temporarily or permanently the statutory standards for several reasons, including cost.

When the EPA set out to implement its congressional mandate to reduce diesel emissions further, it proposed extensive changes in test procedures and instrumentation requirements to make the tests resemble engine-use conditions more closely. The most important development was the EPA's creation of the transient-engine test standard in 1979 to simulate urban driving conditions (U.S. House Committee on Commerce 2000, 5).

The transient system was designed to make the tests more representative of in-use conditions. The EPA selected the specific test conditions based on a survey of trucks and seven buses driven in New York City and Los Angeles. However, the agency did not attempt to validate its new test and even denied that validation was desirable. Engine manufacturers criticized the EPA's proposal for transient testing and expressed concerns about the lead time necessary to implement new standards and procedures.

Over the next decade, the EPA continued to tighten heavy-duty diesel engine emission standards under the transient test. The regulations evolved into a complex and stringent list of regulations covering hydrocarbons, carbon monoxide, NO_x, and particulates. Model year 1988 brought the first particulate standards for heavy-duty diesels, five years after the EPA imposed the first diesel particulate standards in the world on cars and light-duty trucks. Model year 1991 standards tightened the NO_x standard to 5.0 g/bhp-hr and introduced an innovation—allowance for the averaging and trading of emission credits as long as engine families met certain levels.

Truck manufacturers managed to meet the increasingly tight standards through the 1980s by improving combustion rather than by adding postcombustion treatment of exhaust. Indeed, the first particulate standards, effective in 1988, required relatively minor actions to reduce emissions levels.

One important result of the tougher clean air standards was the increasing reliance on electronic controls in mobile-source engines to meet standards and to improve performance. Whereas the first electronic controls were simply add-ons to existing engines, by the mid-1980s engine manufacturers had begun to introduce fully electronic control systems.

The 1990 Clean Air Act Amendments

The next major amendments to the Clean Air Act came in 1990, after more than a decade of political stalemate caused in part by Michigan congressman John Dingell's attempts to weaken mobile-source regulation and by political divisions over acid rain (Morriss 2000, 305–6). Section 201 of the Clean Air Act Amendments of 1990 revised the standards for emissions of hydrocarbons, carbon monoxide, NO_x, and particulates from heavy-duty vehicles or engines. The new standards pushed the envelope of technology and cost. The EPA was given the authority to revise heavy-duty vehicle or engine standards. The amendments also required that regulations remain fixed for at least three model years and set a lead time of no earlier than the model year commencing four years after promulgation of a revised standard. After the 1990 amendments, the EPA added regulations that forbade the use of defeat devices that would interfere with emissions controls in automobiles and light trucks.

By 1997, the EPA was reporting that heavy-duty diesels were the largest sources of particulates and NO_x among mobile sources. This assessment was made on a per engine basis, not on the basis of comparing gasoline and diesel fleets. Although the proportion of diesels was rising, gasoline fleets were still larger by far.

Regulation by Negotiation

While tightening standards during the 1980s, the EPA also attempted to accommodate the need for greater flexibility in engine and truck manufacture. First, the agency introduced delays in implementation because of poor economic conditions in the industry. Second, it delayed tighter standards to give manufacturers more lead time. Third, it introduced "noncompliance penalties" that allowed engines to be sold even if they exceeded the standards, as long as they did not pollute beyond an "upper limit" of acceptable pollution (U.S. EPA 1985, 35374). The companies could pay these penalties rather than keep their products off the market. These noncompliance penalties were intended to accommodate the difficulties manufacturers had in meeting the technology-forcing regulations (U.S. EPA 1985, 35375). This innovation arose from the EPA's first negotiated rulemaking exercise. Agreement on the rule was reached in four months.

REGULATION BY LITIGATION

In 1998, the EPA sued the major diesel engine manufacturers, makers of more than 95 percent of U.S. heavy-duty diesel engines. According to the EPA, diesel emissions were not declining—as the agency had predicted—but were instead increasing (Bowman 1997). In its suit, the EPA argued that the use of electronic controllers to increase fuel economy under nonurban driving conditions amounted to the employment of illegal "defeat devices" under the Clean Air Act (France n.d., 19). In essence, it contended that although the engines passed the agency's test, the electronic controller then adjusted to increase long-haul fuel economy, causing the emission of higher levels of pollutants. EPA administrator Carol Browner claimed that the engine manufacturers "programmed the engine so that it knew when it was being tested and when it was on the road" (qtd. in Johnson 1998, 1).

The engine manufacturers denied the EPA's claim that use of the controller was illegal and alleged that the EPA had known about their use of electronic controllers from the beginning and at least tacitly had approved it. A highly critical U.S. House Committee on Commerce staff report, *Asleep at the Wheel*, also concluded that the agency was aware of the engine manufacturers' use of electronic controllers as early as 1991 (2000, 14). Nevertheless, on October 22, 1998, seven U.S. heavy-duty engine manufacturers settled the enforcement actions by agreeing to pay substantial fines and to devote resources to approved environmental activities. The total resources committed amounted to $1 billion.

WHY REGULATION BY LITIGATION?

The EPA had a number of incentives to adopt regulation by litigation over other options in the case of diesel engines. Because of the legislated lead-time provisions, it could not tighten diesel emissions standards before model year 2007 through traditional regulation. Moreover, the EPA, the California Air Resources Board, and the diesel engine manufacturers had negotiated a Statement of Principles in 1995 to stabilize regulatory initiatives (U.S. Environmental Protection Agency 1995, 45602–4). Any movement toward revisions of past rules, such as those related to emission testing and controllers, would have been seen as a violation of the Statement of Principles and therefore resisted by the manufacturers.

The EPA had other incentives to adopt regulation by litigation. First, the gap between predicted and actual diesel emissions was contributing to some regions' failure to comply with NAAQS. Inspection and maintenance programs around the country, part of the effort to come into compliance, were arousing popular unrest in several states. Second, the EPA and the Clinton administration could reap immediate political rewards by appearing "tough on polluters" during the run-up to the 2000 presidential election (Wilson 1999, 3268–9). Third, the EPA faced relatively low risks of losing the litigation because the enormous leverage it had over the engine manufacturers made a settlement all but assured.

This leverage arose from the requirement for annual certification of engines. Mack's vice president of engineering and product planning, for example, told a reporter that the EPA "held a gun to our head by threatening to withhold certification for 1999" (qtd. in Galligan 1998, 2). Other companies echoed this concern. The settlement negotiations took place after the EPA had issued "conditional" certificates of conformity for model year 1998 engines. These conditional certificates did not apply to engines that employed defeat devices, which is to say that the EPA created considerable uncertainty about whether the current family of engines could be sold. At the same time, the engine manufacturers were seeking certificates for model year 1999 engines, and a related "show cause" order from the EPA to the engine manufacturers was pending (U.S. House Committee on Commerce 2000, 16). Although the EPA almost certainly could not have rejected all the heavy-duty engine manufacturers' model year 1999 engines, inasmuch as doing so would have shut down the industry, the agency did effectively create a "prisoners' dilemma" for the engine manufacturers. If they all refused to settle, the EPA would have to relent, but if any company settled while the others did not, that company would reap increased market share at the others' expense. Because the companies could not rely on each other not to settle (and antitrust law precluded their contracting to enforce such a position), each was better off settling. Thus, the companies settled.

Regardless of the merits of the EPA's case, the agency and the Clinton administration reaped a publicity windfall from the settlement. Attorney General Janet Reno, for example, was quoted as saying, "Every polluter in America had better take note of these record penalties—if you pollute America's air, you are going to pay a very high price" (qtd. in Johnson 1998, 120).

Since this settlement, the EPA has continued to call for tighter emission standards for heavy-duty diesel engines, proposing in 2001 a significant reduction for model year 2007. These standards incorporate elements based on the settlement. Japanese and European regulators also are tightening heavy-duty diesel engine emissions standards, but the standards proposed for model year 2007 in the United States are significantly tougher than those proposed for Europe and Japan.

ANTICIPATION AND RESULTS

We can now compare what EPA litigators may have anticipated when they obtained the 1998 consent and what has resulted. Taking a public-interest theory approach, one might argue that the EPA chose litigation in order to obtain large reductions of NO_x and other emissions sooner than traditional regulation would have provided them and to force technical solutions to eliminate the whole matter of defeat devices. These expectations also assume that engine producers somehow would meet the stricter emission standards on time and that truck builders and fleet operators would be eager to purchase the new engines when they emerged from the factories—that is, engine buyers would not change their normal buying patterns dramatically. Indeed, the consent agreement required that engine producers not engage in marketing to subvert the intentions of the agreement. The expectations also might assume that engine producers would leapfrog the electronic engine-controller technology and adopt a new technology that could not be operated as a defeat device.

As it turns out, these expectations have not been realized so far and may not be ever. One result of the litigation has been a huge trucking industry prepurchase response to the consent mandates, which suggests that the new engines that meet the tighter standards will not sell very well later on, so their impact in reducing emissions will be less than expected. This prebuy of trucks occurred despite the settlements' explicit ban on advertising to promote it, and it was driven entirely by customer unease over the new engine designs that the settlements required and by the lack of time for customer quality testing of the new engines. One manufacturer, Detroit Diesel Corp., received so many preorders that its manufacturing plant operated at capacity twenty-four hours per day using three shifts. By contrast, the demand for its new, post–October 2002 engine dropped precipitously. Detroit Diesel's sales of model year 2002 engines have been more than double the normal sales volume, and it has had to reject more than one thousand orders owing to lack of capacity (Allran 2002, 2).

Consumers apparently do not want to buy the new engines because they cost more and their performance is untested. By some estimates, more than 70 percent of the clean air benefits are lost because of these prepurchase effects. Thus, a larger stock of preconsent-decree engines will be operating than might have been the case had the consent decree not been negotiated.

In short, what EPA litigators reasonably might have anticipated if they were acting according to the public-interest theory has not been achieved. At the same time, trucking companies, truck producers, and diesel engine producers are bearing massive costs as they grapple with the EPA's efforts to short-circuit the route to cleaner engines. Consumers will eventually bear these costs.

REGULATORY CHOICE

In the diesel engine controversy, the EPA had a choice of regulatory strategies. The agency faced different costs and benefits from regulation by rulemaking, regulation by negotiation, and regulation by litigation.

Environmentalists might criticize a decision to continue using regulation by rulemaking because of the delay in getting results. In contrast, regulation by litigation offers the attraction of bringing about a timely outcome in a politically visible way. Environmental groups, accustomed to using the courts in pursuit of their interests, predictably would support bringing suits against violators and penalizing them.

The diesel engine producers and their constituencies apparently lacked the political clout to deflect the action against them. At the same time, they were big and sufficiently sound financially to pay the fines. They had an economically strong product. The increase in cost associated with litigation, which would be spread across hundreds of thousands of units, was not likely to be large enough to capsize the product in the market.

Timing is another consideration. A new administration is not likely to interrupt actions under litigation, whereas it can easily stop regulations passing through the pipeline. Selection of regulation by litigation places the action out of the reach of an opposing political party should a change of administration occur.

The dynamics of the marketplace constrained the EPA when it sought to increase quickly the reductions in NO_x emissions from mobile sources. This fact, too, encouraged the choice of litigation. Only a small part of the stock of diesel engine vehicles is replaced annually. Because of the long life of heavy-duty diesel engines, older engines continue to emit pollutants long after comparable car engines have been scrapped. Furthermore, if newer engines cost more and use more fuel than older ones, then more older engines will be operated longer, and replacements will be purchased before the more costly engines arrive. These facts reduce the potential improvement from tighter standards.

In a search for potentially large reductions in NO_x, the EPA previously had identified diesel engine producers as the most readily targeted component of the industry. Regulation would eventually bring the emission reductions the agency desired, and it cost less to impose regulation on engines made by a few manufacturers and destined to be installed in trucks and other heavy equipment than to sue all those who might install or operate diesel engines in their equipment. The logic that supported initial regulation became the logic that supported litigation. With a favorable settlement, the agency would be able to trumpet the prospects of earlier improvements, no matter what the final accounting might render.

Changing technology provides a rationale for the litigation strategy. The Clean Air Act sometimes forces technology for mobile sources. Manufacturers then must invest in developing technology to which their customers are indifferent or even hostile.

One of the technologies that earlier regulatory efforts had "forced" was sophisticated electronic control of combustion, making possible different modes of operation under different conditions sensed by the controller. Such controllers allow engine manufacturers to offer customers enhanced performance in dimensions other than those the regulators' tests examine directly. At times, the customer may assign higher value to fuel efficiency than to emission reductions. Trade-offs exist between regulator-desired engine characteristics (such as low emissions) and customer-desired engine characteristics (such as low cost and high fuel economy).

As diesel engine emission regulation evolved, the EPA took a performance-standard approach in setting emission standards. It did not specify the technologies to be applied to engines, allowing the standards to be met in a variety of ways, and it spurred producers to compete in the development of engines that would meet the standard.

However, the EPA used a technology standard in developing test procedures. Although producers competed in designing engines, ultimately they also competed with the details of the test procedure itself in developing controls for operating the clean engine technologies that would sell in the marketplace. A characteristic of diesel engine technology causes a trade-off between NO_X (and other) emissions and fuel economy. (NO_X emissions and carbon monoxide emissions present another trade-off.) At the margin, cleaner emissions come at the expense of fuel efficiency, and vice versa. The controller developed by the industry and recognized by the EPA generates fewer NO_X emissions in urban areas, where population density is high; it generates more NO_X emissions, fewer carbon monoxide emissions, and better fuel economy in long stretches of open highway, where population density is low.

Seeking to minimize the cost induced by regulation, diesel engine producers have an incentive to design engine controllers that enable engines to satisfy EPA-dictated tests for urban air quality standards, but also that improve fuel efficiency when the engine operates outside urban environments. In terms of the relevant private interests, one part of the technology mattered little to trucking companies and their customers; the other part, fuel efficiency, mattered very much. Apparent decisions to trade private benefits (fuel-cost savings) for public benefits (lower NO_X emissions) played into a successful litigation strategy for the EPA. The engine controller, which could be described as a performance maximizer subject to an urban emission control constraint, could now be demonized as being simply a "defeat device."

SERVING THE PUBLIC INTEREST

Before the EPA engaged in regulation by litigation, it employed traditional regulation. Such regulation can be justified on the grounds that air pollution causes some harm to many individuals, but the harm to each individual is so small and the number of individuals is so large that no private action is feasible. Assuming such to be the case, the government can act on behalf of the harmed individuals.

In the case of air pollution, the passage of the Clean Air Act and its amendments exemplified such government action. Regulations affecting diesel engine emissions evolved from the statute. When Congress debated the statutes, the affected industries and all other interested parties had access to the debate. When the EPA engaged in regulation by rulemaking and regulation by negotiation, the industry and all other interested parties had access to the regulatory process and to the courts if they regarded the regulatory process as improper. Everyone had the same number of bites at the apple. In the process, some modicum of regulatory certainty was assured for the industry and for all who favored stricter standards. The process was transparent to the participants and to the monitors of the regulatory process in the legislative and executive branch.

The EPA's decision to litigate did not necessarily represent a second bite at the apple for those who support cleaner air. Rather, it was a fresh bite by the regulator for the regulator. In our view, the EPA as a regulator faced a political challenge. The agency was confronted by northeastern states that faced the potential costs of nonattainment status; it recognized that past estimates of improvements in air quality were faulty; it was part of an administration that wished to be considered tough on polluters; and the regulatory process constrained fast action. By employing regulation by litigation, the EPA took its own bite from the apple.

Unfortunately, this regulatory shift has caused uncertainty to increase in the diesel engine industry. The cost of this uncertainty may far exceed the amount of penalties imposed by courts or through settlements.

The EPA's recent extensive employment of regulation by litigation has set a new precedent in the already controversial annals of federal regulation. In time, we shall see whether regulation by litigation becomes a dominant form of regulation or whether the EPA's expansive use of the process will bring about the demise of the process. We have no reason to predict that regulation by litigation will end soon. Indeed, the various theories we have employed to explain this episode suggest that when the conditions that triggered it arise again, then regulation by litigation will be employed again.

NOTE

1. The emission reduction data given in this section are taken from various EPA regulations published in the *Code of Federal Regulation*, various years.

REFERENCES

Allran, Robert. 2002. Appendix C: Declaration of Robert Allran in *Detroit Diesel Corporation's Motion in the Alternative to Stay Compliance with the Consent Decree Pull-Ahead Requirements.* (Copy on file with authors.)

Bernstein, Marver. 1955. *Regulating Business by Independent Commission.* Princeton, N.J.: Princeton University Press.

Bielaczyc, Piotr, and Kerzy Merkisz. 1999. Euro III/Euro IV Emissions: A Study of Cold Start and Warm Up Phases with SI (Spark Ignition) Engine. In *Emissions: Technology, Measurement, and Testing,* 117–27. Warrendale, Pa.: Society of Automotive Engineers.

Bowman, Chris. 1997. EPA Off on Diesel Rigs' Emissions? Clean Air Goals May be Tougher to Meet. *Sacramento Bee,* October 18. [Reprinted in U.S. House of Representatives 2000.]

Domike, Julie R., and Alec C. Zacaroli. 2000. Reinterpretation of NSR Regulations Could Have Costly Implications for Business. *Daily Environment Report* 45 (March): B1.

France, Chet. n.d. *Accounting for Off-Cycle NO_x Emissions from Heavy-duty Diesels.* Washington, D.C.: U.S. Environmental Protection Agency, Engine Programs and Compliance Division, EPA Office of Mobile Sources.

Galligan, Jim. 1998. EPA Betrayed Us. *Transport Topics,* November 2, 2 and 30.

Harter, Philip J. 1982. Negotiating Regulations: A Cure for Malaise. *Georgetown Law Journal* 71: 1–117.

Johnson, Jeff. 1998. EPA Fines Engine Makers. *Transport Topics,* October 26, 1.

Kolko, Gabriel. 1963. *The Triumph of Conservatism: A Reinterpretation of American History,* 1900–1916. New York: Free Press.

McChesney, Fred. 1991. Rent Extraction and Interest Group Organization in a Coasean Model of Regulation. *Journal of Legal Studies* 20: 73–90.

Morriss, Andrew P. 2000. The Politics of the Clean Air Act. In *Political Environmentalism: Going Behind the Green Curtain,* edited by Terry L. Anderson, 263–318. Stanford, Calif.: Hoover Institution Press.

Parker, Larry B., and John E. Blodgett. 2001. *Air Quality and Electricity: Initiatives to Increase Pollution Controls.* CRS Report for Congress. Washington, D.C.: Congressional Research Service, Library of Congress.

Peltzman, Sam. 1976. Toward a More General Theory of Regulation. *Journal of Law and Economics* (August): 211–40.

Reich, Robert B. 1999. Regulation Is Out, Litigation Is In. *American Prospect Online,* February. 11.

Rhodes, Karl. 2003. Up in Smoke. *Region Focus* (spring): 21–25.

Stagg, Michael K. 2001. The EPA's New Source Review Enforcement Actions: Will They Proceed? *Trends* (November–December).

Stigler, George. 1974. The Economic Theory of Regulation. *Bell Journal of Economics and Management Science* 5 (spring): 3–21.

Susskind, Lawrence, and Gerard McMahon. 1985. The Theory and Practice of Negotiated Rulemaking. *Yale Journal of Regulation* 3 (fall): 133–65.

Tierney, Gene. 2002. Mobile6 and Ngm. Presentation.

U.S. Environmental Protection Agency. 1985. Control of Air Pollution from New Motor Vehicles and New Motor Vehicle Engines; Nonconformance Penalties for Heavy-Duty Engines and Heavy-Duty Vehicles, Including Light-Duty Trucks. *Federal Register* 50 (August 30): 35374–401.

———. Office of Air and Radiation, Office of Mobile Sources. 1995. EPA, *California Air Resources Board, and Manufacturers of Heavy-Duty Engines Sign "Statement of Principles."* EPA420-F-95-010a. Washington, D.C.: U.S. Environmental Protection Agency, July.

U.S. House Committee on Commerce. 2000. *Asleep at the Wheel: The Environmental Protection Agency's Failure to Enforce Pollution Standards for Heavy-Duty Diesel Trucks.* March.

Viscuzi, W. Kip. 2002. Overview. In *Regulation Through Litigation*, edited by W. Kip Viscuzi, 1–21. Washington, D.C.: AEI-Brookings Joint Center.

Wilson, George C. 1999. Clinton: Build Trucks That Save Bucks. *National Journal*, June 6, 3268–69.

Yandle, Bruce. 2001. Public Choice and the Environment. In *The Elgar Companion to Public Choice*, edited by William F. Shughart II and Laura Razzolini, 590–610. Cheltenham, England: Edgar Elgar.

Yandle, Bruce, and Stuart Buck. 2002. Bootleggers, Baptists, and the Global Warming Battle. *Harvard Environmental Law Review* 26: 177–229.

Yandle, Bruce, Andrew P. Morriss, and Lea-Rachel Kosnik. 2002. *Regulating Air Quality Through Litigation*. Bozeman, Mont.: PERC.

Acknowledgments: The authors express appreciation to PERC for encouraging the use of materials developed in a PERC research program. Special appreciation goes to Jane Shaw for her editorial assistance. Bruce Yandle is a professor of economics emeritus at Clemson University. Andrew P. Morriss is the Galen J. Roush Professor of Business Law and Regulation and the director of the Center for Business Law and Regulation at Case Western Reserve University School of Law. Both are senior associates at PERC—The Center for Free Market Environmentalism, Bozeman, Montana.

18

The Environmental Propaganda Agency

CRAIG S. MARXSEN

In 1990 the U.S. Environmental Protection Agency (EPA) published *Environmental Investments: The Cost of a Clean Environment.* That work became the most widely cited source of compliance-cost data used in studies reported in economic journals. An impressed Congress, in the 1990 revision of the Clean Air Act, required the EPA to conduct a thorough cost-benefit analysis of its clean air requirements imposed since 1970. Finally, in October 1997, the EPA submitted to Congress *The Benefits and Costs of the Clean Air Act, 1970 to 1990.* (The whole study has been published on the Internet.) Although the EPA acknowledges that it has imposed clean-air-compliance costs totaling $523 billion from 1970 to 1990, it claims that its rules have produced benefits worth $22.2 trillion (measured in dollars of 1990 purchasing power). For comparison, the U.S. Department of Commerce estimates that in the early 1990s all U.S. fixed reproducible tangible wealth totaled less than $20 trillion.

To estimate the benefits and costs of the Clean Air Act, the EPA compared the differences in economic, health, and environmental outcomes under two alternative scenarios: a "control scenario" and a "no-control scenario" (U.S. Environmental Protection Agency [U.S. EPA] 1997, ES-1). The "control scenario" is based on actual historical data. The "no-control scenario" is the EPA's hypothetical description of what the United States would have experienced had no air-pollution controls been established beyond those in place prior to enactment of the 1970 amendments (ES-1). The EPA claims to have, by 1990, eliminated 99 percent of lead emissions. It also claims to have effected reductions of 40 percent in average atmospheric levels of sulfur dioxide, 30 percent in nitrogen oxides, 50 percent in carbon monoxide, and 15 percent in ground-level ozone (ES-3). The "control scenario" involves reduced "acid rain," a 45 percent reduction in total suspended particulate matter, and a 45 percent reduction in smaller particles (PM10 and PM2.5) (ES-4). The particulate matter is the most important pollutant in the EPA's benefit assessment.

The costs of the Clean Air Act include expenditures due to requirements to install, operate, and maintain pollution-abatement equipment, costs incurred in monitoring and reporting regulatory compliance, and other costs of maintaining the regulatory bureaucracy. Many of these costs showed up as higher prices for goods and services. The benefits of the Clean Air Act were estimated by aggregating a dollar-value estimate of the entire harm that would have been done by the greater levels of air pollution that would have occurred in the "no-control scenario." The EPA concluded that many premature deaths would have resulted from the air pollution. Other projected health problems included chronic bronchitis, hypertension, hospital admissions, respiratory-related symptoms, restricted activity, decreased productivity of workers, soiling damage, lost visibility, and agricultural output reductions (U.S. EPA 1997, ES-7). The EPA estimated dollar values for each of the negative effects and concluded that total monetized benefits ranged between $5.6 trillion and $49.4 trillion, with a central estimate of $22.2 trillion, whereas EPA-estimated costs were $0.5 trillion (ES-8). When the EPA's Science Advisory Board criticized the study, the EPA prepared an alternative estimate using a modified methodology and reported benefits ranging between $4.8 trillion and $28.7 trillion, with a central estimate of $14.3 trillion (ES-9).

The EPA's conclusions: "First and foremost, these results indicate that the benefits of the Clean Air Act and associated control programs substantially exceeded costs." The EPA has produced such a large benefit estimate that it goes on to claim, "Even considering the large number of important uncertainties permeating each step of the analysis, it is extremely unlikely that the converse could be true" (U.S. EPA 1997, ES-9). The EPA, therefore, has produced a cost-benefit study that, if true, should forever silence opponents who contend that our air-pollution regulations have cost more than they were worth.

But the truth has eluded the authors of the EPA report. Their study actually represents a milestone in bureaucratic propaganda. Like junk science in a courtroom, the study seemingly attempts to obtain the largest possible benefit figure rather than to come as close as possible to the truth.

VALUING LIVES ALLEGEDLY SAVED

The EPA valued each life saved as a $4.8 million benefit. The underlying concept is sensible enough. Environmental regulations are forcing people to spend money and divert valuable resources from other uses in order to reduce certain environmental health risks. Citing respectable studies, the EPA concludes that workers will accept lower-wage jobs to reduce the probability of being accidentally killed on the job. Installing windows in tall buildings pays better than installing them in one-story houses partly because of the difference in the worker's risk. The claim is not that a human life is worth only $4.8 million but rather that improving safety is worth $4.8 million in the sense that people are spending that much per life saved in other contexts.

Government is largely specialized in the business of saving human lives. This is part of the duty of the military, the police department, the fire department, the public health agency, the Federal Aviation Administration, the highway department, and various other agencies of the various levels of government. Government passes laws and spends money to improve public safety. Cost-benefit analysis is simply a method of assisting in saving the greatest number of lives per dollar spent. If one government activity saves lives at a lower cost per life saved than another, then the same total money expenditure could save more lives if some of the spending were reallocated in favor of the former at the expense of the latter. The Department of Transportation spends no more than $2.7 million to save a statistical life (Hopkins 1997, 6 [March 1996 figure]). If the EPA balanced clean-air costs against benefits, then the Department of Transportation could still save almost two lives for every one life the EPA would save with the same mandated spending. Unlike the Department of Transportation, which values saving the average motorist at $2.7 million, the EPA is theoretically valuing at $4.8 million the saving of some elderly person to live a short time longer on his deathbed.

Valuation of the saving of a life allegedly lost to small amounts of smoke and dust in the outdoor air is critical in generating the EPA's large benefit total. A recent chapter in *Risk in Perspective* (July 1999) deals with "Valuing the Health Effects of Air Pollution." The chapter explains that the EPA's figure of $4.8 million per life lost comes from the idea that an average worker might be paid $480 per year to accept an added fatality risk of one in ten thousand on the job. But a healthy middle-aged worker is quite different from an elderly person with a serious cardiac or respiratory disease. The EPA, the chapter reports, demonstrates that more realistic estimates of what a particulate-pollution casualty has actually lost result in dramatic reductions in the estimated value of life-years saved.

The EPA is notorious for overzealous mandates ostensibly intended to save a few life-years. A study by Tammy O. Tengs and others has become a widely cited source of information on what various agencies spend to save a statistical life-year. The Harvard Center for Risk Analysis features some estimates of the median value of cost per life-year saved for various regulatory agencies (Tengs and others 1995, 369–89):

Federal Aviation Administration	$23,000
Consumer Product Safety Commission	$68,000
National Highway Traffic Safety Administration	$78,000
Occupational Safety and Health Administration	$88,000
Environmental Protection Agency	$7,600,000

Indeed, the Clean Air Act seems to be one of the EPA's better bargains.

Gains in life expectancy from medical interventions are very similar, in principle, to the alleged gains from cleaner air. The Harvard Center for Risk Analysis refers readers of another issue of its newsletter to estimates of such gains (*Risk in Perspective*, November 1998, 2). If all smokers were to quit smoking, the life expectancy of the general population would increase by about nine months. Gains for smokers alone, as a group, are also listed. A thirty-five-year-old smoker gains only fourteen months of life expectancy by cutting smoking by 50 percent (Wright and Weinstein 1998, 383, table 2), whereas quitting altogether increases life expectancy only twenty-eight months. The EPA's alleged 45 percent reduction in outdoor particulate pollution would not seem likely to have a greater effect on an individual's life expectancy than quitting smoking.

About 90 percent of the benefits of the Clean Air Act supposedly come from reductions in mortality (75 percent) and chronic bronchitis (15 percent) that particulate matter (that is, dust and smoke) allegedly would otherwise have caused. Another 6 percent of the $22.2 trillion in benefits is related to preventing people from dying of lead poisoning. During 1990, in the "no-control scenario," the EPA figures that 184,000 additional people thirty years old and older would have died prematurely because of particulate matter suspended in the air (U.S. EPA 1997, ES-4). At $4.8 million per death, this life-saving adds $883.2 billion to the alleged benefits. Summing over the whole twenty-year period, particulate matter alone causes loss of life in the "no-control scenario" that the EPA tallies to be worth $16.632 trillion.

The EPA Science Advisory Board Council on Clean Air Act Compliance Analysis found fault. The council asked the EPA to provide an alternative estimate of benefits based on the value of statistical life-years lost rather than lives lost per se. In response, the EPA revised the value of particulate-matter mortality to $9.1 trillion and came up with a new central estimate of $14.3 trillion for the total benefits of the Clean Air Act (U.S. EPA 1997, ES-9). Although the Federal Aviation Administration mandates spending at $23,000 per life-year gained (Bolch and Pendley 1998, 2), the EPA valued a life-year lost at $293,000 (U.S. EPA 1997, 58). Whenever a moribund seventy-year-old dies, the EPA figured, the fourteen years remaining of actuarial life expectancy for an average seventy-year-old represents the life-years lost (37, 44). Indeed, the average alleged mortal victim of particulate matter was assumed to have lost fourteen years of life expectancy *because* the average victim was an elderly person (44). This means the seventy- year-old lost fourteen years times $293,000 per year, or $2.9 million worth of life-years (I-23). In fact, the EPA treated every dying elderly person as if he would have lived out the full number of years remaining for an average person of his age. The EPA acknowledges doubts about the validity of this procedure, conceding that L. Cifuentes and L. B. Lave had found that 37 to 87 percent of deaths from short-term exposure to air pollution could have been premature by only a few days (I-25).

In summary, one must suspect a pervasive EPA tendency to exaggerate. To arrive at its $9.1 trillion worth of life-years that allegedly would have been lost to particulate- matter mortality, the EPA figures that the average victim loses fourteen years of remaining life expectancy (I-25). If, in fact, the average victim really suffers death only three days prematurely, then the total would be reduced by a factor equal to three days divided by fourteen years. The EPA's $9.1 *trillion* figure drops to $5.3 *billion*. Reducing the estimated value of a life-year saved to the same order of magnitude presumed by other government agencies brings the total down below $1 billion. But we are still ignoring the deeper issue. Air pollution may never have caused those hypothetical deaths in the first place. The EPA Science Advisory Committee (nineteen of the committee's twenty-one members) believed that no causal mechanism linking mortality with particulate matter had been established (34).

DISCOVERING THE VICTIMS

The EPA claims to base its prevented-deaths estimates on a number of studies that attempted to correlate mortality rates with levels of particulate matter polluting the outdoor air. The EPA picked the best one (or so it claims), the study by C. A. Pope and others published in 1995. It also cites twelve studies that report finding significant relationships between daily PM10 (particles 10 microns or smaller) concentrations and daily mortality (EPA 1997, D-15). The main cause of daily variations in these particulate-matter levels, however, is the weather. On calm, hot, summer days, pollution levels outdoors tend to be high. Smog is almost exclusively a summer phenomenon. Is it a shocking surprise if statistical studies discover that the "dog days of summer" don't kill just old dogs?

The study by Pope and others (1995) compares death rates for the over-thirty population to annual median PM2.5 exposure in fifty cities (PM2.5 is particulate matter 2.5 microns or smaller). Temperature differences are ignored (Fumento 1997, 1). Pope and others found a 17 percent difference in death rates over eight years between the cleanest and dirtiest cities (Kaiser 1997, 468). The study did not control for wealth or lifestyle but did control for schooling and obesity (Merline 1997, 13). Differences in annual pollutant emissions accounted for far less of the measured ambient pollutant concentrations than did variations in meteorological conditions (Crandall, Rueter, and Steger 1996, 41–42). As a model for predicting mortality, the 1995 model of Pope and others was seriously misspecified. It omitted many important variables. This misspecification results in mistakenly crediting causality to any included variable (such as PM2.5) that happens to be correlated with an important left-out variable. For instance, towns where everyone works in steel mills might tend to be towns with more particulate-matter pollution. And the differential mortality may arise from working in the steel mill rather than from breathing the outdoor air. Maybe the steel-town residents all smoke high-tar cigarettes, get drunk on Saturday night, and fail to attend church on Sunday. Likewise, towns where people drive older cars might tend to be smokier. Towns located in more densely populated industrial regions might tend to have above-average levels of fine particles. Particulate matter may blow onto places where beef is cheap and people eat too much fat. The possibilities are endless.

Recent studies do not consistently find a relationship between outdoor air pollution and mortality. The EPA cites a 1993, six-city study by Douglas Dockery and others. Yet three of the six cities showed no correlation between death and air pollution (Fumento 1997). A study that followed 6,000 Seventh-Day Adventists for ten years found no connection between particulate matter and mortality, nor between a variety of other air pollutants and mortality (Abbey and others 1993, 35, 45). Another study, by S. H. Moolgavkar, E. G. Luebeck, and E. L. Anderson (1997), adjusts for temperature and finds no correlation between air pollution and hospital admissions for lung problems in Birmingham, Alabama, although such a correlation was found in Minneapolis–St. Paul, Minnesota. In Athens, Greece, a study found mortality correlated with temperature but not with an air-pollution index (Katsouyanni 1994, 264). Another study in Belgium found temperature and ozone significant but not other pollutants (Sartor and others 1997, 116). Laurence S. Kalkstein (1993) evaluated the differential impact of stressful weather and air pollution in several cities and concluded that high pollution concentrations seemed a much less important predictor of acute mortality than stressful weather.

WHY DO PEOPLE DIE?

Some people might have died from indoor pollution coming from indoor sources rather than from outdoor pollution coming from sources regulated by the EPA (Crandall, Rueter, and Steger 1996, 42). The same weather conditions that result in urban smog may impede the airing out of houses and buildings. Indoor pollution coming from indoor cooking, cleaners, dust, pets, mold, and other sources would then rise to high levels indoors. Pollution levels are much higher inside people's houses than outside, according to a recent chapter in *Scientific American* (Ott and Roberts 1998, 90). Americans spend 95 percent of their time indoors (91).

Most of the people who die on high-pollution days are elderly and suffer from circulatory or lung diseases. They spend even more of their time indoors than other people. Outdoor pollution levels might be little more than an indicator of weather conditions promoting indoor pollution coming from indoor sources. Indeed, toxicological studies indicate that particulate matter, even at more than twenty times outdoor levels, causes no measurable effect on the breathing of people with chronic obstructive pulmonary disease (Crandall, Rueter, and Steger 1996, 43). Allergies to indoor pollutants do cause breathing problems for some people. Most people, for whatever reason, die gradually over a period of several weeks or months (American Medical Association 1987, 768). In the end, breathing often becomes labored, partly because the lungs become waterlogged due to heart failure, partly because of saliva that trickles down the windpipe. Pneumonia is not so much the cause of death as the mode of dying from some disease of the heart, liver, or kidneys. The condition is exacerbated when the dying person ceases to rise from the bed. Life may be prolonged by helping such a person get up for a short walk or a ride in a wheelchair. Kenneth Chilton (1997), testifying before a Senate subcommittee, said: "It is curious, to say the least, that the statistical link that has been demonstrated is between fine particles and cardiopulmonary deaths, and not deaths due to respiratory disease or lung cancer alone" (4).

Particulate matter might be a killer largely because it is visible. People die as a result of averting behavior. The dying person's caregiver fears the "deadly pollution" on a hazy day and therefore refrains from getting Grandma up from the bed. Grandma dies of pulmonary edema exacerbated by her failure to get up. The death certificate reports cardiovascular disease as the cause of death because it is the underlying cause. In Israel, Iraqi missile attacks caused the deaths of several elderly people who, fearing poison gas, smothered themselves by wearing government-issued gas masks that had not been properly unpacked—the filter material was still sealed in its plastic wrapper! The missiles that struck Israel delivered no poison gas. Similarly, the EPA may have unwittingly encouraged people to, in effect, smother their dying relatives in an effort to keep them from breathing nearly harmless pollution.

Epidemiological studies that correlate asthma attacks or hospital admissions with particulate-matter pollution may actually be measuring a psychological phenomenon. People, having been persuaded that the haze in the air is deadly, then react accordingly. Children complain that they can't breathe; they put on a big show similar to the one some have now been taught to perform whenever they are exposed to secondhand cigarette smoke. Adults likewise rush to the hospital in fear that the pollution might be the last straw in their struggle against chronic bronchitis or some cardiopulmonary disorder. The environmental movement has encouraged phobic behavior by the more gullible segment of the population. Maybe some people just tend to close their windows to keep the pollution out and then are afflicted with asthma or bronchitis caused by the much more irritating indoor pollution coming from indoor sources.

Any correlation between daily ambient pollution levels and mortality may simply be a misidentification of modest timing effects. Caregivers may simply tend to neglect a dying elder when spring turns to summer and the windows must be closed and the air conditioner turned on. Grandma may get less attention simply because her room smells bad and is repugnant to enter. Maybe no one checked on Grandma after the windows were closed to keep the outdoor pollution from coming in.

REGRESSIONS TO SAVE THE WORLD!

One must appreciate the questionable significance of weak statistical regression. Once a set of observations has been amassed, many experimental regression equations are typically calculated. By adding and deleting variables or observations, experimenting with transformations, and so forth, the analyst can often "find" a relationship between variables that are actually not related. The pressures to experiment until the desired or expected "fit" is obtained are considerable. The weaker the hypothesized relationship, the easier it is to find it spuriously.

Suppose we examine a sample of fifty-three cities, including New Orleans, Pasadena, Pittsburgh, and New York. Regression analysis reveals no statistically significant relationship between mortality and PM2.5 (Jones, Gough, and Van Doren 1997, 3). Next, we eliminate the four named cities from the sample and add Ashland, Kentucky, even though its PM2.5 measurements seem to be 50 percent above their actual values when we compare it with neighboring cities and look at its sulfate measurements. Voila! PM2.5 suddenly appears to cause a 17 percent increase in mortality in the dirtiest city, compared with the cleanest. We have arrived at the sample of Pope and others (1995) and found those analysts' result. We have found what we expected, and we publish that finding. We do not report our first finding, which showed no relationship.

Establishing a true connection between PM2.5 and mortality may be impossible with regression analysis. Multicolinearity cannot be overcome because ambient levels of particulate matter correlate too strongly with other variables, such as temperature, other pollutants, day of the week, season, averting behavior, characteristics of the local economy, transportation patterns, and so forth. Researchers must cheat and misspecify the model by omitting some important variables; otherwise, the results fail to point to one specific culprit as the cause of differential mortality. Moolgavkar and Luebeck (1996), Edward Calthrop and David Maddison (1996), and others have testified to the unsuitability of regression analysis for testing the relationship between PM and mortality. A false positive result is, theoretically, almost assured for anyone who cares to seek it.

As a general rule, epidemiological evidence is rejected when it establishes a relative risk of less than three, or 300 percent (Taubes 1995, 167). Indeed, Marcia Angell, an editor of the *New England Journal of Medicine*, says that, as a rule of thumb, papers are not accepted for publication unless they find a relative risk of three or more, especially if a finding is biologically implausible or unprecedented (Taubes 1995, 167).

But, with regard to the PM–mortality link, the print media exhibit an exceptional bias in favor of the environmentalist agenda. The peer review process is driven by a sort of political correctness. Papers that support the established orthodoxy are more likely to be published. Special-interest groups trying to further an environmental agenda dominate the funding of such research. The EPA itself funds many studies. A great many zealots and hired-gun statisticians are getting grants, tenure, promotion, and fame for "proving" that pollution is destroying the earth. Zeal for the environment drives many investigators to pursue what would ordinarily be regarded as insignificant research. It increases the number of such studies published, and it magnifies the rewards of researchers such as Joel Schwartz, who received the MacArthur Fellowship—a no-strings-attached $275,000 grant—for his work on lead and particulate pollution (Skelton 1997, 29). Schwartz told an interviewer that his strategy was one of "working very hard to dump out a lot of papers very fast, on the ground that if I just kept on pounding this out in the literature, people could not ignore it" (30).

People seem to forget that regressions once "established" a connection between overhead power lines and leukemia. The electromagnetic field (EMF) findings were stronger than those for PM2.5 and mortality and yet were later disproved (Merline 1997, 12). The EMF threat lacked the bandwagon appeal that might have motivated researchers to construct a solid wall of scientific evidence such as the EPA and the environmentalist groups claim exists for PM2.5 and mortality (10). Many epidemics of anxiety have come and gone in recent years as a result of weak epidemiologic findings (Taubes 1995, 164). Radon gas and lung cancer were associated via mismeasurement of actual individual exposure; now the findings are doubted. Cancer and pesticide residues were linked; DDT and breast cancer were linked and then unlinked; electromagnetic fields from power lines seemed to cause brain cancer; hair dryers seemed to cause cancer; coffee seemed to cause cancer, so did saccharin. For particulate matter, so much "good" can be done by the "citizen of the world" statistician. He can "prove" that fossil fuels are killing people. For him, it may be a matter of the "earth in the balance." He can help fight global warming. He can prevent the collapse prophesied in that sensational book, *The Limits to Growth.*

Once, in the late nineteenth century, William Stanley Jevons became enthusiastic about a theoretical relationship between sunspots and the business cycle. He became convinced that a relationship existed, and the more he looked, the stronger seemed to be the correlation he believed he had discovered. Because his theory fell out of favor, it became disreputable for subsequent investigators to seek such correlation. Seemingly good correlations were subsequently found, however, and a few even gained publication. Jevons's sunspot theory might still make a comeback if the environmentalists discover that it somehow serves their cause.

LEAD PHOBIA

Lead is assumed to have been causing many deaths. The EPA's mean estimate for lead-related mortality prevented by the Clean Air Act is $1.339 trillion (1990 dollars) (U.S. EPA 1997, ES-7). The EPA relied on published estimates of a relationship between blood lead levels and high blood pressure (G-9). It then multiplied by other published coefficients distilled from degrees of correlation between high blood pressure and mortality (G-13). The result is a far cry from making inferences based on death certificates that list lead poisoning as the cause of death. Applying the EPA's reasoning, we could save even more lives by eliminating final exams from all colleges and universities. Final exams also tend to elevate blood pressure, and elevated blood pressure is correlated with mortality. The fallacy of such reasoning lies in the source of the correlation between blood pressure and mortality. As Crandall, Rueter, and Steger (1996, 44) have observed, high blood pressure is a predictor of mortality because it indicates heart disease. Heart disease has been the leading cause of death in recent years. Moderately elevated blood pressure itself does not cause nearly the amount of mortality with which it is statistically associated via its connection with heart disease. Maybe getting the lead out of gasoline was worth only a few billion dollars or even less. In any event, the EPA's method results in a huge exaggeration.

Although preventing mortality and chronic bronchitis generates 96 percent of the benefits claimed by the EPA, the largest part of the remainder is attributed to preventing IQ reductions that the EPA says lead would have caused. Drawing on amalgamated regression studies (Schwartz 1993), the EPA relies on a statistical correlation between blood-lead levels in children and low IQ (U.S. EPA 1997, G-2). Supposing that lead is reducing IQs, the EPA values the loss at $3,000 per IQ point per child.

But the EPA might be getting it backward (Juberg, Kleiman, and Kwon 1997, 13). Cross-sectional studies are confounded by the fact that children with lower IQs tend to descend from parents with lower IQs. Stupid parents tend to expose their children to more opportunities for ingesting lead, and the stupid children contribute by eating more dirt. Stupid parents tend to achieve lower socioeconomic status and live in residential areas where lead is more prevalent. Moreover, they tend to fail to provide good diets for their children. Iron-deficient children will actually absorb more of a given amount of ingested lead, and the iron deficiency will further contribute to lower IQ measurements.

Longitudinal studies, tracking the same children over time, persuade some otherwise skeptical scientists of a lead–IQ connection (Powell 1999, 166). However, the relationship between iron deficiency, IQ loss, and lead uptake illustrates the problem of confounding factors that are not evaded by switching from cross-sectional to longitudinal studies. Children living under conditions that expose them to more lead may seem to suffer IQ loss from lead when the loss is actually due to other problems associated with low socioeconomic surroundings. IQ test results are supposedly influenced by the cultural deprivation common to children in low socioeconomic surroundings. The same children who are exposed to more lead may also be exposed to illegal drug use, domestic violence, poor schools, bad diets, and a large number of other factors detrimental to the enhancement of their intellect. Combining many regression studies is almost sure to pick up a false relationship if confounding factors can produce one. Maybe moderate blood lead levels don't really reduce IQ after all, but IQ is just a good predictor of lead ingestion and absorption (Juberg, Kleiman, and

Kwon 1997, 13). It seems likely, at least, that the EPA's estimate of the benefits from lead control is exaggerated.

The possibility of synthesizing epidemiological findings has long plagued the lead-IQ controversy. In 1979 Herbert Needleman and his collaborators reported finding a three- or four-point IQ drop associated with modestly elevated lead levels measured in children's teeth (Needleman and others 1979, 689–95). Mark Powell (1999, 163) reports that Claire Ernhart and her coauthors discredited Needleman's research by charging that confounding variables had been inadequately controlled and that Needleman had performed so many regression comparisons that a few statistically significant outcomes were assured by chance alone (Ernhart, Landa, and Schell 1981, 911–19). Later, in a 1990 Superfund court case, Needleman served as an expert witness. Sandra Wood Scarr, a Virginia psychologist who had been an EPA panelist reviewing Needleman's work, testified against Needleman (Powell 1999, 164). Scarr explained that Needleman's first set of analyses found no lead–IQ relationship and that only by eliminating other IQ-affecting variables did Needleman finally get the results he sought (165). Unfortunately, Scarr and Ernhart were condemning practices probably hidden in most of the epidemiological evidence underlying the EPA's claims of benefits of every kind from the Clean Air Act.

THE COST

The EPA provides an estimate of the cumulative costs of twenty years of compliance with the Clean Air Act—roughly half a trillion dollars. This estimate is not very far from the total one gets by adding up the relevant figures in the EPA's previous (1990) study. After all, the previous study was the EPA's prime source of data for the current cost study. However, the EPA has now added its own version of a general equilibrium analysis. The EPA undercuts previous findings by Michael Hazilla and Raymond J. Kopp by concluding that ripple effects reduced annual GDP by about 1 percent, or $55 billion 1990 dollars by 1990 (U.S. EPA 1997, 9). Calculating the present value of the GDP reductions, the EPA concludes that the aggregate impact of clean-air regulation on production was $1,005 billion 1990 dollars. Hazilla and Kopp had found that, by 1990, environmental regulation of all kinds reduced GNP by 5.85 percent. The Clean Air Act, accounting for 30.46 percent of pollution spending, would, therefore, presumably have reduced 1990 GDP by 1.78 percent, according to Hazilla and Kopp. The EPA concedes that some economists identify a stifling effect of environmental regulation on technological innovation, which hinders productivity growth (U.S. EPA 1997, 11). The agency concedes missing some of that effect, although it claims that its general equilibrium model incorporated "endogenous productivity growth" that results from factor-price changes within the model. These are internal ripple effects, however, and do not represent the aggregate changes in multifactor productivity (or total factor productivity) that economists equate with technological advance.

Focusing on the best private study of the productivity effects of the whole of environmental regulation—the study by James Robinson (1995)—leads one to quite a different conclusion. Robinson, whose work was supported by a grant from the Office of Technology Assessment, concludes: "Overall, the U.S. manufacturing sector attained a level of multifactor productivity in 1986 that was 11.4 percent lower than it would have attained, absent the growth in environmental and occupational health regulation since 1974" (414). Moreover, the contribution of occupational health regulation was negligible, leaving the EPA as the sole cause of this entire effect (411). Robinson's finding that actual manufacturing output was 11.4 percent lower than its potential means that, by 1986, potential manufacturing output was 12.86 percent higher than actual output. This implies that, over the twelve years from 1974 to 1986, manufacturing output would have grown by a factor of 1.01 percent per year in addition to its actual growth, had environmental regulations not existed. Had the manufacturing sector enjoyed this addition to annual growth from 1970 to 1990, then manufacturing output in 1990 would have been 22.35 percent higher than it actually was in 1990.

Michael Hazilla and Raymond J. Kopp (1990) developed a model that ignores technological progress but emphasizes the ripple effects produced as disturbances of production in one sector alter factor inputs into other sectors. In their computable general equilibrium model, environmental regulation reduced manufacturing output about 6.33 percent by 1990, largely by reducing capital accumulation and diverting other inputs. (The figure of 6.33 percent is approximate and is obtained by averaging all sectors of the manufacturing industry from Hazilla and Kopp 1990, table 4, 868–69.) The ripple effects from manufacturing and several other directly affected sectors resulted in a reduction of real GNP of 5.85 percent by 1990 (Hazilla and Kopp 1990, 867). In other words, Hazilla and Kopp found that, by 1990, environmental regulation reduced real GNP (and, therefore, real GDP) by 92.41 percent as much as it reduced manufacturing output alone. If manufacturing output would have been 22.35 percent higher in 1990 without EPA regulation, then real GDP probably could have been 20.65 percent higher than it actually was in 1990. Annual growth of real GDP from 1970 to 1990 appears to have suffered; the absence of environmental regulation would have added 0.94 percent to the annual growth rate of real GDP from 1970 to 1990. This figure is consistent with conclusions reached by Richard Vedder (1996, 16), who concluded that annual productivity growth would have averaged one percentage point higher had regulation in general remained at 1963 levels. Thomas Gale Moore (1997, 2) also infers a reduction in total factor productivity in the neighborhood of $2 trillion, based on findings by Gray and Shadbegian (1993) that every dollar spent on compliance costs in the paper, oil, and steel industries reduced total factor productivity by $3 to $4. Moore applies this ratio to Hopkins's $667 billion estimate of overall regulatory compliance cost. (For a discussion of the EPA's role in causing post-1970 American wage stagnation, see Marxsen 1999.)

By comparing actual real GDP each year with estimated potential real GDP (which would have grown larger than actual real GDP by an annually compounding factor of 1.009), we find a difference that totals $9,951.7 billion (1990 dollars) over the period from 1970 to 1990. Because compliance costs for air-pollution control alone constituted 30.46 percent of total 1970–1990 pollution-control compliance costs (U.S. EPA 1990, 8–20, 8–21), the lost GDP attributable to air pollution control would total $3.03 trillion (1990 dollars). This figure needs to be added to the EPA-estimated half a trillion dollars of direct compliance cost because the compliance cost represents a sinkhole for some of the GDP that actually was produced. To put the figures into perspective, the entire U.S. stock of residential structures in 1990 was worth roughly $6 trillion; the stock of consumer durables (all the durable goods you own besides your house) was worth almost $2 trillion. The implied cost of the Clean Air Act from 1970 to 1990 is more than six times larger than the EPA's official compliance cost estimate. It seems reasonable to suppose that the true cost must be somewhere between $1.5 trillion (the EPA's total, including the EPA-estimated lost GDP due to productivity effects) and $3.5 trillion ($0.5 trillion plus $3 trillion, inferred from Robinson's findings plus those of Hazilla and Kopp). The most likely value is probably near $3.5 trillion because, unlike the EPA, Robinson, Hazilla, and Kopp had little to gain by inflating their estimate. Apart from the dubious health benefits, mean EPA-estimated benefits from visibility, soiling damage avoided, and agriculture add up to just $151 billion and are probably also exaggerated. Quite opposite to the conclusion reached by the EPA, it seems very unlikely that the benefits of the Clean Air Act exceeded the costs.

CONCLUSION

The EPA's cost-benefit study appears to be a bureaucratic cover-up. In reality, the EPA has probably squandered a substantial portion of America's resources. The stifling effect on productivity may account for a big part of America's post-1973 wage stagnation (Marxsen 1999). Without the illusory benefit of all the lives saved, the actual benefits of the Clean Air Act were very modest and probably could have been achieved nearly as well with far less sacrifice. The Clean Air Act and its amendments force the EPA to mandate reduction of air pollution to levels that would have no adverse health effects on even the most sensitive person in the population (Powell 1999, 91). The EPA relentlessly presses forward in its absurd quest, like a madman setting fire to his house in an insane determination to eliminate the last of the insects infesting it.

The EPA has very recently issued a new study in which it attempts to measure the benefits and costs of the Clean Air Act Amendments of 1990 (U.S. EPA 1999). The EPA projects that annual benefits will total $110 billion by the year 2010, $100 billion of which will allegedly come from a 5 to 10 percent reduction in fine particulate matter (iv, 75). Without the amendments, PM10 and PM2.5 supposedly would kill 23,000 people, based on the mid-1990s study by Pope and others (U.S. EPA 1999, 60). The EPA once again has valued each of these fatalities at $4.8 million (1990 dollars) (71). The total comes up short of the product of 23,000 times $4.8 because the EPA decided to add a five-year lag structure to death after exposure to particulates and it used discounting in its calculations (75). Estimating costs to be just $27 billion per year by 2010, the EPA arrives at a four to one ratio of benefits to costs (71). Needless to say, the agency continues to exaggerate.

REFERENCES

Abbey, David E., F. Peterson, P. K. Mills, and W. L. Beeson. 1993. Long-Term Ambient Concentrations of Total Suspended Particulates, Ozone, and Sulfur Dioxide and Respiratory Symptoms in a Nonsmoking Population. *Archives of Environmental Health* 48: 33–46.

American Medical Association. 1987. *Family Medical Guide*. New York: Random House.

Bolch, Ben, and Bradford Pendley. 1998. How the EPA Says It Makes Us Rich. *Liberty* 11 (May): 1–2.

Calthrop, Edward, and David Maddison. 1996. The Dose-Response Function Approach to Modelling the Health Effects of Air Pollution. *Energy Policy* 24: 599–607.

Chilton, Kenneth W. 1997. *Has the Case Been Made for New Air Quality Standards?* Policy Brief 181 (April). St. Louis: Center for the Study of American Business, Washington University.

Crandall, Robert W., Frederick H. Rueter, and Wilbur A. Steger. 1996. Clearing the Air: EPA's Self-Assessment of Clean-Air Policy. *Regulation* no. 4: 35–46.

Dockery, D. W, F. E. Speizer, D. O. Stram, J. H. Ware, J. D. Spengler, and B. B. Ferris, Jr. 1993. An Association between Air Pollution and Mortality in Six U.S. Cities. *New England Journal of Medicine* 329 (24): 1753–59.

Ernhart, C., B. Landa, and N. Schell. 1981. Subclinical Levels of Lead and Developmental Deficit: A Multivariate Follow-Up Reassessment. *Pediatrics* 67 (6): 911–19.

Fumento, Michael. 1997. The EPA's Killer Air Pollution Proposals. *American Enterprise Institute: On the Issues*, October.

Gray, Wayne B., and Ronald J. Shadbegian. 1993. *Environmental Regulation and Manufacturing Productivity at the Plant Level.* Working paper no. 4321 (April). National Bureau of Economic Research.

Harvard Center for Risk Analysis. 1998. Gains in Life Expectancy from Medical Interventions. *Risk in Perspective* 6 (November): 1–4.

———. 1999. Valuing the Health Effects of Air Pollution. Risk in Perspective 7 (July): 1–6.

Hazilla, Michael, and Raymond J. Kopp. 1990. Social Cost of Environmental Quality Regulations: A General Equilibrium Analysis. *Journal of Political Economy* 98 (August): 853–73.

Hopkins, Thomas D. 1997. U.S. Environmental Protection Agency's Rule on National Ambient Air Quality Standards for Particulate Matter. *Economists Incorporated: Papers Available Online* (March 12). Retrieved June 18, 1998, from http://www.mercatus.com/pdf/materials/39.pdf.

Jones, Kay, Michael Gough, and Peter VanDoren. 1997. Addendum to the CSE Foundation study *Is the EPA Misleading the Public about the Health Risks of PM2.5?* Washington, D.C.: Citizens for a Sound Economy Foundation, May 12, 1997. Retrieved December 9, 1999, from http://www.csef.org/csefhome/kjonesaddendum.htm.

Juberg, Daland R., Cindy F. Kleiman, and Simona C. Kwon. 1997. *Lead and Human Health.* New York: American Council on Science and Health, December. Retrieved December 9, 1999, from http://www.acsh.org/publications/booklets/lead.html.

Kaiser, Jocelyn. 1997. Showdown over Clean Air Science. *Science* 275 (July): 446–50.

Kalkstein, Laurence S. 1993. Direct Impacts in Cities. *Lancet* 342 (December 4): 1397–1400.

Katsouyanni, K. 1993. Evidence for Interaction between Air Pollution and High Temperature in the Causation of Excess Mortality. *Archives of Environmental Health* 48: 235–42.

———. 1994. Evidence for Interaction between Air Pollution and High Temperature in the Causation of Excess Mortality (abstract). *Journal of the American Medical Association* 271 (January 26): 264.

Marxsen, Craig S. 1999. Why Stagnation? *B>Quest.* Retrieved December 9, 1999, from http://www.westga.edu/~bquest/1999/stag.html.

Merline, John. 1997. How Deadly Is Air Pollution? *Consumers' Research Magazine* 80 (February): 10–15.

Moolgavkar, S. H., and E. G. Luebeck. 1996. A Critical Review of the Evidence on Particulate Air Pollution and Mortality. *Epidemiology* 7 (July): 420–28.

Moolgavkar, S. H., E. G. Luebeck, and E. L. Anderson. 1997. Air Pollution and Hospital Admissions for Respiratory Causes in Minneapolis St. Paul and Birmingham. *Epidemiology* 8 (July): 364–70.

Moore, Thomas Gale. 1997. Issues in Regulatory Policy. Hoover Institution: World Bank Paper prepared for the Conference on Economic Reform in Korea, January 15–16, 1997. Retrieved November 19, 1999, from http://www.stanford.edu/~moore/RegPolicy.html.

Needleman, H., and others. 1979. Deficits in Psychological and Classroom Performance in Children with Elevated Dentine Lead Levels. *New England Journal of Medicine* 300 (March 29): 689–95.

Ott, Wayne R., and John W. Roberts. 1998. Everyday Exposure to Toxic Pollutants. *Scientific American* 278 (February): 86–91.

Pope, C. A. III, M. J. Thun, M. M. Namboodiri, D. W. Dockery, J. S. Evans, F. E. Speizer, and C. W. Heath, Jr. 1995. Particulate Air Pollution as a Predictor of Mortality in a Prospective Study of U.S. Adults. *American Journal of Respiratory Critical Care Medicine* 151: 669–74.

Powell, Mark R. 1999. *Science at EPA*. Washington, D.C.: Resources for the Future.

Robinson, James C. 1995. The Impact of Environmental and Occupational Health Regulation on Productivity Growth in U.S. Manufacturing. *Yale Journal on Regulation* 12 (Summer): 387–434.

Sartor, F., C. Demuth, R. Snacken, and D. Walckiers. 1997. Mortality in the Elderly and Ambient Ozone Concentration During the Hot Summer, 1994, in Belgium. *Environmental Research* 72 (February): 109–17.

Schwartz, J. 1993. Beyond LOEL's, p Values, and Vote Counting: Methods for Looking at the Shapes and Strengths of Associations. *Neurotoxicology* 14 (2–3): 237–48.

Skelton, Renee. 1997. Clearing the Air: An Epidemiologist Takes on the Worst Air Pollution Problems of Our Times. *Amicus Journal* 19 (Summer): 27–31.

Taubes, Gary. 1995. Epidemiology Faces Its Limits. *Science* 269 (July 14): 164–69.

Tengs, Tammy O., and others. 1995. Five-hundred Life-Saving Interventions and Their Cost-Effectiveness. *Risk Analysis* 15 (June): 369–89.

U.S. Department of Commerce. 1997. *Survey of Current Business*, May, table 15.

U.S. Environmental Protection Agency. 1990. *Environmental Investments: The Cost of a Clean Environment.* Washington D.C.: Environmental Protection Agency.

———. 1997. *The Benefits and Costs of the Clean Air Act*, 1970 to 1990. Prepared for U.S. Congress by U.S. Environmental Protection Agency, October, 1997. Retrieved November 24, 1999, from http://www.epa.gov/airprogm/oar/sect812/index.html.

———. 1999. *The Benefits and Costs of the Clean Air Act, 1990 to 2010.* Prepared for U.S. Congress by U.S. Environmental Protection Agency, November, 1997. Retrieved November 24, 1999, from http://www.epa.gov/airprogm/oar/sect812/index.html.

Vedder, Richard K. 1996. *Federal Regulation's Impact on the Productivity Slowdown: A Trillion-Dollar Drag.* Study no. 131 (July). St. Louis: Center for the Study of American Business, Washington University.

Wright, Janice C., and Milton C. Weinstein. 1998. Gains in Life Expectancy from Medical Interventions: Standardizing Outcomes Data. *New England Journal of Medicine* 339 (August 6): 380–86.

PART VII

Debating Market-Based Environmentalism

19

Market-Based Environmentalism and the Free Market—They're Not the Same

ROY E. CORDATO

Since the collapse of communism in Eastern Europe, overtly socialist solutions to public policy problems have fallen into disrepute, even among socialists. It now seems to be widely accepted among policy analysts of both the Left and the Right that direct government control of market activities and market outcomes—the so-called command-and-control approach to public policy—is an excessively costly way to achieve public policy goals.

Yet despite widespread rejection of outright socialism and command-and-control policies, there is little appreciation of truly free markets and the outcomes they are likely to generate. Policy makers do not value market exchange because it maximizes liberty and personal satisfaction of wants. Instead, policy makers value the market because they can manipulate it to produce a centrally planned outcome. This approach describes so-called market-based environmental policy.

All approaches to market-based environmentalism (MBE) tend to follow the same pattern. As MBE advocates Robert Stavins and Bradley Whitehead (1992) point out, "There are two steps in formulating environmental policy: the choice of the overall goal, and the selection of a means or 'instrument' to achieve that goal" (3). Specifically, government authorities first select a particular outcome (e.g., level of sulfur dioxide emissions or amount of recycled paper used in grocery bags) as a desirable goal. Viewing the behavior of individuals making exchanges in the relevant markets as something to be manipulated through public policies that create incentives to "do the right thing," policy makers then select an appropriate means for this purpose. In environmental policy, the two most highly touted instruments are excise taxes and tradable permits. These market-based approaches, advocated by professional economists and think-tank policy analysts on both the Left and the Right, actually use markets against themselves. In reality they are often meant to thwart the outcomes of true free-market activity.

I shall criticize the arguments advanced for market-based environmentalism, the most important of which have their roots in Pigovian welfare economics.[1] My criticism relies on arguments advanced by F. A. Hayek to demonstrate the impossibility of efficient central planning and by James Buchanan regarding the subjective nature of costs and benefits. The economic arguments for MBE have given it widespread appeal across the political spectrum. I shall argue that MBE has the same defects as full-blown socialism: it is inconsistent with individual liberty and, in practice, impossible to implement successfully.

ENVIRONMENTAL PROBLEMS: MARKET FAILURE OR GOVERNMENT FAILURE?

The reigning view of environmental problems considers them as inherent in a free society. If people are free to pursue their own self-interest—to produce and consume whatever they want, how and when they want it—polluted air and waterways, littered streets, and depleted natural resources will result. The typical characterization in most of the social science literature is that such problems represent "market failure." Pollution and environmental degradation are cited as evidence that Adam Smith was wrong, or at least naïve. People pursuing their self-interest do not necessarily advance the well-being of society as a whole. Therefore, it is not only appropriate for, but incumbent on, government to correct the market's failings.

This view constitutes a misunderstanding of the nature of a free society and a free-market economy. Contrary to the standard view, environmental problems are not an unavoidable side effect of a free-market economy. Instead, they occur because the institutional setting—the property rights structure—required for the operation of a free market is not fully in place. Because, in all modern societies, government has taken nearly complete responsibility for the establishment and maintenance of this institutional setting, environmental problems are more appropriately viewed as manifestations of government failure, not market failure.

ENVIRONMENTAL PROBLEMS AS CONFLICTS OVER THE USE OF PROPERTY

In current debates over environmental issues, it has become common to abstract from individual decision makers in society and to view certain uses of resources as inherently problematical. Traditionally, conditions such as air and water pollution aroused concern to the extent that they harmed people. More recently this view has been abandoned. Now many argue that certain uses of resources should be regulated or proscribed not because they harm third parties but because they degrade "the environment." For example, for many, strip mining, the use of landfills for the disposal of trash, and the cutting down of old-growth forests do not constitute problems because of harm to humans. Indeed, the fact that humans usually benefit from these practices is viewed with disdain. The traditional view of environmental problems, that they should concern policy makers because these problems involve harm to human beings, has been turned on its head. The modern view, adopted by many who advocate market-based "solutions" is that harming humans is justified in the pursuit of "saving" some aspect of the nonhuman environment.[2]

In a free society, concern for human beings must take center stage. In assessing environmental problems, the core question is how and why such problems interfere with individual decision making, construed as the formulation and execution of plans. As all formulation and execution of plans involve the use of physical resources, and such plans can legitimately employ only resources to which one has rights, any environmental analysis focused on the individual decision maker must pay attention to property rights.

For example, air pollution creates a problem to the extent that it interferes with individuals as they formulate and execute plans. This can happen only if the pollution somehow interferes with an individual's exercise of rights to his or her property or if uncertainty prevails concerning who actually has the rights to a particular resource. Viewed in this way, all environmental problems involve conflict over the use of property. Person A and person B are attempting to use resource X for conflicting purposes. Either A or B clearly has the relevant rights to X but these rights are not being enforced, or the rights to X have not been clearly defined, that is, neither A nor B nor anyone else has the relevant rights to X. In the former case, the environmental problem is one of property-rights enforcement. In the latter case, an authoritative decision must be made regarding who should have the rights. The foregoing conditions establish the relevant parameters of environmental problems from a humanist, as opposed to an environmentalist, perspective.

Two simple examples can highlight each of these possibilities. Imagine a community with a cement factory that emits dust into the air without the consent of people nearby. Because of the dust, people in the community must wash their cars and house windows more frequently. The dust also soils clothing hung out to dry and creates respiratory problems for those who breathe it. This problem is clearly one of property-rights enforcement. The problem arises not because the dust is emitted into the air but because it has direct contact with what is indisputably people's property—their cars, houses, laundry, and lungs—and thereby interferes with their planned use of it. Here the conflict concerns the use of property to which ownership is clearly defined but regarding which some rights are not being strictly enforced.

An example of the second type of problem involves the use of a public waterway such as a river. A factory uses the river as a receptacle for waste generated by its production process. Downstream, homeowners use the river for fishing and swimming. Suppose factory waste renders the river unfit or a t least less fit for these purposes. The central problem here is not simply that the river is being polluted, but that plans for its use are in conflict. Unlike the rights in the cement-dust case, the rights to the river are not clearly defined, so the public policy issue involves who should have what rights.

Property rights must be clearly defined and enforced in order for a free market to exist. If problems arise because these institutional requirements are not met, it is wrong to blame the free market for the problem. In each of the examples just presented, a problem arises because the institutional prerequisites of a free market are, in one way or another, not fulfilled. The problems should not be blamed on market failure when a free market is prevented from coming into existence. The problem actually represents institutional failure, as the institution of private property itself is not being sustained. From a public policy perspective, a "crack" exists in the property-rights structure. In general, a genuine free-market policy would first identify the specific interpersonal conflicts that have emerged, then identify and repair the flaws in the property-rights structure that have given rise to the conflicts.

This assessment suggests that "free-market environmentalism" is simply an attempt to advance the free market in areas plagued by conflicts over portions of the physical environment. It should be noted, however, that the focus is on human freedom and welfare and not the nonhuman environment itself. The result of free-market policies in dealing with such issues may not always be consistent with goals of environmentalism as typically construed, although often they will be.[3] The defining characteristic of free-market activity is the institutional setting in which it occurs, not the outcomes it generates.

In the ideal institutional arrangement, all resources are privately owned, and all owners can employ their property in any way they wish. The only legal constraint is that no one be allowed to infringe the equal rights of others.[4] Once these conditions are established, the market process is open ended. The actual results reflect the interaction of individuals pursuing their own objectives, often by making exchange contracts with others. True free-market public policy should not focus on particular outcomes with regard to the environment or anything else, including prices, costs, and output levels. Instead, it should focus on correcting flaws in the institutional setting that are giving rise to human conflict and thereby preventing the efficient pursuit and attainment of goals. From this perspective, what have come to be called environmental problems are indeed problems in many cases, because they are rooted in deviations from the optimal property-rights structure.

On occasion, however, establishing a free market will conflict with the goals of environmentalism as usually construed. An example pertains to the treatment of endangered species. It has become common for the U.S. Fish and Wildlife Service (FWS), in enforcing the Endangered Species Act, to place controls on the use of privately owned land deemed either an actual or a potential habitat for an animal listed as endangered. From a free-market perspective, such restrictions become the source of institutional failure rather than the solution of market failure. In such cases the policy generates a conflict between the actual owners of the land and certain nonowners who use the state to gain decision-making power over the use of the property. A free-market advocate, environmentalist or not, would oppose such policies and favor the individual whose property rights are being transgressed. Free-market environmental policy cannot be unbiased or even democratic. It must be distinctly biased in favor of whoever has title to the portion of the environment in dispute.

Of course, the free-market environmentalist would oppose the FWS regulations, but might be willing to attempt to raise funds to purchase rights to the desired property for the purpose of preserving the habitat. This kind of activity distinguishes a general supporter of free markets from a free-market environmentalist. They differ in their private activities, not in their public policy stance.

In general, free-market advocates argue that nature will fare better under a regime of private property and free exchange than it will under other institutional arrangements, because the profit motive, coupled with the obligation not to violate the property rights of others, leads to the conscientious stewardship of natural resources. As stated in one study in the growing literature:

> Unfortunately, under current institutional arrangements, too many people find that environmental destruction rather than conservation is in their self interest. Most of our environmental problems arise because resources

> such as air, water, forests, and many species of birds, fish, and other wildlife are owned in common. Because these resources have no owners, they have few protectors and defenders. Because there is no market for these resources, people have poor incentives to maintain their value. . . . The institutions that have worked well for us in other areas of economic life include private property, markets, a price system and methods for punishing people who violate the rights of others. . . . Government is needed to create the legal framework. Within that framework, people should be free to experiment and innovate to solve problems which large bureaucracies are unlikely to solve. (Task Force Report 1991)

Free-market environmentalists have done a good job of demonstrating how the institution of private property has led in the past, and could lead in the future, to attaining many of the goals espoused by the environmental movement.

Still, the end result of a free-market process may not coincide with the goals of environmentalists, particularly if those goals require restrictions on the use of resources for their own sake or relate to issues where no physical or economic harm to human beings is involved. Recognizing this fact, most mainstream environmentalists remain antagonistic to truly free markets. But some have also come to recognize that politically controlled markets can be useful tools, in a manipulative sense, for the advancement of their aims. These more mainstream environmentalists, with the intellectual assistance of some segments of the economics profession, are spearheading the recent advocacy of the policies known as market-based environmentalism.

MARKET-BASED ENVIRONMENTALISM: STEALTH SOCIALISM

It is ironic that although the intellectual foundation of MBE has been laid largely by conservative economists, MBE is embraced more or less enthusiastically by liberal environmentalists. The economists have given the environmentalists a means of dispensing with command-and-control regulatory policies while maintaining their command-and-control ends. Whether intentionally or not, many economists, most of them conservative, have become efficiency consultants for the traditionally anti-free-market environmental movement. These economists have demonstrated that the environmentalists can attain their goals more "efficiently" by creating suitable incentives in the market than by setting rigid rules and standards for production processes. Writing as both a conservative and an economist, Murray Weidenbaum, former chairman of President Reagan's Council of Economic Advisers, succinctly states the basic principle of

MBE: "In the various circumstances when government does regulate (as in the case of reducing environmental pollution), conservatives prefer that government policy makers make the maximum use of economic incentives. Thus, to an economist, the environment pollution problem is not the negative task of punishing wrongdoers. Rather, the challenge is a very positive one: to alter people's incentives" (Weidenbaum 1992, 497).

Conservative economists have been quite successful in convincing former regulatory zealots in the environmental movement that command-and-control policies are a cumbersome and unnecessarily costly way to achieve their ends. MBE has won the whole-hearted endorsement of the Progressive Policy Institute (PPI), the think tank most closely associated with the Clinton administration. PPI authors Robert Stavins and Bradley Whitehead (1992) state:

> Command and control regulations were powerful in the early battles against environmental degradation, but they have begun to reveal many of the same limitations that led to the collapse of command and control economies around the globe. Command and control regulations are often economically inefficient—that is excessively costly. . . . Market based policies start with the assumption that the best way to protect the environment is to make it in the daily self-interest of individuals and firms to do so. The key to greater environmental protection, then, is . . . decentralization—by changing the financial incentives that face millions of firms and individuals in their private decisions about what to consume, how to produce, and where to dispose of their wastes. (iii)

Clearly, MBE is outcome driven. Its thrust is to manipulate market incentives in order to achieve a centrally planned outcome with respect to the use of natural resources. As argued by Stavins and Whitehead (1992), "Policies are needed to mobilize and harness the power of market forces on behalf of the environment, making economic and environmental interests compatible and mutually supportive. Policy makers must begin to link the twin forces of government and industry" (3). Clearly, MBE differs fundamentally from public policy meant to promote free markets. Whereas proponents of the latter see free markets as a means of fostering liberty and human well-being, proponents of MBE see it as an instrument for "harnessing" the activities of people "on behalf of the environment."

GREEN TAXES AND ECONOMIC ANALYSIS

Excise taxes and tradable permits are the two major policy instruments advocated most frequently as part of MBE. The justification for using excise taxes to promote environmentalist goals ("green taxes") has its roots in Pigovian welfare economics. Welfare economics is the part of economic analysis that sets the standards by which economists make public policy prescriptions. Mainstream economists have adopted Pigovian welfare economics, the approach expounded in most economics textbooks. Nearly all advocates of MBE, on both the Left and the Right, give lip service to this sort of welfare economics (Repetto and others 1992; Viscusi 1992; Weidenbaum 1992; Stavins and Whitehead 1992).

From this perspective, the success of market activity depends on market outcomes, particularly the prices and quantities that markets generate. Certain price-quantity outcomes are deemed "efficient," contributing positively to social welfare, whereas "inefficient" outcomes reduce social welfare. When efficient outcomes are not generated, free markets are declared a failure, and the primary purpose of public policy is manipulating the choices of market participants to achieve the correct results.

This approach to environmental policy has many flaws, most of which stem from methodological errors in the economic analysis underlying the policy prescriptions.[5] In the standard argument, social welfare is maximized when markets conform to a set of ideal conditions known as "perfect competition." Perfectly competitive markets have many buyers and sellers, all perfectly informed of relevant market conditions; costless entry and exit; and homogeneous product lines. Given these conditions and the attainment of systemic equilibrium, prices will accurately and completely reflect the marginal costs of production and, as a result, the quantity of any good produced and sold will be efficient. All costs of production or consumption are "internalized," that is, borne by the producers or consumers of the product. The market thus "succeeds" in generating the correct, perfectly competitive outcome.

Within this framework, pollution constitutes a problem because it gives rise to incorrect outcomes. In the cement-dust example, the cement company is not bearing all the costs of its production. People in the surrounding community whose plans are disrupted by the pollution also bear some of the costs. In such cases the price of the product is too low, the amount produced too high. The market is said to fail. If the cement company were bearing all the costs, including the pollution costs now borne by others, its costs of production would be higher, it would produce less, and consumers would pay a higher price for cement.

Policy makers use market-based policies to manipulate market incentives ostensibly to obtain the "correct" result, that is, the result that would occur in a perfectly competitive market. In a situation like that of the example, most economists would advocate levying an excise tax. The goal is to impose a tax on each unit of production of the polluting firm that exactly equals the pollution costs borne by the outside (nonconsenting) community. The firm would then have an incentive to behave as if it bore all the costs of production and hence to generate the sought-after "efficient" market outcome. The tax "succeeds" where the free market "fails." As Weidenbaum (1992) argues, "Pollution taxes serve to correct a serious source of market failure: the absence of 'price' needed to prevent the careless and excessive use of scarce environmental resources. Taxation . . . is a basic way of working through the price system" (499).

This approach to dealing with pollution problems necessarily leads to the manipulation of markets for the achievement of political goals. The tax "corrects" the market failure only if imposed exactly as economic theory dictates. But the perfectly competitive conditions that the policy is supposed to induce are so highly stylized and otherworldly that they provide no real-world guidance for imposing the appropriate taxes.[6]

To implement the Pigovian program, policy makers must be able to measure the spillover costs associated with the pollution. But as James Buchanan (1981) has emphasized, "Cost is subjective; it exists only in the mind of the decision-maker or chooser. . . . Cost cannot be measured" (14–15; see also Buchanan 1969). Economically relevant opportunity cost is the satisfaction forgone in choosing to do one thing rather than another. If someone washes his car more frequently because cement dust from a nearby cement plant is soiling it, his cost is the greatest satisfaction he expected to receive by doing something else. Clearly, this is both unmeasurable and unknowable by any outside observer. Therefore the "correct" market outcome also is unknowable. As Buchanan (1969) has concluded, "In order to estimate the size of the corrective tax . . . some objective measurement must be placed on these external costs. But the analyst has no benchmark from which plausible estimates can be made" (72). Anyone defending the use of such taxes as a means of enhancing economic efficiency would first have to explain how Buchanan's objection can, in practice, be overcome.[7]

Advocates of corrective taxes ignore the subjective-cost issue; they proceed with their analysis as if the problem did not exist. They reach conclusions about social costs without a hint that the numbers cited are supposed to measure something conceptually unmeasurable. For example, one analyst confidently proclaims that "the current net tax per gallon [of diesel fuel] is 13 percent of the price, while the environmental cost per gallon is 50 percent of the price. The tax on this fuel could be raised substantially to promote its efficient use" (Viscusi 1992, 18). In light of Buchanan's arguments, one can only wonder what these numbers actually measure, but clearly they are not measures of economically relevant opportunity costs.

In applications, mainstream welfare economics is timeless and does not allow for change, invoking static equilibrium analysis. Analysts assume that information gathered today relates equally to tomorrow. But such constancy would require that input scarcities, technology, population, and people's preferences remain fixed. Once any of these variables changes, current information —intended to shed light on the costs of pollution and therefore on the correct outputs, prices, and taxes—becomes outdated. Inasmuch as these variables are in reality constantly changing, identifying the appropriate corrective tax is an impossibility. Even if actual opportunity costs were being measured, all cost-benefit analysis would necessarily be based on historical evidence, much of it already several years old when gathered. Any corrective tax would be obsolete even before its calculation.

Both the subjective-cost problem and the time-passage problem exemplify a more fundamental problem of information or knowledge. This dooms all attempts at efficient central planning, including Pigovian corrective taxation. F. A. Hayek articulated this argument against socialist planning in the 1930s and 1940s. He emphasized that the information necessary for central planning "never exists in concentrated or integrated form but solely as dispersed bits of . . . knowledge which all the separate individuals possess" (Hayek 1948, 77). Pigovian welfare economics rests on a general-equilibrium analysis of the economy. But in order to know the "correct" price-output combination in any one market where the tax is to be applied, the analyst must have the same information for all markets. As attested by an ardent supporter of such taxes, "The general equilibrium model of resource allocation which underlies formal welfare economics represents . . . a general analysis of the interrelationships of markets throughout the economy . . . it requires knowledge of the structure of preferences of all consumers and the technologies available to all producers" (Kneese 1977, 57).

Because the data required to manipulate markets as prescribed by the theory are impossible to gather, and the efficient outcome therefore impossible to identify, in practice the model serves merely as "cover" for those seeking to manipulate markets for various purposes.[8] Because there is no real standard by which the adequacy of the data can be gauged, the Pigovian approach to dealing with environmental problems has given rise to a frenzy of green-tax proposals, all claiming to promote economic efficiency.

The most famous, or notorious, of the green-tax proposals was President Clinton's ill-fated BTU tax. Although one may suspect that the real purpose of the proposal was to fatten the U.S. Treasury, its supporters touted the tax as a weapon in the battle against global warming. The proposal was a case study in the methods of MBE. An environmental goal was established: reduction of carbon dioxide emissions into the atmosphere. No evidence was presented that human welfare had ever been impaired by CO_2 emissions (Michaels 1992). Further, the global warming hypothesis—that at some unspecified future date such emissions will harm humans—has been accepted by only a small minority of atmospheric scientists. But proponents forged ahead. A 1990 Congressional Budget Office study, which helped to set the stage for the BTU tax proposal, candidly noted that "although there is great uncertainty about the extent to which such global warming is likely to occur, what its effects might be and the costs of efforts to slow the progress of warming, the potential consequences have led to calls for immediate action" (ix). In other words, regardless of the science or the economics, supporters voiced "calls for immediate action." Ultimately, all such tax schemes promote the goals of the politicians and interest groups supporting them.[9] As with the BTU tax, they have the added effect—fortunate in the eyes of many of their advocates—of increasing federal revenues and expanding government control over the use of productive resources.

TRADABLE PERMITS ARE NOT PROPERTY RIGHTS

Taxes are one tool in the kit that advocates of MBE carry to the policy making table. Another is tradable permits. Whereas green taxes alter prices in order to affect resource use in accordance with political desires, tradable permits serve as "property rights" for the same purpose. In advocating tradable permits, environmentalists again have taken their cue from ideas originally put forth by conservative economists. Working along lines laid down by Ronald Coase (1960), these economists have emphasized correctly that many, if not all, genuine pollution problems arise from a lack of property rights in the use of resources. Coase argued that if property rights are clearly defined and well enforced and can be cheaply exchanged, then parties can resolve pollution problems by bargaining. When Coase's conditions are met, there is no need for a cumbersome regulatory apparatus or a government bureaucracy to create an efficient allocation of resources. The "Coase Theorem" does not tell us what the efficient outcome should be. This intellectual modesty distinguishes Coase's argument from the argument of those who invoke his analysis to justify MBE. Although Coase's argument is sound, it says nothing about the nature of property rights, requiring only that once established, they should be freely tradable. Unfortunately, Coase's ideas have been imported into environmental policy debates, not to expand property rights but to justify the rearrangement and restriction of existing rights.

The now-standard approach of the advocates of MBE goes as follows. First, they identify something as excessive, for instance, the amount of waste going into landfills or the amount of fossil fuels used to generate electricity. Then, they arrive at an amount acceptable to the relevant politicians and special interests. To realize this amount, they take away the existing right to engage freely in the activity and then allocate new rights such that the total amount of the activity does not exceed the politically determined target. The new, more restricted "rights" take the form of a specific number of permits, each of which allows an individual or firm to engage in a certain amount of the disfavored activity. Holders of permits may buy or sell them. Holders therefore have a financial incentive to reduce the amount of their own disfavored activity. If a company can reduce its restricted activity sufficiently, it can obtain revenue by selling its unnecessary permits. To reiterate, the objective is not to advance free markets as such but to "harness market forces" to achieve a politically determined goal.

Tradable permits to pollute are the most commonly espoused form. But other forms are also advocated, where no specific pollution or emission is involved, most notably to promote recycling and to create a market for recycled materials where otherwise none would exist. With respect to pollution permits, which were adopted as part of the 1990 Clean Air Act Amendments, Stavins and Whitehead (1992) describe the system as follows:

> [T]he government establishes an overall level of allowable air pollution and then allocates permits among the firms . . . in a relevant geographic area so that each firm is allowed to emit some fraction of the overall total. Firms which keep their emissions below the allotted level may sell or lease their surplus permits to other firms or use them to offset excess emissions in other parts of their own facilities. (6)

One may argue that such schemes simply establish property rights to the use of air. And after all, doesn't air quality suffer because no one owns—and therefore protects—the air? Yes, but the real problem is not air quality as such. The real problem is that emissions eventually land on someone's property. In the cement-dust example, problems arise not because the dust is emitted but because it touches people's cars, houses, laundry, and lungs. Tradable pollution permits that might be issued to the cement factory could, depending on the permissible level of emissions and the prevailing cost conditions, still fail to prevent the factory from impinging on the property rights of people nearby. Robert McGee and Walter Block (1994) have argued forcefully that such permits are simply licenses to violate rights and therefore inconsistent with free markets.

They maintain that "perhaps the major fault with trading permits is that while they allow market forces to allocate resources, they entail a fundamental and pervasive violation of property rights" (57).

A truly free-market approach would allow the damaged parties to sue the offending parties for remedies rooted in a stricter enforcement of property rights. Plaintiffs would seek either compensation for damages or some form of injunctive relief. Another alternative would be a Coasean solution: the cement companies would purchase the relevant rights from the affected parties.

If a tradable permit implicitly grants the polluter the right to disregard the property rights of others, it is clearly inconsistent with a free-market economy. On the other hand, if a polluter's production activities do not violate the property rights of others, no problem exists and no policy action is necessary.[10] Tradable permits have been advocated increasingly in the latter setting, where no one's rights are being violated but where an activity is deemed harmful to the environment per se or otherwise aesthetically displeasing.

One such instance pertains to setting recycled-content standards for production processes, as advocated by the Progressive Policy Institute's Mandate for Change:

> [T]he government would set an industry-wide . . . recycled content standard which individual firms could meet in one of two ways: They could use the required percentage of secondary materials or they can use fewer secondary materials and buy permits from other firms that exceeded their recycling requirements. . . . Recycling credit systems could be . . . used for a variety of products, including newsprint and used lubricating oil. (Stavins and Grumbly 1993, 211)

The sole purpose of this policy is to promote recycling and reductions in landfill usage for their own sake. In the market, producers and consumers are rejecting many recycled materials as less desirable than virgin materials. Recycling advocates complain that "as more states and municipalities have adopted recycling programs, the increased supply of recovered materials has often outpaced demand for recycled or secondary materials. In some instances, this glut has resulted in the subsequent landfilling of separated, recyclable materials" (Stavins and Grumbly 1993, 211). Instead of reaching the obvious conclusion, recycling zealots support tradable permits as a scheme for forcing trash onto a resistant market.

CONCLUSION

Hayek (1967) argued that in a free society the "rules of just conduct" need to be "ends independent" (160–77). Legal arrangements should not favor the goals or purposes of some individuals or groups over others. Instead, such rules should be structured so that they maximize each individual's chances of accomplishing his own goals. Hayek also argued that once it is decided that a liberal social order is desirable, the propensity of policy makers to predetermine specific market and behavioral outcomes must be stifled, as liberty conflicts inherently with deterministic public policies. The rules "should consist solely in prohibitions from infringing the protected domain of each which these rules enable us to determine. Liberalism is, therefore, inseparable from the institution of private property" (165).

Hayek's guidelines point toward a true free-market approach to environmental issues. We must establish rules of conduct that clearly define people's rights, their "protected domain." The primary goal of all public policy, including environmental policy, should be the enforcement of rights once they are clearly defined. There is no proper role for the "ends-dependent" policies of the market-based environmentalists. As evidenced by Stavins and Grumbly's view of industries that fail to incorporate the "right amount" of recycled materials in their production process, the purpose of MBE is to thwart free decision making and the results to which it gives rise.

In assessing environmental issues, alternative policy approaches have been incorrectly categorized. The primary choice is not between command-and-control and market-based policies. Instead, it is between free-market policies, based on clearly defining and protecting property rights, and socialist—or, perhaps more precisely, mercantilist—policies, based on furthering the societal and personal goals of politicians and special-interest groups.[11] The latter includes both command-and-control policies and those labeled "market-based."

NOTES

1. This area of economics is named for economist A. C. Pigou, who originated the analysis. See Pigou ([1927] 1952).
2. For an excellent discussion of this issue and the motives underlying much environmental advocacy, see Kaufman (1994).
3. For an excellent discussion of how respect for private property and free markets often comports with the goals of environmentalism more broadly conceived, see Anderson and Leal (1991).
4. For discussions of the economic efficiency implications of this institutional framework, see

Kirzner (1963) and Cordato (1992). For discussions of some of the philosophical aspects, see Rand (1967) and Rothbard (1973).

5. For extensive discussions of the economics, see Cordato (1992, 1995).
6. For a detailed discussion, see Cordato (1989, 1992).
7. Although economists advocating MBE recognize Buchanan's public-choice analysis in assessing the extent to which government activity is likely to correct for market failure, they ignore the implications of Buchanan's cost theory for the use of social cost-benefit analysis. A consistent application of Buchanan's arguments implies that real-world social cost-benefit analysis, including the type suggested by Coasean property-rights analysis, is a logical impossibility outside a perfectly competitive general-equilibrium world. But any world with externalities is not such a world (Cordato 1989).
8. For example, the same model is invoked to justify increasing taxes on tobacco.
9. Public choice analysis maintains that politicians and bureaucrats, even if they possess the information necessary to improve social welfare, support laws and regulations that help them achieve their own goals and the goals of their political supporters. For a classic case study of environmental policy making along these lines, see Ackerman and Hassler (1981).
10. Arguably this was the case with the 1990 Clean Air Act Amendments, which instituted tradable permits with respect to emissions associated with the generation of acid rain. Prior to passage of the statute, studies demonstrated that many of the harms allegedly caused by acid rain either did not exist or were much less severe than originally thought (Krug 1990). If acid rain harms northeastern lakes as some allege, a free-market remedy would consist of holding the offending utility companies responsible for damages, possibly by requiring them to pay for liming the damaged lakes. This action also would be much less costly.
11. For why mercantilism may be a more precise term, see Ekelund and Tollison (1981).

REFERENCES

Ackerman, Bruce, and William Hassler. 1981. *Clean Coal/Dirty Air* . New Haven: Yale University Press.

Anderson, Terry, and Donald Leal. 1991. *Free Market Environmentalism.* San Francisco: Pacific Research Institute for Public Policy.

Buchanan, James M. 1969. *Cost and Choice.* Chicago: University of Chicago Press.

———. 1981. L.S.E. Cost Theory in Retrospect. Introduction in *L.S.E. Essays on Cost,* edited by J. M. Buchanan and G. F. Thirlby. New York: New York University Press.

Coase, Ronald. 1960. The Problem of Social Cost. *Journal of Law and Economics* 3 (October): 1–44.

Congressional Budget Office. 1990. *Carbon Charges as a Response to Global Warming.* Washington, D.C.: Congress of the United States.

Cordato, Roy E. 1989. Subjective Value, Time Passage, and the Economics of Harmful Effects. *Hamline Law Review* 12: 229–44.

———. 1992. *Welfare Economics and Externalities in an Open Ended Universe.* Boston: Kluwer Academic Publishers.

———. 1995. Pollution Taxes and the Pretense of Efficiency. *Journal of Private Enterprise* 10: 105–18.

Ekelund, Robert, and Robert Tollison. 1981. *Mercantilism as a Rent Seeking Society.* College Station, Tex.: Texas A&M University Press.

Hayek, F. A. 1948. The Use of Knowledge in Society. In *Individualism and Economic Order.* South Bend, Ind.: Gateway Editions.

———. 1967. The Principles of a Liberal Social Order. In *Studies in Philosophy, Politics, and*

Economics. Chicago: University of Chicago Press.

Kaufman, Wallace. 1994. *No Turning Back: Dismantling the Fantasies of Environmental Thinking*. New York: Basic Books.

Kirzner, Israel. 1963. *Market Theory and the Price System*. Princeton: Van Nostrand.

Kneese, Allen. 1977. *Economics and the Environment*. New York: Penguin Books.

Krug, Edward. 1990. The Acid Rain Flimflam. *Policy Review* 52 (Spring): 44–49.

McGee, Robert, and Walter Block. 1994. Pollution Trading Permits as a Form of Market Socialism and the Search for a Real Market Solution to Environmental Pollution. *Fordham Environmental Law Journal* 6: 51–77.

Michaels, Patrick. 1992. *Sound and Fury*. Washington, D.C.: Cato Institute.

Pigou, A. C. [1927] 1952. *Economics of Welfare*. London: Macmillan.

Rand, Ayn. 1967. What Is Capitalism? In *Capitalism: The Unknown Ideal*, edited by A. Rand. New York: New American Library.

Repetto, Robert, Roger Dower, Roger Jenkins, and Jacqueline Geoghegan. 1992. *Green Fees: How a Tax Shift Can Work for the Environment and the Economy*. Washington, D.C.: World Resources Institute.

Rothbard, Murray. 1973. *For a New Liberty*. New York: Macmillan.

Stavins, Robert, and Thomas Grumbly. 1993. *The Greening of the Market: Making the Polluter Pay*. In Mandate for Change. Washington, D.C.: Progressive Policy Institute.

Stavins, Robert, and Bradley Whitehead. 1992. *The Greening of American Taxes: Pollution Charges and Environmental Protection*. Washington, D.C.: Progressive Policy Institute.

Task Force Report. 1991. Executive Summary. *In Progressive Environmentalism: A Pro- Human, Pro-Science, Pro-Free Enterprise Agenda for Change*. Dallas, Tex.: National Center for Policy Analysis.

Viscusi, Kip. 1992. *Pricing Environmental Risks*. St. Louis: Center for the Study of American Business.

Weidenbaum, Murray. 1992. Reforming Government Regulation to Promote Prosperity. *The World & I* (August): 487–503.

Acknowledgments: The original draft of this paper was prepared for the Future of Freedom Foundation conference "Halting the Destruction of American Liberty," October 1994. The author thanks two anonymous referees for their helpful suggestions.

20

Market-Based Environmentalism and the Free Market—Substitutes or Complements?

PETER J. HILL

Roy Cordato has presented an insightful critique of market-based environmentalism. His analysis provides several reasons why one should be suspicious of recent attempts to shape environmental policy through the use of "market-based" rules. Four of Cordato's criticisms are especially telling.

First, environmental problems exemplify not market failure but institutional failure. The lack of well-defined and -enforced private property rights causes the voluntary interaction of individuals to produce undesirable outcomes.

Second, people err by characterizing environmental problems as the overuse of certain resources, independent of whether such use harms people's ability to formulate and execute plans. Thus, many efforts to reform public policy amount to attempts to keep certain resources from being used, whether or not such use represents an infringement of anyone's rights. Humans cannot inhabit this planet without using resources. The relevant question is: What sorts of rules facilitate human cooperation and the satisfaction of human desires? Of course, such desires may include the preservation of certain species or resources, but a well-functioning market allows individuals to satisfy these desires only if they make acceptable payments when they use other people's property for such preservation. Unfortunately, when government coercion is involved, people can use the political process to achieve ends for which they are unwilling to pay.

Third, it is difficult to implement appropriate market-based environmentalism because the political process does a poor job of measuring subjective costs and benefits. Instead, people with their own agendas can pervert the process to achieve ends that impose substantial costs on others. Statists, bureaucrats, or radical environmentalists can use many of the market-based environmental proposals to accomplish purely political ends. Economists often act as if choosing the optimal tax level or deciding how many tradable pollution permits should be created is a straight-forward matter easily accomplished through the political process. They give little consideration to the influence of the special-interest groups who capture the process to achieve their own purposes (Mitchell and Simmons 1994).

Fourth, Cordato correctly identifies the hubris of much environmental policy. Such policy presumes that the solutions are known and need only be implemented, thus ignoring the importance of the discovery process. Having frozen environmental policy into a particular configuration, people have little capacity for adaptation as they learn of and encounter changes in preferences and resource availabilities. Although market-based environmentalism (MBE) can be more adaptive than command-and-control techniques, it remains less flexible than a true market order.

Cordato's insights are valuable, but he does not go far enough. He identifies the solution to environmental problems as well-defined and -enforced private property rights, but he does not provide a theory or evidence of how such property rights come into being. Without a framework for such analysis, we can only make categorical statements that are not very constructive. But we need to assess whether particular policy innovations are helpful or harmful in moving us toward a more effective structure of property rights. Cordato closes his paper by saying,

> The primary choice is not between command-and-control and market-based policies. Instead, it is between free-market policies, based on clearly defining and protecting property rights, and socialist —or, perhaps more precisely, mercantilist—policies, based on furthering the societal and personal goals of politicians and special-interest groups. The latter includes both command-and-control policies and those labeled "market-based."

But exactly what are these free-market policies based on clearly defining and protecting property rights? Do such policies involve any role for government? If so, what is it? Cordato criticizes economists who use the nirvana model of perfect competition to analyze "market failure" and make policy prescriptions. However, Cordato himself may have slipped into nirvana thinking by urging a regime of pure private property rights but giving no attention to the transaction costs inherent in the definition and enforcement of those rights. While recognizing the importance of Cordato's insights into some of the problems of market-based environmental policy, one who takes a transaction-cost approach may view MBE somewhat more favorably.

TRANSACTION COSTS AND THE ROLE OF GOVERNMENT

Market solutions are superior to coercive ones because voluntary exchange offers the assurance that social interactions are mutually advantageous. However, transaction costs prevent some potentially profitable voluntary exchanges from taking place. Through the use of appropriate rules, government can provide feasible alternatives. In the standard examples of roads and national defense, the transaction costs of individual exchange are high and the free-rider problem substantial. Thus, there is at least some potential for using tax-financed provision of these public goods as a corrective mechanism. Of course, government provision of public goods is fraught with numerous problems, and one ought not be overly optimistic that government will get it right. However, we should not automatically rule out all government intervention.[1]

An even more telling example of not relying completely on voluntary contracts under well-defined and -enforced private property rights pertains to common-law proceedings. Although parts of the common law (contract and property law) are designed to facilitate voluntary transactions, another part (tort law) amounts to using the coercive power of government to enforce solutions that simulate, but do not completely capture, the essence of exchange solutions.

When I drive my car on the road, we could be more assured of "correct" solutions to all interaction problems if I had contracted with everyone I might potentially harm in an accident. Such contracts would specify the different types of accidents, the various degrees of harm, and the appropriate compensation of those harmed. Obviously, were such contracts required before entering the public roadways, the transaction costs of identifying and contracting with all who might be affected by one's driving would be so high as to preclude most people from using automobiles. Therefore, we have adopted an alternative solution: tort law holds me responsible if I dent the fender of your car.[2]

Under common-law rules the problem of subjective cost is ignored and a coerced solution mandated. I will have to compensate you for the damage I cause to your car, but the amount of compensation will be determined by a third party who cannot know your subjective costs. Nevertheless, such a "market-based" solution has many advantages. It lowers the transaction costs of certain types of human interaction, allows us to be involved in useful activities and, despite its shortcomings, forces people who harm others to pay damages, even if the payments do not precisely correspond to the true costs of the harm.

One may offer a similar defense of market-based environmentalism. A more benign reading than Cordato's of the efforts to introduce market-based solutions is that the transaction costs of defining and enforcing rights in some cases are very high and that the coercive power of government can be used productively to give us solutions that are better than not taking any action at all.

WHENCE PRIVATE PROPERTY RIGHTS?

It is not very helpful simply to suggest that the answer lies in well-defined and -enforced private property rights without indicating how we get those rights. In the remainder of this paper I shall present and discuss some preliminary observations about how property rights come into being. I shall also suggest how one might assess the involvement of government in that process.

Four hypotheses about the development of private property rights are as follows.

1. Property-rights definition and enforcement are themselves economic activities that respond to perceived benefits and costs (Anderson and Hill 1975). The incidence of benefits and costs is crucial to the process: those who can reap the gains from such activity are more likely to undertake it. In other words, the potential for appropriation of the profits is as important here as elsewhere.
2. The transaction costs of defining and enforcing rights depend on culture, norms, and group size. Small, culturally homogeneous groups have the best chance of success in developing effective rights structures. However, when the resource in question is mobile, the local community may be unable to provide effective governance.
3. Government can facilitate the definition and enforcement of rights. Coercion is a powerful means for reducing free-rider problems; thus, it can improve the institutional framework within which private decision makers act.
4. Government can also interfere with the development of effective property rights. For reasons well stated by Cordato, no one should look upon government action as a panacea for all property-right problems.

Events in the American West illustrate well the influence of benefits and costs on property-rights development. When settlers first came to the West, they devoted few resources to defining and enforcing rights because, given the scarcity of people and the abundance of natural resources, claimants rarely came into conflict. They published announcements in newspapers, directly informed potential neighbors, or used simple devices of demarcation to define their rights to land. However, as more settlers came and property became more valuable, people devoted more resources to the definition and enforcement of rights.

Costs of defining and enforcing rights also changed. The development of barbed wire in the 1870s—an entrepreneurial response to the high costs of wooden or stone fences in the Great Plains—dramatically lowered the cost of defining and enforcing rights to land in that region.

Ability to appropriate profits crucially affected the development of workable rules. In 1862, the first Homestead Act mandated a laborious system of establishing property rights to unclaimed federal land. Under this and later statutes, activities required to establish ownership included living on the land for a specified number of years, plowing acreage unsuited for farming, planting trees, and carrying out irrigation. However, settlers developed alternatives. To govern property rights, land claim clubs established local rules quite different from those mandated by federal legislation (Anderson and Hill 1983). Because those designing the rules for the clubs stood to gain by reducing the dissipation of resources in the process of defining rights, they chose much more sensibly. They did not require inefficient farming practices, and they dramatically reduced residency requirements. As this example shows, local institutions for defining and enforcing rights are more likely to generate productive rules.

Another example pertains to salmon fishing in the Pacific Northwest. Contrary to popular perception, salmon have not always been overexploited. Local Indian tribes evolved excellent rules to govern access to migrating salmon and their spawning grounds (Anderson and Leal 1996). Only with the imposition of modern institutions by the white settlers did these property rights break down (Higgs 1982).

In contrast, the tribe was not an effective decision-making unit for governing the overall rate of use of buffalo, and overexploitation occurred. It is often assumed that white hunters caused the near destruction of buffalo on the Great Plains, but there is evidence that American Indians, once they had the horse and the rifle, overexploited the resource because of the difficulty of securing adequate agreement among different tribes as to the appropriate rate of use (Anderson 1995). Numerous other case studies make the same point: local groups can often devise effective rules for governing the use of resources, but sometimes these communities do fall victim to the tragedy of the commons.

The cultural homogeneity of groups, the existence of moral constraints, and the willingness of individuals to identify with the group affect significantly the development of property rights (Anderson and Simmons 1993). High transaction costs impede social interactions among people who do not share a common cultural or religious heritage or who do not develop trust by interacting repeatedly. In traditional societies of small local communities, established multidimensional relationships increase the likelihood that effective private property rights will evolve to govern the use of most resources and also that institutions will develop for controlling the use of the commons (Leal forthcoming). However, when shocks, such as the arrival of a different cultural group, hit the society, its ability to maintain effective rules suffers.

Because of the difficulties of securing voluntary agreement, government can often reduce transaction costs. There are economies of scale in the production of information, and the use of coercion can overcome certain barriers to the definition and enforcement of rights. One of the most effective ways government can intervene is by measuring and monitoring property rights. For example, although settlers in the West relied a great deal on localized institutions for governing the use of resources, statewide brand registration laws were passed quickly. Because cattle could be moved from community to community, requiring centralized registration of brands as legal proof of ownership reduced the transaction costs of monitoring theft. Likewise, requiring the recording of deeds in the county courthouse lowers the cost of contract enforcement and facilitates using real property as collateral for loans.

But government action is a two-edged sword, with both benefits and costs. The use of coercive power can lower transaction costs, but it can also substantially increase them and obstruct the development of private property rights. For instance, instream flows of water have become more valuable in many western states. Because of an increased demand for recreation, fishing, and other amenities, leaving water in the streams generates far more benefits than it did previously. However, in most western states, private parties cannot own water left in the stream; they must divert their water or lose their rights to it. Therefore, transferring property rights in recognition of shifting demands has been difficult.

Government action has also raised transaction costs in environmental law. The replacement of common law by statute law after the establishment of the Environmental Protection Agency has slowed the evolution of common law (Meiners 1995). Application of the doctrine of nuisance has substantial potential for solving environmental problems, but statute law has preempted much of the common law of nuisance. Again, government has stood in the way of the development of appropriate private property rights.

WHAT SHOULD GOVERNMENT DO?

The transaction-cost approach to environmental problems points to a somewhat different solution than that suggested by Cordato. He sees the choice as between markets and socialism (or mercantilism). Granted, the lack of well-defined and -enforced private property rights lies at the root of environmental problems. But because transaction costs prevent the complete specification of property rights, it is helpful to have an evaluative framework by which to judge government involvement in environmental matters. Government can do three basic things under the rubric of market-based environmentalism to facilitate the definition and enforcement of rights.

Get Out of the Way

In many instances government simply needs to get out of the way of private efforts to define rights. When it makes sense to define and enforce rights, individuals will search for a way to do so. However, when government insists on maintaining control of the relevant resources, it will thwart such efforts. A prime example is big-game hunting on public lands. As hunting pressure has increased, the common-pool problem of hunters taking less-than-trophy-size animals has also increased. Hunters seeking trophies and those seeking meat increasingly come into conflict. Under state ownership of both the game and the land, no residual claimant (profit appropriator) can prosper by superior game management, preventing hunters from engaging in premature harvest.[3] A few large ranches with resident game herds provide strong evidence that such management is profitable. Numerous creative property-rights solutions have also evolved on other private lands, such as the pooling of lands for hunting purposes and the selling of rights to hunt for trophy game. Likewise, private game management in Africa has prospered where private ownership of land and game prevails.

In other cases, higher levels of government should allow local communities the freedom to create and enforce rights. The likelihood that effective agreement can be reached is much greater in smaller, more homogeneous communities. Ellickson (1991) has found that viable communities of people engaged in numerous repeated interactions produce useful rights structures outside the legal system. Other case studies reveal effective local governance of fisheries where higher levels of government don't interfere (Anderson and Simmons 1993).

Government should not restrict trades of property rights. Grazing permits on federal lands and water rights are among the property rights not freely transferable because of legal impediments.

Legal doctrine or federal statutes can also stand in the way of the evolution of effective property rights. For instance, the public trust doctrine implies that government, acting as the agent of society, should have rights to many natural resources. Unfortunately, this doctrine has often been invoked to prevent the effective development of private rights.

Measure and Monitor Property Rights

Another potentially effective government action is to provide—and in some cases to mandate—basic registration functions. Cattle ranchers in the West turned to the states to register their brands. Similarly, government can use its coercive power to require the registration of potential pollutants. In Canyonlands National Park in Utah, effective monitoring and tracing of the pollutants from a particular source have been carried out by the addition of tracer chemicals to a smokestack (Crawford 1990). It would also be feasible to require all users of an herbicide to "brand" the product by adding a small amount of an inert chemical or radioactive isotope. If the brands were placed in a central registry, then, if the herbicide appeared in groundwater, the responsible party could be identified. It might be unnecessary for government to do anything more, particularly if the common law of nuisance were embraced.

Patience is in order. The definition and enforcement of property rights are evolutionary activities. Government can set the stage by removing artificial barriers and, in some cases, providing registration or monitoring facilities. However, if a solution does not ensue immediately, further government action is not necessarily indicated. Conflicts over the use of resources may not be sufficiently serious, or parties may need to develop new technology. Devising workable arrangements takes time. A good general guideline is to permit numerous attempts to solve the problem. Localized solutions hold more promise than a centralized effort aimed at a single national remedy.

Taxes, Subsidies, and Pollution Permits

The most difficult issues of government involvement arise where problems extend across the boundaries of local communities. In such cases, removing government obstacles to the development of property rights, requiring the registration of potential pollutants, or relying on local, community-based solutions may be insufficient. Major problems of global climatic change and of pollution that crosses political boundaries are much more difficult to solve.[4] Taxes, subsidies, tradable pollution permits, and treaties among political units may be appropriate to approximate markets and property rights. However, at this level, Cordato's criticisms of market-based environmentalism are the most telling. Attempts to use government to simulate the solutions of the marketplace will be plagued by both information and incentive problems. Thus, government ought to move with extreme caution in implementing solutions of this type.

IN SUM

In assessing environmental problems and policies, I do not draw the line between government and the marketplace as sharply as Cordato does. Effective environmental policy making must center on an understanding of the process by which private property rights develop. Government can play both positive and negative roles in that process. We need an adequate theory of the creation and change of property rights to assess the government's involvement. Market-based environmentalism can be a smoke screen for statists and radical environmentalists who simply want to force their preferences on an unsuspecting public. However, it can also represent an intelligent effort by people who understand and believe in markets to facilitate the development of private property rights so that markets can operate.

NOTES

1. Some have suggested that public-good problems are not significant enough to warrant government action (Friedman 1973), but a more mainstream position allows at least some positive potential for government involvement in environmental problems. For instance, Anderson and Leal (1991) offer a program of free-market environmentalism that relies on appropriate, limited government actions.
2. Also at issue is the more complicated question of whether the rule of negligence or strict liability should apply. For a discussion, see Epstein (1995, chap. 5).
3. A similar problem exists in fisheries. Where government insists on maintaining the rights of all comers to fish, many useful property-rights solutions to problems with fishery commons are prevented from evolving (Libecap 1989, chap. 5).
4. Of course, one must be reasonably certain that such problems exist at a level worth attempting to ameliorate before plunging headlong into massive state solutions. Many of the supposedly substantial transboundary problems that have received a great deal of media attention have been found not to be serious at present (Bast, Hill, and Rue 1994).

REFERENCES

Anderson, Terry. 1995. *Sovereign Nations or Reservations? An Economic History of American Indians.* San Francisco: Pacific Research Institute for Public Policy.

Anderson, Terry, and Peter J. Hill. 1975. The Evolution of Property Rights: A Study of the

American West. *Journal of Law and Economics* 18:163–79.

———. 1983. Privatizing the Commons: An Improvement? *Southern Economic Journal* 50: 438–50.

Anderson, Terry, and Donald Leal. 1991. F*ree Market Environmentalism*. San Francisco: Pacific Research Institute for Public Policy.

———. 1996. The Salmon in Economics: Back to the Future of Property Rights. Paper presented at Fraser Institute conference "Managing a Wasting Resource," Vancouver, B.C., May 30–31.

Anderson, Terry, and Randy Simmons, eds. 1993. *The Political Economy of Customs and Cultures*. Lanham, Md.: Rowman & Littlefield.

Bast, Joseph, Peter J. Hill, and Richard C. Rue. 1994. *: A Common-Sense Guide to Environmentalism*. Lanham, Md.: Rowman & Littlefield.

Crawford, Mark. 1990. Scientists Battle over Grand Canyon Pollution. Science 247: 911–12.

Ellickson, Robert C. 1991. *Order without Law: How Neighbors Settle Disputes*. Cambridge: Harvard University Press.

Epstein, Richard A. 1995. *Simple Rules for a Complex World*. Cambridge: Harvard University Press.

Friedman, David. 1973. *The Machinery of Freedom*. New York: Harper & Row.

Higgs, Robert. 1982. Legally Induced Technical Regress in the Washington Salmon Fishery. *Research in Economic History* 7: 55–86.

Leal, Donald. 1996. Community-Run Fisheries: Preventing the Tragedy of the Commons. In *Taking Ownership: Property Rights and Fishery Management on the Atlantic Coast*, edited by B. Crowley. Halifax, Nova Scotia: Atlantic Institute for Market Studies.

Libecap, Gary D. 1989. *Contracting for Property Rights*. New York: Cambridge University Press.

Meiners, Roger E. 1995. Elements of Property Rights: The Common Law Alternative. In *Land Rights: The 1990s' Property Rights Rebellion*, edited by B. Yandle. Lanham, Md.: Rowman & Littlefield.

Mitchell, William C., and Randy T. Simmons. 1994. *Beyond Politics: Markets, Welfare, and the Failure of Bureaucracy*. Boulder: Westview Press for The Independent Institute.

PART VIII

Environmental Philosophy

21

Does "Existence Value" Exist?

Environmental Economics Encroaches on Religion

ROBERT H. NELSON

In *Encounters with the Archdruid*, John McPhee (1971) relates a discussion he had with David Brower, regarded by McPhee and many others as the leading environmentalist of our time. Brower is talking about the real meaning of wilderness. He notes that

> I have a friend named Garrett Hardin, who wears leg braces. I have heard him say that he would not want to come to a place like this by road, and that it is enough for him just to know that these mountains exist as they are, and he hopes that they will be like this in the future. (74)

As Brower said of his own views, "I believe in wilderness for itself alone."

Economics as traditionally practiced, however, finds it difficult to accommodate this perspective on the world (Norton 1991). Human beings, the economics way of thinking assumes, live for happiness. Happiness is, moreover, a product of consumption. As the economist Stanley Lebergott (1995) writes, "the goal of every economy is to provide consumption. So economists of all persuasions have agreed, from Smith and Mill to Keynes, Tobin, and Becker" (149). Historically, there has been little or no place in economic thinking for the idea that something that is never seen, touched, or otherwise experienced—that is not consumed in any direct way—can have a value to an individual.

Yet, as McPhee's discussions with Brower indicated, this thinking rooted in economics was deeply at odds with an emerging environmental awareness that in the 1960s and 1970s was spreading widely in American society. Economists, it appeared, might be faced with an awkward choice: either reject their own economic perspective on the world or disagree with a powerful new social movement. Most likely, some economists were themselves drawn personally to the environmental values difficult to express in conventional economic terms. For them, the potential dilemma was also internal: either limit their own commitment to certain environmental goals such as the intrinsic importance of wilderness and endangered species preservation or reject the economic way of thinking in an important area of their life.

However, in a famous 1967 chapter in the *American Economic Review*, John Krutilla (1967) proposed a reconciliation. Krutilla suggested that the scope of economics should be expanded to include a new concept, which has come to be known as "existence value." The enjoyment of life need not have as its limit things that can be seen and touched. Consumption, even as economists think about it, should extend to include the simple fact of knowing that a wilderness, endangered species, or other object in nature exists. Formally, the variables in a person's "utility function" would not only comprise the amounts of food, clothing, and other ordinary goods and services consumed but also the various states of knowledge each person has of the existence of social and physical characteristics in the world. Implicitly at least, consumers would be willing to pay something for this form of consumption; hence the efforts by economists to estimate existence values in dollar terms (Mitchell and Carson 1989; Portney 1994).

By the 1980s, the concept of existence value was coming into use by a number of economists for purposes such as estimating the benefits of government actions or calculating damage assessments against corporations whose actions had harmed the environment. A federal appeals court in 1989 directed the Department of the Interior to give greater weight to existence values in its procedures for assessing damages to public resources under the Superfund law (*State of Ohio v. United States Department of the Interior* [D.C. Circuit. 880 F. 2d 432, 1989]). The concept has even been received favorably in literary circles such as the *New York Review of Books*, where the author of one chapter concluded that it would be central to achieving preservation of tropical forests and other world biodiversity objectives: "But why should citizens of industrialized countries pay to preserve resources that are legally the domain of other countries? An obscure tenet of economics provides a rationale. Certain things have what is known as an 'existence value'" (Terborgh 1992, 6).

The large dollar magnitudes that some economists were attributing to this new source of economic benefit emphasized its potential importance. In 1992, Walter Mead surveyed a variety of estimates of existence value. In one study the value to households across the United States of preserving visibility in the Grand Canyon was estimated to equal $1.90 per household per year, yielding a long-run discounted value to all U.S. households of $6.8 billion. Preservation of the northern spotted owl in the Pacific Northwest was estimated in another study to be much more valuable, having a total existence value for U.S. households of

$8.3 billion per year. Still another existence-value study came up with higher numbers, asserting that, however implausible it might seem, preserving whooping cranes would be worth $32 billion per year for all the people of the United States. Having such impressive dollar estimates to cite raised the prospect for some environmentally concerned economists that the government might alter calculations of the economic merits of various policy proposals.

A GROWING DEBATE

Initially, most of the economic discussion of existence value reflected the views of proponents. Beginning in the 1970s, a small circle of economists sought to introduce a novel concept to the profession and to show that it could be applied successfully in practice. At first, most mainstream economists paid little attention. However, as the potential uses have widened and the policy stakes escalated, an active debate has broken out in recent years within the economics profession concerning the merits of the existence-value concept (Bate 1994; Desvousges 1993a; Diamond and Hausman 1993; Edwards 1992; Kopp 1992; Quiggin 1993; Randall 1993; Rosenthal and Nelson 1992; Stewart 1995). Noneconomists also entered the controversy, in some cases questioning the use of existence value ("Ask a Silly Question" 1992; DiBona 1992).

The Exxon Corporation, facing large potential damage assessments as a result of the Exxon Valdez oil spill, and fearing that these assessments might be based in part on economic estimates of existence value for various states of nature in Prince William Sound, committed large financial resources to the issue. Exxon hired a number of leading economists to examine whether the use of existence value was an appropriate economic method. Their evaluation was on the whole negative (Hausman 1993). The state of Alaska and the federal government hired several leading environmental economists who took a more positive view (Carson and others 1992).

Reflecting the growing controversy inside and outside the economics profession, the National Oceanographic and Atmospheric Administration (NOAA) convened a panel of leading economists, chaired by Nobel Prize winners Kenneth Arrow and Robert Solow, to review the issue. In 1993 the panel declared that although great care must be exercised to prevent misuse, existence value should be included among the economic tools available to government analysts (Arrow and others 1993). However, the NOAA report failed to resolve the matter, and an active debate continues (Portney 1994).

From a technical economic standpoint, existence value raises a number of problems, which a growing literature has been probing. Massachusetts Institute of Technology economists Peter Diamond and Jerry Hausman (1994) conclude that "surveys designed to test for consistency between stated willingness-to-pay and economic theory have found that contingent valuation responses are not consistent with economic theory" (46). Other critics find that in practice existence-value studies often yield implausible estimates (Desvousges 1993b). For example, respondents to survey questionnaires sometimes give similar estimates for saving a group of wild animals from human harm, even when the exact number of animals specified in the surveys may vary by orders of magnitude.

Thus far, those who have actually attempted to measure existence values have studied mostly wilderness areas, threatened species, and other environmental concerns. However, the use of the concept is potentially much broader. People in rich nations may value the existence of tropical forests, but these same people may value the existence of higher incomes for poor people in poor countries, a goal whose fulfillment may depend on cutting the forests.

Indeed, possibilities for the calculation of existence value are endless (Milgrom 1993). Virtually any object invested with symbolic importance will have an existence value. For example, the presence of an abortion clinic in a community will cause some of the residents to feel good, others to feel bad. Burning the American flag will have a large negative existence value for many people. However, the knowledge that freedom of speech, including flag burning, is protected will also have a large positive value for many others. Should survey questionnaires, based on statements of dollar values as a way of communicating views about the desirability of government actions, be used to help resolve such issues? One can pose the same sorts of questions for a large array of issues.

Diamond, Hausman, and several other leading economists have called on the profession to abandon the use of existence value on both theoretical and empirical grounds, such as those noted so far (Diamond and Hausman 1994; Castle and Berrens 1993). But others, although acknowledging the significant difficulties and major potential pitfalls of the theory, argue that Americans care a great deal about the environment even when they are not directly affected and that any decision-making calculus not incorporating such preferences as a benefit would suffer from serious inadequacies (Carson and others 1996; Hanemann 1994).

Despite their importance, these particular issues are not my subject here. I conclude, like other critics, that the use of existence value should be abandoned. My argument, however, is grounded in what might be called "economic theology" (Nelson 1991). To be sure, I mean theology in a broader sense than Christianity or other traditional religions alone. The distinguished theologian Paul Tillich (1967) once said in all seriousness that in terms of actual impact Karl Marx was "the most successful theologian since the Reformation" (476).

Secular religions such as Marxism, it is now common to point out, have been a dominant feature of the modern age, often the decisive force in shaping the course of history (Talmon [1960] 1985). Secular religions do not speak directly of or appeal to God for authority. However, they are religions in the sense that they set a framework of meaning by which a person understands his or her life and the fundamental values that will shape it. Moreover, secular religions are often suffused with the same themes present in Christianity and Judaism (Glover 1984). In all likelihood, that is the explanation for their great appeal (Nelson 1991).

Existence-value methods have thus far been applied mostly to issues such as wilderness and endangered species that, I will show, have a religious basis. To anticipate my conclusion, the theological problem with existence value is that in such cases it attempts to answer a religious question by applying an economic method. Making estimates of existence value then is both as silly and as meaningless as asking how much the knowledge of God is worth.

NATURE AS THE PATH TO KNOWLEDGE OF THE DIVINE

McPhee's discussions with Brower went well beyond the importance of preserving wilderness areas. Indeed, for Brower, wilderness was but one element in an overall worldview. For many years, Brower toured lecture halls on college campuses and other places across the United States, preaching what McPhee (1971) labeled "the sermon" (79). To many people, Brower's great appeal was essentially religious. As McPhee wrote,

> to put it mildly, there is something evangelical about Brower. His approach is in many ways analogous to the Reverend Dr. Billy Graham's exhortations to sinners to come forward and be saved now because if you go away without making a decision for Christ coronary thrombosis may level you before you reach the exit. Brower's crusade, like Graham's, began many years ago, and Brower's may have been more effective. (83)

It was, and Brower enjoyed greatest influence in the segments of secular society where environmentalism was most popular and Graham's voice scarcely heard at all.

Indeed, Brower's approach had its own religious tradition, with its environmental prophets, great texts, and sacred sites. According to McPhee (1971), "throughout the sermon, Brower quotes the gospel—the gospel according to

John Muir, . . . the gospel according to Henry David Thoreau" (84). As a former executive director of the Sierra Club for seventeen years in the 1950s and 1960s, Brower followed directly in the line of Muir, who had founded the Sierra Club in 1892. In the late nineteenth and early twentieth centuries, Muir was the foremost advocate of setting aside wild areas to preserve them for the future as free of human impact as possible.

For Muir, the wilderness had an explicitly religious significance. He referred to primitive forests as "temples" and to trees as "psalm-singing." As Roderick Nash (1973) writes in *Wilderness and the American Mind*, Muir believed that the "wilderness glowed, to be sure, only for those who approached it on a higher spiritual plane. . . . In this condition he believed life's inner harmonies, fundamental truths of existence, stood out in bold relief" (125–26).

For Muir this was one way of saying that he experienced the presence of God in the wilderness. On other occasions he was still more explicit. He believed that in the natural objects of wild areas, one could find "terrestrial manifestations of God." They provided a "window opening into heaven, a mirror reflecting the Creator," making it possible to encounter in nature some true "sparks of the Divine Soul" (Nash 1973, 125).

By creating the world, God had enabled human beings to experience directly a product of divine workmanship. The experience of nature untouched by human hand was as close to a direct encounter with God as possible on this earth. Yet, because of the spread of science and industry in the modern era, this opening to the mind of God was being erased. People were building dams, cutting forests, farming the land, and in many ways imposing a heavy human footprint on the divine Creation. It was only in the limited remaining areas of wilderness, as Nash (1973) relates, that "wild nature provided the best 'conductor of divinity' because it was least associated with man's artificial constructs" (125). If all the wild areas should be lost, future generations would be forever cut off from this main avenue of contact with God.

All this is to say that for Muir a wilderness area was literally a church. A church is a place of spiritual inspiration, where people come to learn about and understand better the meaning of God in their life. In church especially, God communicates his intentions for the world. A wilderness church is, furthermore, in one sense more imposing and spiritual than any church people can build. A wilderness is a church literally built by God.

A SECULAR RELIGION

Today, the religious convictions that motivated Muir still lie behind the demand for wilderness, but with one significant difference. Environmentalism has become a secular religion. As Joseph Sax (1980) has said, in seeking to preserve national parks and other wild areas, he and his fellow preservationists are "secular prophets, preaching a message of secular salvation" (104). Roger Kennedy (1995), the former director of the National Park Service, agrees: "Wilderness is a religious concept," he wrote recently, adding that "we should conceive of wilderness as part of our religious life." Wilderness puts us "in the presence of the unknowable and the uncontrollable before which all humans stand in awe" (28)—that is to say, although Kennedy does not put it in just these words, in wilderness we stand in the presence of God.

In his essay "John Muir and the Roots of American Environmentalism," the distinguished environmental historian Donald Worster (1993) explores the process of secularization. Muir was brought up in Wisconsin and immersed in the doctrines of a strict Protestant sect, Campbellism. These doctrines would play a major role in shaping his thinking for the rest of his life. But like so many others in the modern age, by his twenties he had left the traditional religious forms of his youth well behind. As Muir said, "I take more intense delight from reading the power and goodness of God from 'the things which are made' than from the Bible" (Worster 1993, 193). Worster concludes that while the influence of Muir's youthful piety remained strong, he "invented a new kind of frontier religion; one based on going to the wilderness to experience the loving presence of God." This type of religion would later prove immensely attractive for the "many Americans who have made a similar transition from Judeo-Christianity to modern environmentalism" (195–96).

Although Muir abandoned the established Christian churches of his time, his writings contain frequent references to God. Today, environmentalists such as Brower seldom speak directly of God but do regularly describe experiencing a "spiritual inspiration," "sense of awe," "source of values," "humbleness of spirit," and so forth in the wilderness. These descriptions differ little from the language used by earlier generations to describe the feeling of being in the presence of God.

Many leading environmental thinkers in the United States today explicitly characterize their mission as "religious," if not as Christian. In *The Voice of the Earth*, Theodore Roszak (1992) states that "the emerging worldview of our day will have to address questions of a frankly religious character" (101). Environmentalism, he argues, will have to provide answers to "ethical conduct, moral purpose, and the meaning of life," and thereby help to guide "the soul" to the goal of "salvation" (51). In early 1996, Interior Secretary Bruce Babbitt stated that "religious values are at the very core of the 1973 Endangered Species Act" (ESA). Babbitt and other environmental leaders have sought to enlist Christian religious organizations to support the ESA as a "Modern Noah's Ark" (Steifels 1996, A12).

The present motto of the Wilderness Society, borrowed from Thoreau, is "In wildness is the preservation of the world," that is, the salvation of the world. In its appeals for public support, the Wilderness Society typically asserts of wilderness areas that "destroy them and we destroy our spirit . . . destroy them and we destroy our sense of values." The issue at stake in preserving wilderness is not merely a matter of the aesthetics of a beautiful landscape or the retention of a museum piece of the geologic past. The real objective, as the Wilderness Society declares, is to maintain the moral foundations of the nation.

This declaration might seem outlandish, or mere fund-raising rhetoric, to those who know little of the theological history of the idea of wilderness. However, in a long religious tradition that dates to seventeenth-century New England, "a genuine reading of the book of [wild] nature is an ascension to the mind of God, both theoretical and practical" (Miller 1954, 209). When the Wilderness Society tells us today that our national values depend on preserving the wilderness, it is only expressing in a secularized way what many others have asserted before: that without God, no foundation for values exists. And God, as Muir said explicitly and contemporary secular environmentalists say implicitly, is encountered most closely in the wilderness.

Although some people have seen modern environmentalism as borrowing from Asian, pantheistic, and other sources, the core of the religious conviction for most environmentalists is a secularized Christianity. This should not be surprising in a nation where Christian influence is ingrained in the national psyche, whether recognized explicitly in all cases or not.

A SECULAR PURITANISM

The process of secularization did not begin with Muir. He regarded himself as a follower of Emerson, whose writings he had studied closely. The philosophy of New England transcendentalism represented the critical point at which Christian theology—largely of a Puritan variety—adapted rapidly to the new demands of the modern age. Historian Arthur Ekirch (1963) observes that in the transcendentalist philosophy "nature was the connecting link between God and man"; thus, "God spoke to man through nature" (51–52).

Emerson, Thoreau, and other transcendentalists, in turn, drew much of their inspiration from their Boston forebears. If the transcendentalists saw an empty worship of false economic gods spreading across the land, the Puritans had always said that income and wealth were among the most dangerous corrupters of the souls of men. The Puritans also, as the Harvard historian Perry Miller (1954) commented, were "obsessed with" the "theology of nature." In Puritan theology of the colonial era,

> the creatures . . . are a glass in which we perceive the one art which fashions all the world, they are subordinate arguments and testimonies of the most wise God, pages of the book of nature, ministers and apostles of God, the vehicles and the way by which we are carried to God. (208–9)

The idea of a moral imperative to preserve every species—God's decree that every species has a right to exist—has religious origins deep in Western civilization. In the sixteenth century, Calvin said that human beings should be "instructed by this bare and simple testimony which the [animal] creatures render splendidly to the glory of God" (Kerr 1989, 27). Indeed, according to Calvin, God intends for "the preservation of each species until the Last Day" (41). The Bible, which some current environmental leaders are invoking, gave explicit instructions on this matter in the story of Noah and his Ark.

Jonathan Edwards, by some accounts America's greatest theologian, linked the seventeenth-century Puritans and their nineteenth-century New England intellectual heirs. Edwards said that "the disposition to communicate himself . . . was what moved [God] to create the world" (Miller 1964, 194). As Miller (1964) observed, "what is persistent, from the [Puritan] covenant theology (and from the heretics against the covenant) to Edwards and to Emerson is the Puritan's effort to confront, face to face, the image of a blinding divinity in the physical universe, and to look upon that universe without the intermediacy of ritual, of ceremony, of the Mass and the confessional" (185).

The present environmental movement offers a secular Puritanism in more than its attitudes toward wild nature. As McPhee (1971, 83) relates, Brower commonly referred in his sermon to the human presence in the world as a "cancer." More recently, Dave Foreman, the founder of Earth First!, again said that "humans are a disease, a cancer on nature" (Looney 1991). As Paul Watson (1988), a founder of Greenpeace, put it, "we, the human species, have become a viral epidemic to the earth"—in truth, "the AIDS of the earth" (82). This all harks back to the doom and gloom of a Puritan world of depraved human beings infected with sin, tempted to their own destruction at every step by the devil and his devious tricks. It should be expected, the Puritan ministers said, that a sinful world would soon have to suffer a harsh punishment imposed by God, both on this earth and for most people in a life to come in hell.

In these and still other ways, environmentalism is today a powerful secular embodiment of the Puritan impulse in American life (Nelson 1993). Indeed, the Puritan tradition has had an extraordinary and enduring influence on the entire history of the United States. It is hardly surprising that although it is taking new and most often secular forms today, the Puritan influence is exerting itself once again. As Worster (1993) explains:

> The second legacy [of the environmental movement] from Protestantism is ascetic discipline. In large measure Protestantism began as a reaction against a European culture that seemed to be given over, outside the monastic orders, to sensuous, gratification-seeking behavior. . . . There was from the beginning, and it reappeared with vigor from time to time, a deep suspicion of unrestrained play, extravagant consumption, and self-indulgence, a suspicion that tended to be very skeptical of human nature, to fear that humans were born depraved and were in need of strict management.
>
> The Protestant tradition may someday survive only among the nation's environmentalists. . . . Too often for the public they sound like gloomy echoes of Gilbert Burnet's ringing jeremiad of 1679: "The whole Nation is corrupted . . . and we may justly look for unheard of Calamities." Nonetheless, the environmentalists persist in warning that a return to the disciplined, self denying life may be the only way out for a world heading towards environmental catastrophe.
>
> Surely it cannot be surprising that in a culture deeply rooted in Protestantism, we should find ourselves speaking its language, expressing its temperament, even when we thought we were free of all that. (197–98, 200)

Today the environmental movement is strongest in Germany, Sweden, and Holland—all countries with strong Protestant heritages. By contrast, in France, Spain, and Italy, shaped much more by Catholicism, green parties and environmental groups play a much smaller role. In Latin countries the full body of the Catholic church itself—with all its history and authority—was the means by which God communicated with the world. The Pope was God's agent on earth; in the Catholic church the faithful could encounter the majesty and mystery of God.

Having expelled Catholicism, Protestants had to look elsewhere. They often found their spiritual inspiration in nature, hearing there the voice of God. The Puritans, who most ruthlessly eliminated ceremony and imagery, had a special need to find in nature a substitute for an abandoned mother church.

HOW MUCH IS A CHURCH WORTH?

This brief excursion into theological and environmental history should suffice to show that the existence value of wilderness, endangered species, and other wild objects in nature is as much a theological as an economic subject. Indeed, extending the concept of existence value to the maximum, one might conclude that God has the ultimate existence value. But unlike God, a candidate wilderness area can at least potentially be tangibly visited, even by those who value it most for the very fact of its existence.

Certainly, many people will find any talk of the dollar value of God to be sacrilegious. Not that long ago a person could be burned at the stake for less. Yet, as the previous discussion has indicated, calculating a monetary value for the knowledge of the existence of a wilderness area comes close to valuing God. Nature untouched by human hand, as found in a wilderness, gives access to knowledge of the existence and qualities of God. In secular environmentalism this message comes in an only slightly revised form: wild nature is "the true source of values for the world."

Admittedly, to value a wilderness in this way is to value the instrument of communication of religious truth rather than the actual knowledge itself. Thus, a more precisely analogous question would be: How much is the knowledge of the existence of a church worth?

At least in concept, this is an answerable question. Economists can point out that although leaders of institutional religions may be offended by the question, they do in fact make such calculations. Other things being equal, they regard more churches as better. But more churches also cost money. In making a decision at some point not to build another church, a religious organization is saying in effect that the religious benefit of the additional church is not worth the cost of building and maintaining it.

How would one go about putting a marginal value on the existence of one more church (wilderness)? Answering this question, assuming a person is willing to think about the matter in these terms, would involve multiple concerns. One question would be: How much does a particular new church (wilderness) add to the religious education of the faithful? How many new people might it draw into the faith? A related question would be: How many churches (wildernesses) should a religious denomination ideally maintain, and how many does it have already? The answers obviously depend partly on the total number of faithful, their geographic distribution, and the expected growth of the religious group in the future.

Yet another factor is that the building of a church is not just a way to be spiritually uplifted. It can also serve to publicly and symbolically announce a depth of religious commitment and formally take an action for the glory of God. Building a grand cathedral, such as Notre Dame in Paris, can acquire a special religious significance when it involves a great sacrifice of effort—as religions have historically found meaning in making large sacrifices of many kinds. A wilderness area might become more meaningful in the same way: the more valuable the mineral, timber, and other natural resources given up, the greater the sacrifice and therefore the symbolic statement of allegiance to the faith.

The symbolic value of great sacrifice explains why the Arctic National Wildlife Refuge (ANWR) has become so important to the environmental movement. Its importance springs not just from the on-the-ground environmental features of the area—many other equally desolate and isolated places are also important to some group of wild animals. The truly distinctive feature of ANWR is that by leaving it untouched, so much valuable oil would potentially be sacrificed. Protecting the area offers a rare opportunity to make a powerful religious statement. An analysis of the benefits and costs of ANWR oil development thus becomes in major part an assessment of a trade-off between two alternative "uses" of the oil: (1) as fuel for a modern economy, and (2) as a symbol signifying the willingness of society to commit vast resources to preserve a multibillion-dollar cathedral, a religious edifice requiring such large sacrifice that it would stand as one of the greatest (certainly most expensive) testimonies ever made to the glory of the faith.

From a social point of view extending beyond the immediate members of the denomination, a church may be valued also by others outside the religion. Just as non-Catholics may admire the Vatican as a work of art or as an important part of their history, many people today no doubt value a wilderness as a museum of natural history. Wilderness areas often have beautiful scenery that can be set aside for others in the future to enjoy.

To be sure, the discussion of the potential analytical problems that surround putting a marginal value on the existence of a new church (wilderness) has begged the question of whether a religious body would ever want to do such a thing. Indeed, most religious leaders would likely reject any such suggestion out of hand. A church partakes of the sacred; to assign a money value to it profanes the faith. The very act of regarding the church in economic terms would in itself diminish the value of the church.

Many environmental leaders do in fact react much as other religious leaders would to proposals to measure the existence value of a wilderness. While recognizing a potential for political gain by making their case in economic terms, environmentalists on the whole have reacted coolly if not antagonistically to efforts by economists to calculate existence values for wild objects in nature.

Mark Sagoff (1994), former president of the International Society of Environmental Ethics, writes that "contingent valuation [is] an attempt to expand economic theory to cover environmental values. . . . But what makes environmental values important—what makes them values—often has little or nothing to do with 'preferences,' with perceived well-being, or with the 'satisfaction' people may feel in taking principled positions" (7–8). Aside from the many practical analytical problems, Sagoff rejects the calculation of existence value in principle as an imperialistic attempt by economists to substitute clever techniques for "the role that the public discussion of values should play in formulating environmental policy" (8). In short, the practice attempts to decide religious questions on (pseudo) scientific grounds.

NEGATIVE EXISTENCE VALUE

For Sagoff and many others, the very act of attempting to attach a money value to the existence of an endangered species, a wilderness, or other object of wild nature is itself a source of mental distress. It is like trying to put a money value on God, a sacrilege in any faith. Indeed, "negative existence values" may be almost as common as positive evaluations, because in any diverse society a cultural or religious symbol regarded favorably by one group will often be regarded unfavorably by some other groups. Not surprisingly, the members of the economics profession who advocate use of existence value have largely neglected this possibility.

Indeed, in the specific case of wilderness, some people do regard the existence of a newly created wilderness area as a symbolic affront to their own values. For some, it is offensive in the manner of throwing away good food, a deliberate waste of good timber, mineral, and other natural resources. A leader of the current Wise Use movement, Ron Arnold (1982), writes that wilderness and other curbs on development "have bit by bit impaired our productivity with excessive and unwise restrictions on forest and rangelands, on water and agriculture, on construction and manufacture, on energy and mineral, on every material value upon which our society is built" (123).

Although they might not phrase it precisely this way, other critics sense intuitively the following: the legal designation of a wilderness area represents symbolically a testimony to the glory of one faith, but this may be a faith different from their own, and they may thereby feel their own religious convictions diminished. One analyst has characterized the current fierce policy dispute over the creation of wilderness areas in southern Utah is at heart a clash between the

Mormon theology of many Utah natives and a competing set of secular religious precepts (Williams 1991).

Still others might object that a wilderness is not a church of any institutional Christian religion. Indeed, the rise of environmentalism reflects the increasing secularization of American society—in itself an unpleasant thought to contemplate for some traditional Christians.[1] "Negative utility" may also arise when secular religions borrow Christian messages and values, even though the followers in these secular faiths may no longer be aware of the original inspiration.

ENVIRONMENTAL CREATIONISM

A "secular religion" is, in truth, an awkward construction. Such a religion typically appropriates the values, religious energy, organizational forms, and other features of an earlier established religion, usually in the Judeo-Christian tradition. Yet, frequently it sets all these attributes in what is said to be a naturalistic or scientific context. The dressing of religion in the garb of science may end up as an attempt to blend contradictory elements.

Consider the theology of wilderness as found in the secular faith of much of contemporary environmentalism. The Puritans believed one could go to the wilderness to gain a unique access to the mind of God. In the sixteenth and seventeenth centuries the Puritans could accept easily enough the biblical message of the Creation, identifying nature as a literal work of God untouched by human hand. But geological, biological, and other sciences since that time have determined that the earth is many billions of years old and has undergone countless upheavals and transformations. Perhaps a wilderness can help to reveal natural laws of the universe, and these laws may themselves reflect a divine source. However, a wilderness can no longer in any real sense be said to reveal the original condition of the earth, as created by God.

Wilderness theology, in short, involves a form of creationism: sometimes with an explicit link to the Judeo-Christian story in Genesis; in other cases, with no explicit mention of God—a "secular creationism." Current environmental writings contain many references of both kinds to "the Creation"; two recent books on environmental matters are *Caring for Creation* and *Covenant for a New Creation* (Oelschlaeger 1994; Robb and Casebolt 1991). A magazine chapter on environmental philosophers describes the belief that the current need is for a "spiritual bond between ourselves and the natural world similar to God's covenant with creation" (Borelli 1988, 35). In much the same vein, and perhaps even more commonly, natural environments isolated from much European contact are widely referred to as a newly found, or currently sought after, "Eden" or "paradise" of the earth ("Inside the World's Last Eden" 1992; McCormick 1989).

Such language has begun to invade even mainstream politics: Vice-President Gore has said that we must cease "heaping contempt on God's creation" (Niebuhr 1993, A13). In a December 1995 speech remarkable for its candor in linking his environmental policy making to his religious beliefs, Interior Secretary Bruce Babbitt said that "our covenant" requires that we "protect the whole of Creation." Harking back to John Muir and to earlier Puritan theology, the secretary said that wild areas are a source of our "values" because they are "a manifestation of the presence of our Creator." It is necessary to protect every animal and plant species, he stated, because "the earth is a sacred precinct, designed by and for the purposes of the Creator."[2]

Such new forms of environmental creationism involve as much tension with the canons of scientific knowledge as with the older more familiar forms of Christian creationism. Although Babbitt referred specifically to God, others do not, even though they speak religiously of "the Creation." Some might find the secular version the most objectionable of all: prominent biologists and other physical scientists sally forth to attack Christian creationism as ignorant obscurism, even as some of them actively proselytize their own secular style of environmental creationism.

If awareness of these matters spreads, the designation of a wilderness area could come to represent yet another cultural symbol: the existence of a large element of religious naïveté, if not rank hypocrisy, among portions of the scientific establishment—yet another potential source of negative existence value for at least some people.

Whether the cultural symbol is established through a public or private action also affects the various forms of potential negative existence value. If a private group builds its own church, at least in America (it can be much different in other countries), few people are likely to be greatly upset, even though they may disagree strongly with the church creed. However, if the government builds the church, matters are altogether different. The government is not only spending taxpayers' money but making an official declaration formally affirming a particular set of religious values. A person might object strongly to the establishment of a government-owned and -operated wilderness area but have little or no objection to a private group's undertaking precisely the same mission. My arguments suggest that the present national system of wilderness areas should be privatized and any further wildernesses created privately as well (Nelson 1992).

WHO ASKS THE QUESTION DETERMINES THE ANSWER

The multiple meanings of wilderness are typical of cultural symbols. An X-rated movie provides valued sexual titillation to one person, while its very existence signifies society's moral decay to another. To some, the existence of a government welfare program may represent the compassion of society for the poor, but for others it may symbolize the coercive confiscation of hard-earned money from deserving people in order to give it away to undeserving others.

The proponents of the use of existence-value methods suggest that they can apply their techniques according to the canons of the scientific method. They suggest further that existence-value measurement, as a scientific exercise, will be replicable. Also, they argue that despite some people's suspicions, the results will not reflect the beliefs of the scientific investigators and that the more resources put into the investigation, the more consistent and reliable the estimates of monetary existence value presumably should become.

None of these things, however, is likely to occur in practice. In fact, when economists estimate existence value, they use uncomplicated methods. In essence, the economic researcher solicits answers to a survey questionnaire. The questions and the answers may be oral or written (sometimes with follow-up). For a particular wilderness area, for example, the questionnaire might begin with a brief description of the potential site and then ask how much money the person—who may be a thousand or more miles away—would be willing to pay to know that this place will be preserved for the future as a wilderness with minimal human intrusion.

However, since the respondent often knows essentially nothing about the possible wilderness, it is typically necessary to provide some background for answering the question. This raises many potential difficulties. Consider some of the possible items that might be mentioned:

1. A brief physical description of the wilderness;
2. in order to provide some needed context, a brief description of how many total wilderness areas have already been established in the United States and how the particular potential wilderness area being studied fits into that broader picture;
3. inclusion of some historical context, such as an explanation of how the belief in preserving wilderness has been traced by leading scholars to John Muir and New England transcendentalists, for whom the purpose of visiting wild nature was to experience the presence of God;
4. for the survey respondents with an interest in theological analysis, brief reference to the idea that in light of modern scientific knowledge, the theology of wilderness today represents a kind of secular creationism.

To be sure, existence-value researchers no doubt would object strongly that administering the questionnaire with any such accompanying materials would bias the results, which is probably true. However, there may be no escaping this problem. To say that only "the facts" will be provided is untenable. Presenting all the facts is not feasible; one must make a ruthless selection. Why would a geologic description be factually more appropriate than a historic or theological description? To argue for the exclusion of the theological information may merely affirm the cultural values of a secular society.

Moreover, the more financial resources that are available and thus the more information that can be conveyed to respondents, the better a scientific analysis should be. However, in this case it will also exacerbate the selection problem. Unlike the normal scientific undertaking, the more systematic the effort, the more variable and thus problematic existence-value results may become. The only truly replicable analysis may well be one that conveys little information beyond the simple identification of the natural object under study, the replicability resting on nothing more than the commonality of ignorance.

Even stating such a minimal detail as that the wilderness has "a total area of such and such" would emphasize this feature relative to other potential descriptions. Another person might think that a more important detail is that the potential wild area has, say, "the second highest elevation in Colorado." Who knows? The point is that no one can provide the background in objective terms. When it comes to matters of cultural symbolism, the researcher can supply the information needed by respondents only by knowing in advance the appropriate cultural frame of reference.

Yet, in matters of public policy debate that relate to the creation of cultural (in many cases religious) symbols, the appropriate cultural frame of reference is very often precisely the matter at issue. The economic researcher thus ends up merely translating his or her own value system, or that of the client providing the money, into a more formal and ultimately pseudoscientific economic result.

CONCLUSION: SCIENTIFIC ECONOMICS IN CRISIS

Economists introduced the idea of existence value in an attempt to solve a new problem facing their profession. The problem was real, but the existence-value cure is worse than the disease.

The economics profession emerged in the Progressive Era as part of the design for the scientific management of American life (Nelson 1987). Since then, economists have occupied a privileged position in American professional and intellectual life. The secular religion of America for much of the twentieth century was economic progress, a concept that transcended the mere satisfaction of crass material desires. The faithful believed that economic progress would bring the end of scarcity, and hence—or so it was supposed—eliminate most human conflict. The end result of economic progress therefore would be nothing less than the salvation of humankind, the arrival of heaven on earth (Nelson 1991).

Biblically, moral actions lead to salvation. Now, in Progressive theology, efficient and inefficient became virtually synonymous with good and evil. The efficiency of an action determined whether it contributed to economic progress and thus to the secular salvation of the world. Historians have aptly described Progressivism as "the gospel of efficiency" (Hays 1959).

As the group responsible for judging efficiency, professional economists became more than expert technicians; they became the ultimate judges of the morality of government programs, policies, and other actions. Not by accident were members of the economics profession, not Christian clergy or other social science professionals, designated by law to sit at the door of the president, by the Employment Act of 1946, which created the Council of Economic Advisors.

By the 1960s, however, this priestly role of economists as the dispensers of moral legitimacy in American society was confronting growing challenge. Many factors contributed, but one development probably had the greatest impact: the claims for the redeeming benefits of economic progress were not borne out by the actual history of the twentieth century. As a matter of material gains alone, the economic progress promised had in significant degree taken place in developed countries (rare success, it might be noted, for a theological prophesy). But the concurrent moral transformation also promised had not occurred. Heaven on earth seemed as far off as ever. Indeed, despite immense material advance, the twentieth century brought world warfare, the Holocaust, mass imprisonments in Siberian camps, and many other horrible events.

With belief in economic progress—or, as one might say more formally, "economic theology"—entering a period of crisis, environmentalism offered a new set of cultural symbols (Nelson 1995) and a new religious vocabulary. If a dam taming a raging river, showing human mastery of the wildest forces in nature, had been a cathedral to economic progress, in environmental religion the same dam became a virtual evil. For environmentalists the new cathedral would be a wilderness area. The Wilderness Act of 1964 officially announced the advent of a powerful new religious symbol in American public life.

Progressive religion had looked to the future; constant change signified continuing advance in building heaven on earth. The constant striving for efficiency ensured that progress would take place as rapidly as possible. The status quo, by contrast, was something to be left behind as rapidly as possible. What was "in existence" per se had no value.

All these ideals came into question, however, as the hopes for moral as well as economic progress clashed against the many unhappy events in the twentieth century. Perhaps constant change was not the path to salvation. Perhaps greater attention and value should be placed on what already existed. Indeed, preservation of wilderness took on such cultural significance because wilderness represented the longest existing thing of all: nature as it had existed since the Creation (or at least this image could be the symbolism, if hard to square with modern geologic science).

The economists who promoted a whole new realm of economic valuation, putting a value on "existence" for its own sake, very likely sensed all this. They saw that the vocabulary of economics, grounded as it was in the values of change, efficiency, and progress, faced growing doubts in important parts of American life. Many of these economists themselves probably sympathized in some ways with this trend of events.

But economists using the concept of existence value sought in effect to elevate new environmental values without abandoning the authority of the reigning economic language, as if to say that Christians and Muslims should stop fighting about religion because both are correct. If efficiency had long been a basic term of social legitimacy, why not simply redefine efficiency to encompass the value of preserving the existing state of the world?

This scheme was bound to fail. Theologically, it required that the forward march of progress should be measured by the extent to which people liked the cessation of progress. If belief in progress were to be displaced in the American value system, the accompanying vocabulary of progress would be abandoned. Existence value would lack pertinence because the very framework of efficiency analysis would no longer have much interest. Some new vocabulary and source of moral legitimacy—one can only guess at what it might be—would take the place of professional economics.

That economists continue to be consulted, receiving large payments to make estimates of existence value, merely indicates that the vocabulary of progress remains a contender in the American value arena. Appealing to efficiency still pays, even in cases in which the underlying goal may be to abolish efficiency. The remaining true believers in progress, however, should recognize that the introduction of existence value amounts to a Trojan horse. Seeming for a time to sustain the social role of economics, in the long run it can only help to undermine it.

I am not arguing that the critics of progress are wrong. Surely, they are right, at least in part, insofar as the Progressive gospel promised heaven on earth. Yet, few people seem prepared to abandon the material comforts that modern science and industry have delivered in such abundance. The ultimate future of progress, in any case, lies outside the scope of this chapter. The important point is that existence value has little or nothing to contribute to this particular religious discussion. The fate of progress will have to be resolved the old-fashioned way, through empirical observation, historical awareness, reasoned argument, moral judgment, testimonies of faith, theological analysis, and other traditional means of religious communication.

NOTES

1. The Catholic League for Religious and Civil Rights, for example, reacting to speeches by Interior Secretary Bruce Babbitt that defended the Endangered Species Act in biblical terms, issued a press release "Bruce Babbitt Maligns Catholicism." Babbitt had said in his speech that he found more spiritual inspiration in nature than in the Catholic church of his youth. The president of the league, William Donohue, declared that the secretary's explanation of his religious turn away from Catholicism showed political "stupidity as well as unfairness." See Human Events, 12 January 1996.
2. Bruce Babbitt, "Our Covenant: To Protect the Whole of Creation," circulated to top staff of the Department of the Interior through the e-mail system, 14 December 1995. This speech was delivered on various occasions, including to the League of Conservation Voters in New York City in early December 1995.

REFERENCES

Arnold, Ron. 1982. *At the Eye of the Storm: James Watt and the Environmentalists.* Chicago: Henry Regnery.

Arrow, Kenneth, Robert Solow, Paul R. Portney, Edward Leamer, Roy Radner, and Howard Schuman. 1993. Report of the NOAA Panel on Contingent Valuation. 58 *Federal Register* 4601 (15 January).

"Ask a Silly Question, . . .": Contingent Valuation of Natural Resource Damages. 1992. *Harvard Law Review* 105 (June): 1981–2000.

Bate, Roger. 1994. *Pick a Number: A Critique of Contingent Valuation Methodology and Its Application in Public Policy.* Washington, D.C.: Competitive Enterprise Institute.

Borelli, Peter. 1988. The Ecophilosophers. *The Amicus Journal* (Spring): 30–39.

Carson, Richard T., Robert Cameron Mitchell, W. Michael Hanemann, Raymond J. Kopp, Stanley Presser, and Paul Rudd. 1992. *A Contingent Valuation Study of Lost Passive Use Values Resulting from the Exxon Valdez Oil Spill.* Report to the Attorney General of Alaska. La Jolla, Calif.: Natural Resource Damage Assessment, Inc.

Carson, Richard T., W. Michael Hanemann, Raymond J. Kopp, Jon A. Krosnick, Robert C. Mitchell, Stanley Presser, Paul A. Rudd, and V. Kerry Smith. 1996. *Was the NOAA Panel Correct about Contingent Evaluation?* Discussion Paper 96-21. Washington, D.C.: Resources for the Future (May).

Castle, Emery N., and Robert P. Berrens. 1993. Endangered Species, Economic Analysis, and the

Safe Minimum Standard. *Northwest Environmental Journal* 9:108–30.

Desvousges, William H., Alicia R. Gable, Richard W. Dunford, and Sara P. Hudson.. 1993a. Contingent Valuation: The Wrong Tool to Measure Passive-Use Losses. *Choices* (Second Quarter): 9–11.

Desvousges, William H., F. Reed Johnson, Richard W. Dunford, Kevin J. Boyle, Sara P. Hudson, and K. Nicole Wilson. 1993b. Measuring Natural Resource Damages with Contingent Valuation: Tests of Validity and Reliability. In *Contingent Valuation: A Critical Assessment,* edited by J. A. Hausman. New York: North Holland.

Diamond, Peter A., and Jerry A. Hausman. 1993. On Contingent Valuation Measurement of Nonuse Values. In *Contingent Valuation: A Critical Assessment,* edited by J. A. Hausman. New York: North Holland.

———. 1994. Contingent Valuation: Is Some Number Better than No Number? *Journal of Economic Perspectives* 8 (Fall): 45–64.

DiBona, Charles J. 1992. Assessing Environmental Damage. *Issues in Science and Technology* (Fall): 50–54.

Edwards, Steven F. 1992. Re-Thinking Existence Values. *Land Economics* 68 (February): 120–22.

Ekirch, Arthur A. 1963. *Man and Nature in America.* New York: Columbia University Press.

Glover, Willis B. 1984. Biblical Origins of Modern Secular Culture: An Essay in the Interpretation of Western History. Macon, Ga.: Mercer University Press.

Hanemann, W. Michael. 1994. Valuing the Environment through Contingent Valuation. *Journal of Economic Perspectives* 8 (Fall): 19–43.

Hausman, Jerry A., ed. 1993. *Contingent Valuation: A Critical Assessment.* New York: North Holland.

Hays, Samuel P. 1959. *Conservation and the Gospel of Efficiency: The Progressive Conservation Movement, 1890–1920.* Cambridge, Mass.: Harvard University Press.

Inside the World's Last Eden: A Personal Journal to a Place No Human Has Ever Seen (cover story). 1992. *Time,* 13 July.

Kennedy, Roger G. 1995. The Fish That Will Not Take Our Hooks. *Wilderness* (Spring): 28–30.

Kerr, Hugh T., ed. 1989. *Calvin's Institutes: A New Compend.* Louisville, Ky.: Westminster/ John Knox.

Kopp, Raymond J. 1992. Why Existence Value Should Be Used in Cost-Benefit Analysis. *Journal of Policy Analysis and Management* 11 (Winter): 123–30.

Krutilla, John V. 1967. Conservation Reconsidered. *American Economic Review* 57 (September): 777–86.

Lebergott, Stanley. 1995. Long-Term Trends in the U.S. Standard of Living. In *The State of Humanity,* edited by J. L. Simon. Cambridge, Mass.: Blackwell.

Looney, Douglas S. 1991. Protection or Provocateur? *Sports Illustrated,* 27 May.

McCormick, John. 1989. *Reclaiming Paradise: The Global Environmental Movement.* Bloomington: University of Indiana Press.

McPhee, John. 1971. *Encounters with the Archdruid.* New York: Farrar, Straus and Giroux.

Mead, Walter J. 1992. Review and Analysis of Recent State-of-the-Art Contingent Valuation Studies. In *Contingent Valuation: A Critical Assessment.* Papers from a Symposium in Washington, D.C., 2–3 April. Cambridge, MA: Cambridge Economics.

Milgrom, Paul. 1993. Is Sympathy an Economic Value? Philosophy, Economics, and the Contingent Valuation Method. In *Contingent Valuation: A Critical Assessment,* edited by J. A. Hausman. New York: North Holland.

Miller, Perry. 1954. *The New England Mind: The Seventeenth Century.* Cambridge, Mass.: Harvard University Press.

———. 1964. *Errand into the Wilderness.* New York: Harper and Row.

Mitchell, Robert, and Richard Carson. 1989. *Using Surveys to Value Public Goods: The Contingent Valuation Method.* Washington, D.C.: Resources for the Future.

Nash, Roderick. 1973. *Wilderness and the American Mind.* New Haven: Yale University Press.
Nelson, Robert H. 1987. The Economics Profession and the Making of Public Policy. *Journal of Economic Literature* 25 (March): 49–91.
———. 1991. *Reaching for Heaven on Earth: The Theological Meaning of Economics.* Lanham, Md.: Rowman and Littlefield.
———. 1992. Wilderness, Church, and State. *Liberty* (September): 34–40.
———. 1993. Environmental Calvinism: The Judeo-Christian Roots of Eco-Theology. In *Taking the Environment Seriously,* edited by R. E. Meiners and B. Yandle. Lanham, Md.: Rowman and Littlefield.
———. 1995. Sustainability, Efficiency and God: Economic Values and the Sustainability Debate. *Annual Review of Ecology and Systematics* 26: 135–54.
Niebuhr, Gustav. 1993. Black Churches' Efforts on Environmentalism Praised by Gore. *Washington Post,* 3 December.
Norton, Bryan G. 1991. Thoreau's Insect Analogies: Or Why Environmentalists Hate Mainstream Economists. *Environmental Ethics* 13: 235–51.
Oelschlaeger, Max. 1994. *Caring for Creation: An Ecumenical Approach to the Environmental Crisis.* New Haven: Yale University Press.
Portney, Paul R. 1994. The Contingent Valuation Debate: Why Economists Should Care. *Journal of Economic Perspectives* 8 (Fall): 3–17.
Quiggin, John. 1993. Existence Value and Benefit-Cost Analysis: A Third View. *Journal of Policy Analysis and Management* 12 (Winter): 195–99.
Randall, Alan. 1993. Passive-Use and Contingent Valuation—Valid for Damage Assessment. Choices (Second Quarter): 12–15.
Robb, Carol S., and Carl J. Casebolt, eds. 1991. *Covenant for a New Creation: Ethics, Religion, and Public Policy.* Mary Knoll, N.Y.: Orbis.
Rosenthal, Donald H., and Robert H. Nelson. 1992. Why Existence Values Should Not Be Used in Cost-Benefit Analysis. *Journal of Policy Analysis and Management* 11 (Winter): 116–22.
Roszak, Theodore. 1992. *The Voice of the Earth.* New York: Simon and Schuster.
Sagoff, Mark. 1994. On the Expansion of Economic Theory: A Rejoinder. *Economy and Environment* (Summer): 7–8.
Sax, Joseph L. 1980. *Mountains without Handrails: Reflections on the National Parks.* Ann Arbor: University of Michigan Press.
Steifels, Peter. 1996. Evangelical Group Defends Laws Protecting Endangered Species as a Modern Noah's Ark. *New York Times,* 31 January.
Stewart, Richard, ed. 1995. *Natural Resource Damages: A Legal, Economic, and Policy Analysis.* Washington, D.C.: National Legal Center for the Public Interest.
Talmon, J. L. [1960] 1985. *Political Messianism: The Romantic Phase.* Boulder, Colo.: Westview Press.
Terborgh, John. 1992. A Matter of Life and Death. *New York Review of Books,* 5 November.
Tillich, Paul. 1967. *A History of Christian Thought: From Its Judaic and Hellenistic Origins to Existentialism.* New York: Simon and Schuster.
Watson, Paul. 1988. On the Precedence of Natural Law. *Journal of Environmental Law and Litigation* 3: 82.
Williams, Brooke. 1991. Love or Power? *Northern Lights* 7 (Fall): 17–21.
Worster, Donald. 1993. *The Wealth of Nature: Environmental History and the Ecological Imagination.* New York: Oxford University Press.

Acknowledgment: This paper is adapted from Robert H. Nelson, "How Much Is God Worth? The Problems—Economic and Theological—of Existence Value," Competitive Enterprise Institute Discussion Paper, Washington, D.C., May 1996.

22

Autonomy and Automobility

LOREN E. LOMASKY

Years before the automobile evolved into a transportation necessity, before multilaned asphalt replaced meandering muddy ruts, intrepidly pioneering motorists took to the roads for pleasure. Today tens of millions drive for pleasure, but increasingly it is a guilty pleasure. From a multitude of quarters, motorists are indicted for the harms they leave in their wake. Drivers generate suburban sprawl, exacerbate the trade deficit while imperiling national security, foul lungs and warm the atmosphere with their noxious emissions, give up the ghosts of their vehicles to unsightly graveyards of rubber and steel, leave human roadkill behind them, trap each other in ever vaster mazes of gridlock and, adding insult to injury, commandeer a comfy subsidy from the general public. Only the presence of unconverted cigarette smokers deprives them of the title Public Nuisance Number One.[1]

Barring a radical reengineering of America, we will not soon toss away our car keys. As the primary vehicles for commuting, hauling freight, and general touring, cars (and trucks) are here to stay. But as the automobile enters its second century of transporting Americans from here to there, it is increasingly dubbed a public malefactor, and momentum grows for curbing its depredations. Construction of significant additions to the interstate highway system has ground to a halt. Designated lanes on urban roads are declared off-limits to solo motorists. Federal Corporate Average Fuel Economy (CAFE) standards require automakers to eschew selling vehicles as capacious as motorists may wish to buy and instead to alter their mix of products to emphasize lighter, less gasoline-hungry cars. Taxes on fuel have been increased only modestly, but if critics of the hegemony of the automobile have their way, America will emulate Europe, pushing the tax up by a dollar or more per gallon. Funds thereby generated will not be designated for motorist services—such earmarking is precisely what has exacerbated the current plague of overautomobilization—but will instead be directed toward more mass transit, pollution relief, and research on alternate modes of transportation.[2] Some argue that employer-provided parking should be taxed as income to the employee or disallowed as a business expense to the provider. Others advocate following Amsterdam's lead, barring nearly all automobiles from entry into the center city. Moral suasion supplements policy proposals. In the name of social responsibility, individuals are urged to carpool or avail themselves of public transportation, scrap their older, fuel-intensive vehicles, and eschew unnecessary automobile trips.

Why this assault on the automobile? I have no wish to deny that it occurs at least in part because some of the critics' charges are true. Automobile carnage is indeed dreadful. The number of people killed each year on our roadways far exceeds the total who succumb to AIDS. Automobiles do pollute, all to some extent, some much worse than others. The cost of petroleum imports into this country exceeds the amount of the entire national trade deficit. And anyone who has ever been trapped in rush-hour gridlock, fuming inside at the delay while being engulfed by the fumes outside spewing from ten thousand tailpipes, knows that the simple job of getting from here to there in one's automobile can be the most stressful part of the day. Cars are not always "user-friendly."

But all these criticisms seem insufficient for explaining the intensity of opposition directed toward the automobile. Any large-scale enterprise entails costs, and so a critique that merely reminds us of the nature and extent of these costs is only half useful. Also required, of course, is a statement of the benefits derived from the enterprise, and a plausible accounting of whether the benefits exceed the costs. Identifying and measuring the costs and benefits of automobile usage pose very difficult methodological problems that I shall not consider here. I do note that the overwhelming popularity of the automobile is itself prima facie evidence that from the perspective of ordinary American motorists, the benefits of operating a motor vehicle exceed the concomitant costs. Just as theorists speak of people "voting with their feet," we can count those who vote with their tires. And this vote is overwhelmingly proautomobile.

Critics may contend, though, that the election has been rigged. They can maintain that the absence of public transportation and compact neighborhoods in which commerce, industry, and housing are integrated forces us so often into our cars. People might like to be able to purchase a loaf of bread without buckling their seat belts, but in many parts of the country they cannot. And even if each of us values the options and mobility that automobile transport affords, we might devalue yet more the stress, delay, and pollution imposed on us by others. Private use of automobiles so understood would approximate game theory's

Prisoner's Dilemma, an interaction in which each player acts in his own rational self-interest but all parties are worse off than they would have been had someone impelled them to choose otherwise. And the critic contends that some such requirement, in the form of regulation or increased taxes or outright prohibitions, is needed to escape the tyranny of the automobile (see Hensher 1993, and Freund and Martin 1993).

The critic's case has at least this much merit: a purely behavioristic appraisal of automobile usage is insufficient for evaluating its normative status. We need also to think more intently about how to classify and understand as a distinctive human practice the action of *driving a car.* Opponents of the automobile argue that the most telling way to understand this is by equating the act with *creating a public bad.* I shall dispute that appraisal. My focus will not be on the many and varied instrumental uses to which the automobile is put (driving to work, carpooling the kids, buying groceries), though in no way do I mean to disparage these. Rather, I shall concentrate on automobility's intrinsic capacity to move a person from place to place. As such, automobility complements *autonomy*: the distinctively human capacity to be self-directing. An autonomous being is not simply a locus at which forces collide and which then is moved by them. Rather, to be autonomous is, minimally, to be a value with ends taken to be good as such and to have the capacity to direct oneself to the realization or furtherance of these ends through actions expressly chosen for that purpose. Motorists fit this description. Therefore, insofar as we have reason to regard self-directedness as a valuable human trait, we have reason to think well of driving automobiles.

I am not maintaining, of course, that all and only motorists are autonomous, that someone persuaded by the slogan "Take the bus and leave the driving to us" thereby displays some human deficiency. A liberal society is one in which people pursue a vast diversity of goods in myriad ways, and this variety accounts for a considerable share of that society's attractiveness. So even if driving a car is an intrinsically worthwhile action, it does not follow that declining to drive is suspect.

But neither am I claiming that automobiles are simply one among thousands of other products that individuals might, and do, happen to find attractive in a cornucopia of consumer goods. The claim is stronger. Automobility is not just something for which people in their ingenuity or idiosyncrasy might happen to hanker—as they have for Nehru jackets, disco music, hula hoops, pet rocks, pink flamingo lawn ornaments, Madonna, and "How many . . . does it take to change a lightbulb?" jokes. Rather, automobile transport is a good for people in virtue of its intrinsic features. Automobility has value because it extends the scope and magnitude of self-direction.

Moreover, the value of automobility strongly complements other core values of our culture, such as freedom of association, pursuit of knowledge, economic advancement, privacy, and even the expression of religious commitments and affectional preference. If these contentions have even partial cogency, then opponents of the automobile must take on and surmount a stronger burden of proof than they have heretofore acknowledged. For not only must they show that instrumental costs of marginal automobile usage outweigh the corresponding benefits, but they must also establish that these costs outweigh the *inherent good of the exercise of free mobility.*

WHEELS OF FORTUNE: MOVEMENT, CHOICE AND HUMAN POTENTIAL

Concern about automobiles may be a modern phenomenon, but analysis of the distinctive nature of automobility is not. For Aristotle, being a self-mover was the crucial feature distinguishing animals from plants and, thus, higher forms of life from lower. A more basic distinction separates the organic realm from that which is lifeless. Living things have an internal animating force, *psyche.*[3] The customary translation is "soul," but in the context of Greek biology that is misleading. For us, "soul" tends to carry a theological and therefore elevated sense, but in classical Greek thought it marks the divide between inert things and those imbued with a vital principle.[4] Psyche appears at three levels. The lowest is vegetative soul. Plants are more than just things insofar as they are not merely acted on but also do something. Specifically, they ingest food, metabolize, and reproduce. At the highest level is the rational soul, the intelligence exhibited among the animals only by humans. Between, and crucial to this discussion, is animal or sensitive soul. Level-2 psyche has the capacities of level-1 psyche (and level-3 psyche those of level-2) plus two further features. Unlike plants, animals perceive and they move themselves.

Perception and movement are enumerated as two qualities but, as set out in *De Anima,* they are to be understood as strongly complementary. Because plants are stationary (or, if mobile, as are the seedpods of some species, carried where they go by external forces), they have no need to perceive. If the wheat is not going anywhere, then it cannot benefit from seeing the swarm of locusts about to descend on it. Aristotle expressed this idea in the teleological language of *purpose* and *natural function* that pervades his metaphysical awareness, but essentially the same point could be made in contemporary terms of inclusive evolutionary fitness. Plants do not perceive because (a) no purpose would be fulfilled by their perceiving or (b) evolution does not select at that biological level for perception. The locusts do perceive, however, as their survival depends on becoming aware of and directing themselves toward potential items of food. We can also state the connection in reverse order: if a being does not perceive the difference between *here* and *there*, then its having the capacity to direct itself there rather than here serves no purpose.

Plants are alive, but their "quality of life" is low (thus the comatose individual referred to as a "human vegetable" and the inert TV-watching "couch potato"). They function in the world but in complete obliviousness to it. Lacking consciousness, the cucumber has no perspective from which there is a "what it is like to be a cucumber." Plants *are*, and in a restricted sense do, but in terms of nearly all that we take to be of value in life, they are nullities.

Animal life differs, and the difference lifts the organism above nullity status. To perceive is to assimilate in some measure the world to oneself. And to be a self-mover is to situate oneself in the world in accordance with one's own desires. Perception plus mobility are prerequisites of *agency*. Patients are beings to whom things happen, but agents act. At some level of awareness agents distinguish between goods and bads and endeavor to direct themselves toward the former and away from the latter. For animals, this direction involves instinctive or acquired responses to pleasure and pain. For human beings action takes on additional complexity. We do not merely react to stimuli in our environment. Instead, we deliberate among available alternatives conceived of not only as pleasing or displeasing but also in terms such as "dishonorable," "what justice demands," "liable to make me famous," "chic," and so on. At this level it is proper to speak in a nonmetaphorical sense of *choice*. Aristotle maintains that animals or young children do no genuine choosing. In choosing, we act to give expression to our settled conceptions of how we want to direct ourselves. Our choices flow from and redound upon our virtues and vices. We do not offer moral appraisals of beings incapable of choice; unlike normal adult human beings, neither infants nor animals can be brave or wicked or temperate.

The conception of motion has a wider scope than traveling from place to place. We retain residual traces of this broader meaning in expressions such as "a moving experience" and in the etymological history of "emotion," but in the philosophical language of the Greeks the more inclusive sense is primary. Any transformation of a subject from a state of potentiality with regard to some quality to the actual realization of that quality is deemed motion.[5] So going from here to there constitutes movement, but so also do an organism's growth, someone's coming to know something, the development of a faculty, and so on. In an Aristotelian universe, motion is ubiquitous because everything tends to progress toward the highest possible self-realization. For simple inorganic forms like a rock, this potential is correspondingly simple, involving only the capacity to fall when unsupported. In organisms the transition from potency to act is more complex. The oak, for example, moves to its actuality through the complex chain of maturation that commences from the acorn stage. For animals, such self-realization incorporates consciousness and self-propulsion. Human actualization adds deliberation and choice. Only for a completely actualized being would movement be otiose (or counterproductive). And indeed, Aristotle hypothesizes that a god dubbed the "Unmoved Mover" occupies the pinnacle of the metaphysical hierarchy because in its enduring perfection it has transcended all reason to change, whereas anything else in the universe, insofar as it realizes any of its potential, is approaching to some greater or lesser degree, consciously or unconsciously, this state of full actualization. Encountering Greek philosophical thought, Christians applied this concept of an unchanging perfection to the Book of Genesis's Creator of Heaven and Earth.

Movement, therefore, does not simply describe getting from here to there; it has normative richness. To move is to progress—though, of course, it can also be to backslide. Only stasis is morally neutral, and ours is a dynamic universe. The greater the variety of dimensions through which an individual transforms itself and things it encounters, the greater the scope for evaluative concerns. The grounds on which human beings appraise themselves and their fellows will be much richer than, say, the standards applied to horses or bottles of wine or the performance of machines. For people, there is not only a better or worse but a *chosen* better or worse toward which we deliberately direct ourselves. Intelligent automobility is crucial to the elevated status of human beings vis-à-vis other beings.

A PHILOSOPHICAL DETOUR

If you bump into me and cause me to lurch from my path, it is clear that my behavior is not that of a self-mover. Less clear, though, is the case in which you glower menacingly at me as you approach down the sidewalk, thereby "persuading" me to step aside. Or suppose that yesterday when you hypnotized me you implanted within me a suggestion that I always make way for you, and so today when I see you approaching I not only defer but am pleased to do so out of concern for your well-being. In the latter two instances I have, in a sense, moved myself. Not only are the muscle contractions that impel my legs the contractions of *my* muscles, but they are preceded by mental activities that can be characterized as *my decision* to move in that way. But that characterization demands qualification. The action is mine, but in its initiation it is also yours by virtue of the threat or hypnotic implantation. It is at least as much a *being done to* as it is a *doing*, and so it qualifies as agency only in a restricted sense.

The many species of such qualified action—or "action"—raise notoriously vexing problems of moral responsibility. Aristotle considers them with regard to the dichotomy voluntary-nonvoluntary and concludes, not all that helpfully, that they are "mixed," though perhaps to be classified closer to the voluntary than the nonvoluntary.[6] The issue is not only theoretical but also sharply practical: Do we blame (or praise) those who act under duress, extraordinary fear, rage, naïve suggestibility, exhaustion, ignorance, or similar other conditions that call into question their full authorship of an action? Lawyers and moralists wrestle with such issues. For purposes of this discussion it is not necessary to resolve these conundrums. Note however that the more qualified the action is with regard to the performer's agency, the less it redounds as either asset or liability to the individual's moral account.

Accountability enters crucially into human dignity. Insane or incompetent persons are not accountable for their doings, and that is symptomatic of their misfortune. We value full authorship of our own actions (or, noncircularly, authorship of the behavior of one's body) and fear conditions—manipulation, coercion, intimidation—that impede such authorship. Those who exercise such control over their actions are said to be *autonomous*.

Autonomy, literally "self-legislating," originated as a term applied to political units, distinguishing the independent ones from those governed by the laws of some other polity. In moral philosophy, autonomy acquired importance as an attribute of individuals in the writings of the eighteenth-century German philosopher Immanuel Kant.[7] Like Aristotle, Kant inquired into the conditions required for the existence of moral responsibility, but for him the universe in which human beings act had a significantly different look than the teleologically structured world of Aristotle, which was thoroughly hospitable to normativity. Newton and the new physics had depicted a deterministic order in which each event is the inevitable consequence of the nexus between universal causal laws and the antecedent conditions to which they apply. Whatever happens does so of necessity rather than caprice or randomness. But if that necessity holds for events in general, it applies to human actions in particular. We are as subject to the physical laws governing the cosmos as are galaxies and atoms. Therefore, our doings are in principle entirely explainable and predictable (depending on whether one is viewing them retrospectively or prospectively) in terms of these laws. But if conditions that obtained five minutes—or five hours, or five years, or five millennia—ago made it inevitable that at this precise moment I would perform Action A, it would seem that I am not free with respect to performing A. It *had to happen* and, thus, I *had to do it*. This determinism may seem to deliver a crushing blow to conceptions of human agency and moral responsibility. If the doing of A was sealed long ago, if even before I was born it was inscribed in the history of the cosmos as an inevitability, then my participation in its unfolding would seem to be purely passive. I can no more be genuinely responsible for its occurrence than I can be for my eye color or an eclipse of the sun. In none of these cases can I change the course of events.

This problem of free will and determinism is one of the most vexing in philosophy. In Kant's day, it loomed very large. If the whole universe is one giant machine obeying its own internal laws, how can we be other than machine cogs ourselves? Kant's way out was drastic. He salvaged human freedom by imposing on persons a metaphysical schizophrenia. We simultaneously belong to the phenomenal universe subject to cause and effect and to a purely intelligible realm, the *noumenal order*, regulated not by mechanical laws of physics but by the normative laws of reason. In the former realm we are self-movers only in a relative and incomplete sense; every action has a cause that necessitated it, and that cause has a cause, and so on ad infinitum. As phenomenal beings we are no more than protoplasmic machines in a thoroughly mechanistic universe. But as noumenal beings we can determine ourselves in accord with self-imposed dictates of reason, thereby achieving autonomy. Insofar as we enjoy autonomy, we are free beings and possess a worth and dignity that set us apart from the realm of necessity.[8]

Ironically, although almost no contemporary moral philosophers accept Kant's complex "two worlds" metaphysics, his guiding idea that autonomy is central to our special moral status as persons informs much modern moral thinking. Its effect appears in quarters as disparate as the existentialist insistence that we are beings with no predetermined essence and thus privileged —or condemned—to define ourselves through our own free choices,[9] John Rawls's (1971) influential conception of justice as principles that would be autonomously chosen by free and equal rational beings deliberating behind a veil of ignorance, and the doctrine of informed consent that dominates contemporary medical ethics.[10] I shall not attempt to sort out these and other variants on the theme of autonomy. It is worth noting, though, that much of the contemporary concern for autonomy is continuous with and indeed has tended to replace the earlier emphasis in moral philosophy on the centrality of liberty in human affairs. The writings of John Stuart Mill provide the locus for much of this transformation.

In his classic *On Liberty*, Mill sought to provide a principled basis for opposition to the imposition of conformity via law and social custom. He trotted out a whole array of arguments to demonstrate that restrictions on liberty are inimical to scientific advance, accumulation of wealth, and other requisites of human happiness. Most of these appeals invoke instrumental considerations of the sort familiar from standard economic analysis. But in perhaps the most important section, the chapter entitled "Of Individuality, as One of the Elements of Well-being," he presented a different sort of argument, predicated on the intrinsic worth of what I have called full authorship of one's actions, Kant called autonomy, and Mill referred to as individuality:

> He who lets the world, or his own portion of it, choose his plan of life for him, has no need of any other faculty than the ape-like one of imitation. He who chooses his plan for himself, employs all his faculties. He must use observation to see, reasoning and judgment to foresee, activity to gather materials for decision, discrimination to decide, and when he has decided, firmness and self-control to hold to his deliberate decision. And these qualities he requires and exercises exactly in proportion as the part of his conduct which he determines according to his own judgment and feelings is a large one. It is possible that he might be guided in some good path, and kept out of harm's way, without any of these things. But what will be his comparative worth as a human being? It really is of importance not only what men do, but also what manner of men they are that do it. (Mill [1859], 1989, 59)

One cannot reduce "what manner of men they are that do it" to statistics of net wealth per household or GDP growth from one year to the next; such achievements matter less than the character one creates in and for oneself. Retaining captaincy of one's soul (if not always mastery of one's fate) is essential to authenticity and a self genuinely deserving of esteem. Conversely, to be prodded by others along paths they have cleared toward goals they have set is servile. It demeans the dignity of the individual. To live well is to live in a manner that one has made distinctively one's own.

Autonomy so understood incorporates Aristotelian self-moving but goes beyond it. A self-mover can be one participant among thousands in a lengthy parade, each following in lockstep the one who goes before, not knowing or caring where he is headed just so long as he ends up in the same place as all the others. But an autonomous individual is not content to leave the course of the march to the determinations of others (or to chance). He has a conception of a good-for-him that he may not have created ex nihilo but which he actively endorses. And in its service he prioritizes, deliberates, and selects means judged appropriate to ends. He acknowledges personal responsibility for those ends and means. If he succeeds, the outcome is in a full sense his success rather than the vagaries of fate playing kindly with him; and if he fails, that outcome is also lodged at his doorstep rather than that of the parents who toilet-trained him, the teachers who instructed him, the community that socialized him, the politicians who competed for his allegiance, or the preachers who offered him slide shows of heaven. Any or all of these persons may have provided elements of value that he has incorporated into his projects, but the compound he concocts from them is his.

It would be overly contentious to maintain, as some exponents do, that without autonomy one fails to lead a fully human life. Countervailing virtues grace traditional modes of life. Individuals do not so much craft these virtues for themselves as they receive and don them as hand-me-downs from others. The monk's life of humility and abasement and the traditionally female roles of nurturance and support within the family display their own quiet dignity. Still, no mode of nonautonomous living fully expresses individuated human agency or so firmly opposes servile conformism. To cite Mill again:

> In our times, from the highest class of society down to the lowest, every one lives as under the eye of a hostile and dreaded censorship. Not only in what concerns others, but in what concerns only themselves, the individual or the family do not ask themselves—what do I prefer? or, what would suit my character and disposition? or, what would allow the best and highest in me to have fair play, and enable it to grow and thrive? They ask themselves, what is suitable to my position? what is usually done by persons of my station and pecuniary circumstances? or (worse still) what is usually done by persons of a station and circumstances superior to mine? . . . It does not occur to them to have any inclination, except for what is customary. Thus the mind itself is bowed to the yoke. ([1859] 1989, 61)

Autonomous people "Just Say No" to the yoke.

COMMUTING AND COMMUNITY

The automobile, definitionally, promotes automobility. The complementarity of autonomy and automobility is only slightly less evident. In the latter part of the twentieth century, being a self-mover entails, to a significant extent, being a motorist. Because we have cars we can, more than any other people in history, choose where we will live and where we will work, and separate these two choices from each other. We can more easily avail ourselves of near and distant pleasures, at a schedule tailored to individual preference. In our choice of friends and associates, we are less constrained by accidents of geographical proximity. In our comings and goings, we depend less on the concurrence of others. We have more capacity to gain observational experience of an extended immediate environment. And for all of the preceding options, access is far more open and democratic than it was in preautomobile eras. Arguably, only the printing press (and perhaps within a few more years the microchip) rivals the automobile as an autonomy-enhancing contrivance of technology.

No one who has been caught in rush-hour gridlock will maintain that commuting to and from work is an unalloyed joy. Competing with tens of thousands of other motorists for scarce expanses of asphalt reminds one of the Hobbesian war of all against all. For critics of the automobile this complaint is not a negligible point. But neither are its implications entirely clear-cut. Just as worthy of notice as the unpleasantness of stop-and-go commuting is how many people voluntarily subject themselves to it. Have they not realized how much time they are wasting in overly close proximity to their steering wheels? Such inadvertence is not plausible. Evidently, people who, individually and collectively, could have devised for themselves residential and occupational patterns not incorporating lengthy commutes chose to do otherwise. In their judgment, the costs of commuting are compensated by the benefits thereby derived. The more the critics emphasize the magnitude of the costs, the more these critics underscore, often unwittingly, the extent of the benefits.

Commentators from the Greek philosophers to Adam Smith to Karl Marx have noted that the nature of the work one does largely shapes the quality of life one enjoys. For nearly all of us, to do work suited to oneself in a satisfactory environment is a great good, whereas to perform alienating labor under unfriendly and unhealthy conditions is a correspondingly great evil. Similarly, to reside in a comfortable and functional dwelling situated in a neighborhood one finds hospitable is also a considerable good. For most people throughout human history, neither occupation nor place of residence has afforded more than a negligible range of choice. One did the work one's father or mother did, or to which one had been apprenticed, or the kind of work available in that place. And one lived near the workplace.

The increased affluence and openness of liberal capitalist society vastly expanded the range of choice. But the coming of the automobile essentially separated the choices. Previously one lived either near one's work or else on a commuter rail line. But the geography of the New York, New Haven, and Hartford tracks did not bind motorists. Depending on how much time they cared to invest in transit, they could live at a considerable distance from their workplaces, yet emancipated from the rigidities of mass transit. Cultured despisers of the idiocy of suburban existence can and do decry this circumstance, but millions of Americans (and, increasingly, the rest of the world) disagree. Even if one believes for aesthetic or other reasons that row upon row of bungalows or ersatz Tudor houses miles distant from the city or industrial area to which they are connected by roadways represent unattractive neighborhoods, one cannot deny that they are genuine objects of choice for those who live there. People, we might say, have a right to banality. To respect the autonomy of persons is to acknowledge that expanding their options for combining work and place of residence is as such a plus.

Nineteenth-century socialist reformers decried the enhanced ability of industrial capitalism's factory system to exploit workers. Human labor, they charged, had become no more than an appendage of mill or machine. Although one could reasonably respond (as Friedrich Hayek [1954] famously did in *Capitalism and the Historians*) that workers who voluntarily abandoned their rural domiciles for the factory town did so only because they themselves regarded the move as a net improvement, one must nonetheless concede that their situation was not enviable. They may have enjoyed a higher standard of living than that available to them on the farm, but their work was grueling and their opportunities for self-directed choice minimal. Against the perceived oppression of industrial society, the reformers contrived various nostrums, one family of which, now mercifully defunct, oppressed millions of unfortunate souls throughout most of this century. No syndicalist scheme or string of workers' cooperatives remotely approaches the automobile as an emancipatory instrument. Insofar as it extended the feasible range of commuting between residence and labor, the coming of the motorcar augmented the bargaining power enjoyed by workers. A company town offers little scope for alternate employment opportunities. Changing jobs very likely requires changing place of residence, and exit costs of both pecuniary and nonpecuniary sorts may render that prohibitive. However, widespread automobile ownership dramatically extended the geographical radius of possible employment venues. Hence, the market for labor came more closely to approximate the economists' model of many sellers and many buyers. In theory, under a legal regime of free contract, workers always enjoyed the right to terminate their employment when they wished to do so, but in practice the exercise of this liberty often proved discouragingly costly. Automobility significantly lowered those costs. The country music song "Take This Job and Shove It" became something of an anthem for the disaffected at a time when car ownership had become almost universal. Musical aesthetics aside, those who value choice not only formalistically but as the existence of genuine live options must appreciate this alternative. Detroit has done more for the liberation and dignity of labor than all the Socialist Internationals combined.

One can also observe liberation by viewing the employment-residence nexus from the other direction. The ability to choose where one will live makes a considerable difference in the exercise of self-determination. Life in the suburbs is not inherently better than life in the central city, but it is different. To the extent that one possesses a real opportunity to choose between them, one can give effect to significant values that shape the contours of a life. A city may offer ready access to arts and education, a succession of ethnically diverse neighborhoods, a feeling of drive and vitality, an ambience that "swings." But cities are often dirty, expensive, and dangerous. Exurban life may provide peaceful neighborliness, gardens and green spots, family-oriented activities that take place in the home or the mall. But exurbs are often antiseptic, provincial, and stultifying. To choose the one is to relinquish (some of) what the other affords. So which is the better alternative? People must answer for themselves based on their own conceptions of what matters most. To the extent that one has geographical mobility, the question is answered by an act of positive choice rather than through inertia or extraneous constraints such as the location of one's place of employment.

Choice of residence serves as a major avenue for Americans to exercise their right to free association. Choosing a neighborhood is the macrolevel correlate to choosing one's friends. One thereby decides with whom one will live. And perhaps even more important, one decides with whom one will not live. In contemporary society, "leaving home" signifies a full coming of age and the concomitant entitlement to direct one's own projects as an adult. But then comes the necessity of finding and making a home in a neighborhood to which one has a tie at least in part because one has freely chosen to live there rather than somewhere else. This choice too signifies and gives effect to one's values. Some people prize a high degree of homogeneity of race or religion or age or economic class among those with whom they will most frequently associate. Others prefer a heterogeneous diversity of different ages, skin tones, and backgrounds from which casual acquaintances and intimate friendships will emerge. Considering whether one of these preferences deserves more admiration than the other carries us away from the theme of this essay, but even if one regrets that some people choose to segregate themselves from those who somehow differ—or conversely, that some defect from tightly knit ethnic communities—an ethic that endorses autonomy must acknowledge that, the content of individual choices aside, it is good that people can make up their own minds and then act on their decision about where to live.

More flexibly and more frequently than anything else, cars get us from one place to another. If we can conveniently drive to a place consistent with work and other commitments, then it passes the first test of eligibility as a possible place of domicile. (Thanks to that other great choice-enhancing device, the microchip, this situation may change as more and more individuals telecommute.) Although critics of the automobile also frequently criticize what they take to be a dreary suburban sameness, within reasonable commuting distance of virtually every urban center in this country are dozens of towns and neighborhoods that differ significantly one from another—perhaps not in factors these critics take to be momentous but certainly along dimensions that the men and women behind the steering wheels consider important. From the perspective of autonomy, their criteria deserve respect.

MOBILITY AND KNOWLEDGE

For much the same reasons that automobility and autonomy are good things, so too is knowledge. Like self-moving, knowing affords us a firmer grip on our world. Indeed, choice and knowledge complement one another. A simple example will help illustrate their relationship.

Consider a shopper in a supermarket deciding whether to buy the can on the left or the can on the right. Neither can has a label, so it is anyone's guess whether one of the cans holds tuna fish or shoe polish or bamboo shoots. How much would a shopper value the freedom to choose between them? The obvious answer is "not much." The minimal ability to distinguish them as "left can" and "right can" does not afford enough information for individuals to judge which is more likely to serve their ends. The "choice" is otiose.

Now suppose that the label is restored to one of the cans. The shopper now knows it to contain mushrooms. The value of choosing has gone up. The magnitude of the increase depends on how this added bit of knowledge relates to the shopper's preferences. If he either strongly likes or dislikes mushrooms, then he has a basis for picking between the cans, but not as good a reason as he would have if the other can were labeled, too. And further knowledge concerning particulars of taste, nutrition, quantity, and so on renders the choice one in which the shopper can give effect to his own distinctive values. Choice without knowledge is blind; knowledge without choice is impotent.

Automobiles enhance mobility, and mobility enhances knowledge. Recall the discussion of the relationship between self-moving and perception in Aristotle's biological theory. As the area in which people can direct their self-aware movements increases, so too does the range of their knowledge-gathering capacities. The knowledge in question is, in the first instance, local knowledge. By traveling through, around, and within a place, one comes to know it in its particularity. This kind of knowledge has no very close substitute. I may have read a score of books about Paris, but if I have never visited the City of Lights, if I have never traversed its streets and bridges and marketplaces, then I could not truly claim, "I know Paris." One can no more reduce knowledge of a place to possessing many facts about that place than one can reduce knowing another person to having read a very detailed resume. Philosophers often distinguish between knowledge by *description* and *knowledge by acquaintance*. To acquire the latter, one often needs mobility.

Of course automobiles are not the only form of transportation that serves to increase local knowledge, and for some types of local knowledge they may serve poorly. One such case is that described in the preceding paragraph: for acquiring up-close knowledge of a city like Paris, shoes serve better than tires. All forms of transportation—from walking to bicycling to trains, buses, ships, and airplanes—enhance knowledge. But with the possible exception of the motorcycle, another means of transportation assailed by no shortage of critics, none combines local maneuverability with extended range to the degree that the automobile does. The train can move me from one city to another at intermediate distance and afford me the opportunity of viewing the terrain in between. But it allows only a limited number of stops along the way, the speed may be slower or faster than one would wish for optimal information gathering, and the route will be exactly the same on the thousandth trip as on the first. Airplanes excel for speed, but everything between points of departure and destination is indistinct. Walking is a wonderful way to observe a neighborhood, but inadequate to take in even the opposite end of a village, let alone a state or country. For genuine exploration at long or intermediate range, the car dominates all alternatives.

How much weight should one give this sort of knowledge? The question deserves an answer. Few of the automobile's critics have a word to say about the knowledge-enhancing aspects of automobility, either because they have never considered the automobile from the perspective of information gathering or because they implicitly suppose that what one learns while behind a steering wheel is trivial. But these critics do not represent the population at large. They are intellectuals and information processors of one stripe or another, most comfortable with information that can be synthesized in books or graphs or computerized databases. They tend to depreciate information that can't be measured, quantified, and represented symbolically. But the information to be gained from reading a history book or running a regression is not the only sort that individuals can use effectively in their pursuits. Knowledge need not be grand or profound to have value in itself and to complement choice. By driving north along the lake to see how the autumn leaves have turned and whether the Canadian geese are still milling or have flown, I may gain an inherently worthwhile experience. Driving through the various neighborhoods of a city reveals where the bakeries, hairdressers, and Thai restaurants are located; who is having a garage sale this week; and which parts of town are becoming distinctly seedier. Teenagers cruising the "main drag" are conducting an epistemological mission motivated by the hope of sniffing out the whereabouts of others of a desirable age and gender. And even the stereotypically boorish Bermuda-shorts-clad tourists with their vans, videocams, and surly children in tow may actually be uplifted by the sights of the Civil War battlefield or seaside to which they have driven.

When the range within which one moves about becomes extended, so too does the range of one's potential knowledge. The automobile is the quintessential range extender, not only by lengthening the trips one can take but also by multiplying the number of available routes. Knowledge by acquaintance has been emphasized in the preceding discussion, but automobility also extends one's ability to acquire other kinds of knowledge. Cars go not only to malls and theme parks but to libraries, universities, and museums. Cars provide regular access to urban centers of learning to those who live many miles distant. The traditional derogatory image of the unlettered "country bumpkin" has been rendered increasingly obsolete by new technologies—telephone, television, computer and, not least, the automobile.

THE WHEELS OF PRIVACY

Privacy complements autonomy. Someone who is private has a life of his own. That is, he is not entirely defined and constrained by a public persona. The capacity to be self-determining requires some quantum of privacy, whereas being an adjunct to a greater whole or an organic part of an organism does not. Individuals are private only to the extent that some part of their personas belongs primarily to them and not to the world at large. Being inappropriately viewed during a moment of intimacy or vulnerability constitutes one of the most basic encroachments on privacy. In an extended sense, privacy incorporates limitations not only on perceptual access but also on the knowledge or control others may have over oneself.

What constitutes an invasion of privacy is not fixed by our nature as human beings but is relative both to more or less arbitrary convention and to the far-from-arbitrary conditions that govern the possibility of forging an identity that is distinctively one's own. "A man's home is his castle" expresses one early manifestation of this impulse. The king is powerful and the king reigns, but in one little corner of the realm the commoner, not the king, enjoys (quasi) regal prerogatives. A right not to be subject to search and seizure without due process of law and a right not to be obliged to incriminate oneself are further manifestations. They express the conviction that personal dignity imposes limits on mandatory subjection to the scrutiny of others.

Some ancient conceptions of privacy endorsed a radical withdrawal from one's fellows. We should view the hermit or anchorite not as essentially a misanthrope but rather as someone who by separating himself from other human beings thereby draws closer to his God. (For Christians, Jesus in the wilderness provides the paradigmatic instance; there are many others.) Monasticism constitutes a slightly less radical version: voluntary sequestration with a few like-minded others away from the main crossroads of urban life. From Qumran by the Dead Sea to David Koresh at Waco, sectarians have acted on the belief that they could achieve a greater inner and external freedom by isolating themselves from the majority culture. When that majority culture nonetheless forcibly impinges on them, results typically are tragic.

Previously I have focused on the value to individuals of the capacity to approach and enjoy particular goods. The concern for privacy underscores the concomitant importance of the capacity to distance oneself from threats. If too many eyes are on me where I am, then I shall enhance my privacy by moving out of the spotlight of public scrutiny. For most of us the relevant degree of privacy rarely involves isolation from all others but usually does require the ability to exercise a significant degree of discretionary control over who will have access to one's body and mind. Adolescents who go out to "do nothing" thereby claim a measure of privacy vis-à-vis their parents; a fishing trip may have less to do with baiting fishhooks than with taking oneself off invasive social hooks.

For twentieth-century American society, the automobile serves as the quintessential bastion of privacy. For many of us the Honda, not the home, is the castle. Ironically or not, those minutes between home and office on a freeway clogged past capacity with multitudes of other cars may be one's most private time of the day. (I do not mean to slight the benefits of the other great solitude-enhancing device of our culture, the bathroom.) Even those who love their spouse and children, delight in the company of friends, and work compatibly alongside colleagues may nonetheless relish a short time each day to be alone. Such interludes do not indicate an antisocial impulse. Intermediate periods of solitude can fuel bouts of gregariousness and sociality just as an astringent serves to clean the palate between sumptuous courses.

Social planners are wont to gnash their teeth at the number of motorists who could arrange to commute by car pool but instead "inefficiently" take up roadway space with solitary-occupant cars. Diamond lanes and other inducements have only a limited effect on the average occupancy. This outcome may be viewed as a failure of policy, but it can also be seen as a reasonable and in some ways estimable response to the valid human desire for privacy. "It is not good for the man to be alone," says Scripture, but for those who live among a surfeit of others, it is sometimes very good indeed to be alone. The closing of the car door can provide a welcome shutting out of the rest of the world, allowing a recapture of the self by the self—as opposed to its usual embeddedness in an array of intersecting public spaces. Car pools are not necessarily a bad thing; in demonstrable respects, we might be better off if more people doubled and tripled up before taking to the roads. Privacy in virtually all its forms, including that afforded by the automobile has significant costs. (Think of the private room versus the hospital ward.) I shall not inquire here whether the costs of automotive privacy exceed the benefits; my point is simply that driving solo has genuine benefits that go beyond merely instrumental facility in getting from here to there. Any unbiased cost-benefit analysis must acknowledge that privacy has a positive value and proceed from there.

Being alone is one aspect of privacy but not, I believe, the most central. More salient to privacy than the distancing of oneself from others is a (re)gaining of control over one's immediate environment. I may be surrounded by other people, but if I can determine to a significant degree what they shall be allowed to perceive of me and know about me and impose on me, then to that extent I have retained a private self. Surely one reason for people's fondness for their cars and for automobility in general is the control afforded over one's immediate environment. Drivers make choices by turning the wheel clockwise and counterclockwise, determining the external environment to which they will move themselves; by other manipulations they arrange the internal environment to their liking. Pushing one button turns on the radio. Pushing others changes the station, lowers the volume, turns off the radio and switches to the tape player. Individuals choose for themselves whether to listen to news reports, Beethoven, the Beatles, or nothing at all. Next to the switches for the stereo are those for climate control, windshield washing, blinking one's lights, and perhaps a cellular phone. (Because the last item supplies incoming as well as outgoing calls, an assessment of whether it extends or diminishes privacy is double edged.) The vehicle's make, model, style, color, and options are more permanent objects of one-time choice. Automobile reviewers write about "responsiveness." This has a limited meaning in the context of evaluating how a vehicle performs, but automobiles, unique among all forms of personal transportation, have a larger responsiveness. Individuals exercise control over the internal environment of their cars in a manner not possible with any alternate mode of getting around.

Contrast the privacy-enhancing features of the automobile with a typical (typical, that is, based on the author's recent experience) commute by public transportation. As one walks down the stairs to the subway, one's nostrils are greeted by a subtle aroma of urine and garbage. If it is rush hour many milling people clog the platform, and so one tries to be careful neither to knock nor be knocked into. When traveling will actually commence is not in one's own hands; it depends on whether the train is on time or delayed. Being able to sit is a matter of luck. So, too, is the company one will keep. A man of indeterminate years holding a hat in his hand treks through the train car by car. He begs the attention of the passengers, tells them that he has no job, no place to sleep, no money. Dope, he announces, has scrambled his brain. That confession probably is true; he twitches, smells bad, looks unhealthy. Some people drop a quarter into the hat, most don't. A few minutes later three kids come through, break into song for a mercifully brief period, smile, wait to get paid. The singing displays few aesthetic gifts, but the boys' smiles are rather sharklike. Maybe another quarter is dropped in another hat, maybe not. Between the bumpings of the car and the performances of these itinerants one may manage to read a few *New York Times* column inches. Eventually one arrives at one's destination.

Again, I am not arguing against mass transportation. In some urban settings it is the only realistic way to move a large number of people through small spaces in a reasonable amount of time. My point rather is that public transportation necessarily encroaches on privacy. On a New York City subway the encroachment tends to be great; with other modalities it may be considerably smaller. However great or small, though, it belongs on the debit side if one counts privacy as a credit. Working out the magnitudes is the tricky part, an exercise that will vary according to differences in individual temperament and preferences. But once we focus attentively on privacy, it will no longer appear obvious to us that rush-hour gridlock on highways is an unacceptably high price to pay for the opportunity to be one's own man or woman behind the wheel of one's own car. Appealing to popular practice is not decisive in these matters, if only because some extraneous force may perversely shape such practice, but it does adduce evidence. That millions of people who bear no obvious marks of incompetence elect to drive when they might otherwise at equal or lower financial cost to themselves employ some means of public transportation indicates that for them automobility is a positive good rather than a necessary evil.

THE ROAD FROM SERFDOM

I have argued that the automobile does not merit the opprobrium its critics have showered on it. My reflections have considered some very general features of automobile usage, which obtain across nearly the whole range of interactions between motorists and their machine. I could have discussed more specialized enjoyments of automobiles: exhilarating in the speed of a high-powered sports vehicle taken flat out, the enthusiast's loving application of wax to a cherished collector car, the teenage boy half buried under the hood of the beat-up Ford whose engine he is tweaking for one last little bit of extra performance. These are automobile dividends, too, but because they appeal to special tastes, I judged that their inclusion might distract from the primary normative significance of automobility. Even with regard to only general considerations, however, one has ample reason to maintain that the ethical status of automobility stands quite high.

Why, then, has motoring fallen under such a cloud? Why does ostensibly enlightened opinion regard it as a bane and a nuisance? Three possible reasons suggest themselves. First, although the critics acknowledge the range of goods afforded by automobility, they have identified accompanying evils that drastically outweigh the goods. Second, the critics may have been oblivious to the various autonomy-enhancing features of automobility. Third, they may have recognized these features but regarded them as having a much lesser status than I have claimed on their behalf or, indeed, even as negatively valued.

Critics have driven home the case against the automobile with lengthy recitations of the social ills it fosters. I listed several of them in the opening section of this essay: polluting the air and littering the landscape with rusting steel cadavers, dependence on foreign oil suppliers, gridlock, the multitude of bodies mangled each year in road accidents, and so on. Let us grant that each is an evil. Still, they are not intrinsic to automobility as such but undesirable side effects. In a proper accounting, one will balance them against the various goods for whose attainment the automobile is instrumental. The overwhelming popularity of automobility among ordinary shoppers, commuters, suburbanites schlepping around the kids, and Sunday drivers out for a spin offers presumptive evidence that people value these goods highly. Precise measures can be left to the econometricians and their professional kin. I shall confine myself to making two different points.

First, the cited ills do not support a general indictment of the automobile and attempts to roll back its use. Rather, the indicated remedy is to adopt policies that reduce spillover costs. Legislators should aim taxes and regulatory controls at the vehicles that pollute excessively or present more than normal dangers to others; differential pricing for peak and off-peak access to highways lies well within the capabilities of currently available technology; and so on. Well-aimed attentiveness to particular avoidable costs is commendable; wholesale denunciations of automobility are not.

Second, the balance sheet of instrumental values and disvalues ignores the intrinsic goodness of automobility in promoting autonomy and complements of autonomy—such as free association and privacy. Even if purely instrumental calculations did not unambiguously display a positive balance in favor of automobility, its autonomy-enhancing aspects are so pronounced both qualitatively and quantitatively that any plausibly adequate normative evaluation of the status of automobile usage must give them primary attention.

Could the automobile's critics have failed to observe that cars support autonomy? If these effects were slight and subtle, that supposition might be reasonable. But when compared with alternate means of transportation the automobile stands out as the vehicle of self-directedness par excellence. To overlook this fact would be like visiting the mammal area at the zoo and failing to notice that the elephants are larger than the zebras, camels, and warthogs.

I am convinced that the automobile's most strident critics appreciate that automobility promotes autonomy—and that is precisely why they are so wary of it. Public policymakers have a professional predisposition to consider people as so many knights, rooks, and pawns to be moved around on the social chessboard in the service of one's grand strategy. Not all analysts succumb to this temptation, but many do. Their patron saint is the philosopher Plato, the utopian architect of the ideal Republic, who embraces propaganda campaigns ("Noble Lie"), eugenic breeding, radical property redistribution schemes and—most tellingly—rule exercised by people just like himself, the philosopher-kings. If one sincerely believes that one knows what is best, and if one benevolently desires to gift one's fellows with this treasure, their obdurate insistence on continuing to do things in their own preferred way can be maddening. "I'll give you what's good for you," the policy specialist vows, first in the soft tones of a promise and then, after experiencing rejection, in the clipped cadences of a threat.

People who drive automobiles upset the patterns spun from the policy intellectual's brain. The precise urban design that he has concocted loses out to suburban sprawl; neat integration of work, residence, and shopping within compact, multipurpose developments gives way to bedroom communities here, industrial parks there, and malls everywhere in between. If people rode buses and trains whenever they could, less oil would be burned and fewer acres of countryside would be paved over. Perhaps the races and classes would mix more. Perhaps communities of an old-fashioned sort, where everyone knew his neighbor, would return. Perhaps the central city would come alive again in the evenings. Perhaps . . . but why go on? These lovely visions give way before the free choices of men and women who resist all blandishments to leave their cars in the garage. They wish to drive, and by doing so they powerfully express their autonomy, but their exercises of choice also have the effect of rendering the planners' conceptions moot. So the intellectuals sulk in their tents and grumpily call to mind utopias that might have been.

Although this essay was stimulated in the first instance by a conviction that the critics of the automobile had, at best, offered distinctly one-sided appraisals, my aim here has been to develop the positive case for the value of automobility, not to respond point by point to the items in the brief against the automobile. (And, of course, I staunchly agree with some of these points.) Many of the argumentative missiles launched at the automobile become more fully intelligible if one understands them as motivated at least as much by a disinclination to tolerate individual autonomy as by any particular facet of automobile technology.

Consider an example. If the critics love anything less than cars, it is the roads they are driven on. If existing highways are too congested to support the quantity of traffic that squeezes along them, would it not be desirable to build more roads to relieve that gridlock? No! respond the critics. They oppose the construction of more highways on the grounds that as soon as a spanking new road opens to divert some of the flow from overused arteries, it too becomes engorged with traffic. The ultimate consequence is yet another venue for tedious stop-and-go automotive crawling. Better, then, not to waste any more dollars on futile freeway building. And at this point the subject usually turns to mass-transportation subsidies and new imposts on automobiles.

Most readers, I am sure, have heard the argument. But consider how odd it would sound in another context. I am in the business of teaching philosophy classes. Suppose that my class in the moral philosophy of Immanuel Kant were very popular, with every seat filled and a waiting list for admission. (Alas, the supposition is counterfactual.) And suppose further that when the philosophy department opens a new section of the class, it too becomes quickly oversubscribed. And the same for a third and then a fourth section. Should we conclude that continuing to pump resources into Kant pedagogy is futile, that instead we ought to use the money for a Nautilus machine in the football training room? That conclusion would be preposterous. Instead, my colleagues and I would rejoice in a renaissance of philosophy in northwestern Ohio.

No renaissance of Kant instruction is occurring, at least not yet. But for other items, one can observe such overflowing demand. McDonald's enjoys success at selling hamburgers. The company has thousands of establishments, many of them filled at rush hour with lines of people in pursuit of Big Macs and Chicken McNuggets. When McDonald's opens a new franchise, it too soon becomes congested with consumers waiting in lines to place their orders. Should we conclude that investing resources in more Golden Arches is futile?

No matter how many millions of instructions per second microprocessors perform, people keep demanding more and faster CPUs. Intel gives them the new generation top-of-the-line chip, and almost immediately people start impatiently clamoring for its successor. Should we conclude from this observed insatiability that investing in computing power wastes resources?

Big Macs and Pentium processors improve people's lives. Similarly, millions of people demand the use of highways because driving enhances their well-being. The striking feature of the critique of highway-building programs is that what should be taken as a sign of great, indeed overwhelming, success is presented as a mark of failure. But the only failure has been the critics' attempt to talk people out of their cars and out of the neighborhoods and workplaces their cars have rendered accessible. If my argument is sound, it shows that the critics' persuasive appeals deserved to fail. Automobile motoring is a good because people wish to engage in it, and they wish to engage in it because it is inherently good.

NOTES

1. Among the more comprehensive critiques are Mumford (1964), and Freund and Martin (1993). The campaign against the automobile is not confined to the United States; see "The Car Trap" (1996).
2. See, for example, Hensher (1993).
3. Aristotle's most extended discussion of *psyche* is in *De Anima*. The secondary literature is vast. A concise and comprehensible overview of Aristotelian philosophy is Ackrill (1981).
4. See "psyche" entry in Edwards (1967).
5. *Physics* is the Aristotelian work that addresses the investigation of being qua subject to change or motion.
6. See *Nicomachean Ethics*, bk. 3.
7. See his *Groundwork of the Metaphysics of Morals* (available in numerous editions). A useful discussion of the nature and importance of autonomy is Dworkin (1988).
8. Immanuel Kant, *Groundwork of the Metaphysics of Morals*, Sec. 3.
9. See, for example, Sartre ([1943] 1956).
10. An illuminating discussion is Childress (1982).
11. See, for example, Adler (1993), Cameron (1995), Calvert (1993), Harrington and others (1994).

REFERENCES

Ackrill, J. L. 1981. *Aristotle the Philosopher*. New York: Oxford University Press.

Adler, Jonathan. 1993. *Reforming Arizona's Air Pollution Policy*. Arizona Issue Analysis Report. no. 127, January, Goldwater Institute.

Calvert, J. G., and others. 1993. Achieving Acceptable Air Quality: Some Reflections on Controlling Vehicle Emissions. *Science*. 2 July.

Cameron, Michael. 1995. *Efficiency and Fairness on the Road*. Environmental Defense Fund.
The Car Trap. 1996. *World Press Review*. December, 6–11.
Childress, James F. 1982. *Who Should Decide? Paternalism in Health Care*. New York: Oxford University Press.
Dworkin, Gerald. 1988. *The Theory and Practice of Autonomy*. Cambridge: Cambridge University Press.
Edwards, Paul, ed. 1967. *The Encyclopedia of Philosophy*. New York: Macmillan.
Freund, Peter, and George Martin. 1993. The Ecology of the Automobile. In *Car Trouble*, edited by S. Nadis and J. MacKenzie. Boston: Beacon.
Harrington, Winston, and others. 1994. *Shifting Gears: New Directions for Cars and Clean Air. Discussion Paper*, no. 94–26, Resources for the Future.
Hayek, Friedrich August von. 1954. *Capitalism and the Historians*. Chicago: University of Chicago Press.
Hensher, David. 1993. Socially and Environmentally Appropriate Urban Futures for the Motor Car. *Transportation* 20:1–19.
Mill, Stuart. [1859] 1989. On Liberty. In *On Liberty and Other Writings*. Edited by Stefan Collini. Cambridge: Cambridge University Press.
Mumford, Lewis. 1963. *The Highway and the City*. New York: Harcourt, Brace & World.
Rawls, John, 1971. *A Theory of Justice*. Cambridge, Mass.: Harvard University Press.
Sartre, Jean-Paul. [1943] 1956. *Being and Nothingness: An Essay on Phenomenological Ontology*. Translated by Hazel E. Barnes. New York: Washington Square Press.

Acknowledgments: This essay was originally produced as a working paper for the Competitive Enterprise Institute (CEI). It was under CEI's auspices that my interest in the ethical dimensions of automobility was first piqued. I am grateful to CEI and, in particular, Sam Kazman for invaluable stimulation and assistance throughout the writing process.

About the Editors

ROBERT HIGGS is Senior Fellow in Political Economy at The Independent Institute and Editor of the Institute's quarterly journal, *The Independent Review: A Journal of Political Economy.* He received his Ph.D. in economics from the Johns Hopkins University, and he has taught at the University of Washington, Lafayette College, and Seattle University. He has been a visiting scholar at Oxford University and Stanford University.

Dr. Higgs is the editor of two Independent Institute books, *Arms, Politics, and the Economy: Historical and Contemporary Perspectives* (1990), *Hazardous to Our Health? FDA Regulation of Health Care Products* (1995), and of the volume *Emergence of the Modern Political Economy* (1985). His authored books include *The Transformation of the American Economy 1865–1914: An Essay in Interpretation* (1971), *Competition and Coercion: Blacks in the American Economy, 1865–1914* (1977), *Crisis and Leviathan: Critical Episodes in the Growth of American Government* (1987), and *Against Leviathan: Government Power and a Free Society* (2004). A contributor to numerous scholarly volumes, he is the author of more than 100 articles and reviews in academic journals of economics, demography, history, and public policy.

His popular articles have appeared in the *Wall Street Journal, Los Angeles Times, Providence Journal, Chicago Tribune, San Francisco Examiner, San Francisco Chronicle, Society, Reason, AlterNet,* and many other newspapers, magazines, and Web sites, and he has appeared on NPR, NBC, ABC, C-SPAN, CBN, CNBC, PBS, America's Talking Television, Radio America Network, Radio Free Europe, Talk Radio Network, Voice of America, Newstalk TV, the Organization of American Historians' public radio program, and scores of local radio and television stations. He has also been interviewed for articles in the *New York Times, Washington Post, Al-Ahram Weekly, Terra Libera, Investor's Business Daily,* UPI, *Congressional Quarterly, Orlando Sentinel, Seattle Times, Chicago Tribune, National Journal, Reason, Washington Times,* WorldNetDaily,

Folha de São Paulo, Newsmax, *Financial Times*, Creators Syndicate, *Insight*, *Christian Science Monitor*, and many other news media.

Dr. Higgs has spoken at more than 100 colleges and universities and to such professional organizations as the Economic History Association, Western Economic Association, Population Association of America, Southern Economic Association, International Economic History Congress, Public Choice Society, International Studies Association, Cliometric Society, Allied Social Sciences Association, American Political Science Association, American Historical Association, and many others.

CARL P. CLOSE is Academic Affairs Director of the Independent Institute and Assistant Editor of *The Independent Review*. His research interests include environmental policy, the history of economic and political thought, and the political economy of propaganda. He received his masters degree in economics (emphasis in business economics) from the University of California, Santa Barbara.

About the Contributors

LORD PETER T. BAUER was emeritus professor of economics at the London School of Economics and author of numerous books on economic development, including *West African Trade*; *The Economics of Under-developed Countries* (co-authored by Basil S. Yamey); *Dissent on Development*; *Equality, the Third World and Economic Delusion*; and *From Subsistence to Exchange.* Before passing away in 2002, he was awarded the Milton Friedman Prize for Advancing Liberty.

JOHN BRÄTLAND is an economist with the U.S. Department of the Interior in Washington, D.C. His articles have appeared in *Natural Resources Journal* and the *Quarterly Journal of Austrian Economics.*

J. R. CLARK is a professor of economics and occupies the Probasco Chair of Free Enterprise at the University of Tennessee, Chattanooga. He is the author of six books and numerous academic articles published in the United States, Japan, Italy, Canada, and Russia. He earned his Ph.D. in economics from Virginia Polytechnic Institute under Nobel laureate James M. Buchanan.

ROY E. CORDATO is Vice President for Research and Resident Scholar at the John Locke Foundation. From 1993 to 2000 he served as the Lundy Professor of Business Philosophy at Campbell University, Buies Creek, North Carolina. He has also taught economics at the University of Hartford, Auburn University, and Johns Hopkins University. He is the author of *Welfare Economics and Externalities in an Open Ended Universe.*

PIERRE DESROCHERS is Assistant Professor of geography at the University of Toronto at Mississauga. He has been a research director of the Montreal Economic Institute and a post-doctoral fellow at the Whiting School of Engineering, Johns Hopkins University, Baltimore, and has published articles the *Journal of Cleaner Production*, *Environmental Politics*, *Journal of Industrial Ecology* and elsewhere. His article "Industrial Ecology and the Rediscovery of Inter-Firm Recycling Linkages: Some Historical Perspective and Policy Implications" (*Industrial and Corporate Change*, vol. 11, no. 5) was selected as top environmental management paper of 2002 by Emerald Management Reviews.

PETER J. HILL is George F. Benett Professor of Economics at Wheaton (Ill.) College and a senior associate of PERC—the Property and Environment Research Center, Bozeman, Montana. His writings on the environment and property rights have appeared in such publications as the *Journal of Legal Studies*, *Agricultural Policy and the Environment*, and *Markets and Morality*. He is the co-editor (with Terry L. Anderson) of the books *Who Owns the Environment?* and *The Technology of Property Rights*.

JACQUELINE R. KASUN is a professor emeritus of economics at Humboldt State University, Arcata, California, and editorial director of the Center for Economic Education, Bayside, California. She is the author of *The War Against Population: The Economics and Ideology of Population Control*.

WILLIAM H. KAEMPFER is a professor of economics at the University of Colorado and the university's Vice Provost and Associate Vice Chancellor for Budget and Planning. He is co-author (with Anton D. Lowenberg) of *The Origins and Demise of South African Apartheid: A Public Choice Analysis*.

DWIGHT R. LEE is the Ramsey Professor of Economics in the Terry College of Business at the University of Georgia. A prolific author, his writings have appeared in the *Journal of Labor Research*, *Environmental and Resource Economics*, *Economic Inquiry*, *Southern Economic Journal*, *Public Choice*, *The Freeman*, and the *Wall Street Journal*, and in the books *Environmental Contaminants, Ecosystems and Human Health*, and *Buying a Better Environment: Cost-Effective Regulation Through Permit Trading* and many others.

LOREN E. LOMASKY is Cory Professor of Political Philosophy, Policy and Law, and Director of the Political Philosophy, Policy and Law Program at the University of Virginia. A recipient of the 1991 Matchette Prize for his book *Persons, Rights, and the Moral Community*, he is also co-author (with Geoffrey Brennan) of *Democracy and Decision: The Pure Theory of Electoral Preference* and co-editor (with Geoffrey Brennan) of *Politics and Process: New Essays in Democratic Theory*. He has published in such journals as *The American Journal* *of Economics and Sociology* and the *Journal of Ethics* and has held research appointments sponsored by the National Endowment for the Humanities, the Center for the Study of Public Choice, the Australian National University, and Bowling Green University's Social Philosophy and Policy Center.

ANTON D. LOWENBERG is a professor of economics at California State University, Northridge. With William H. Kaempfer he has co-authored the books *The Origins and Demise of South African Apartheid: A Public Choice Analysis* and *International Economic Sanctions: A Public Choice Perspective.* He received his Ph.D. in economics from Simon Frazer University.

CRAIG S. MARXSEN is an associate professor of economics at the University of Nebraska, Kearney. His research interests include economic growth and productivity, economic analysis of environmental policy, and the economic effects of government regulation. He received his Ph.D. in economics from Georgia State University.

ANDREW P. MORRISS is the Galen J. Roush Professor of Business Law and Regulation and the director of the Center for Business Law and Regulation at Case Western Reserve University School of Law. He is also a senior associate at PERC—the Property and Environment Research Center, Bozeman, Montana. His recent articles have been published in *Ecology Law Quarterly*, *Environmental Law*, and *Oregon Law Review*.

ROBERT H. NELSON is a professor at the School of Public Affairs of the University of Maryland, College Park, and Senior Fellow of the Competitive Enterprise Institute. He is the author of *Zoning and Property Rights*, *The Making of Federal Coal Policy*, *Reaching for Heaven on Earth: The Theological Meaning of Economics*, *Public Lands and Private Rights: The Failure of Scientific Management*, A *Burning Issue: A Case for Abolishing the U.S. Forest Service*, and *Economics as Religion: From Samuelson to Chicago and Beyond*, as well as numerous articles for scholarly and popular publications.

RANDAL O'TOOLE is senior economist at the Thoreau Institute and author of *The Vanishing Automobile and Other Urban Myths: How Smart Growth Will Harm American Cities* and *Reforming the Forest Service*. He has taught at Yale, the University of California (Berkeley), and Utah State University. He is also the recipient of the Oregon Environmental Council's Neuberger Award for Service to the Conservation Movement and the Oregon Natural Resources Council's David Simons Award for Vision.

JEFFREY J. POMPE is an associate professor of economics at Francis Marion University in Florence, South Carolina. He has co-authored with James R. Rinehart articles and book chapters on the economics of coastal housing and property rights, as well as the book *Environmental Conflict: In Search of Common Ground.*

JAMES R. RINEHART is the Phillip N. Truluck Professor of Economics and Public Policy at Francis Marion University and the co-author (with Jackson F. Lee, Jr.) of *American Education and the Dynamics of Choice.*

RANDY T. SIMMONS is a professor of political science and director of the Institute of Political Economy at Utah State University. He is the author of the forthcoming book *Political Ecology: Politics, Economics and the Endangered Species Act*, co-author (with William C. Mitchell) of *Beyond Politics: Markets, Welfare and the Failure of Bureaucracy*, and co-editor (with Terry T. Anderson) of *The Political Economy of Culture and Norms: Informal Solutions to the Commons Problem.*

RICHARD L. STROUP is a professor of economics at Montana State University and a senior associate at PERC—the Property and Environment Research Center, Bozeman, Montana. He is the author of numerous books on economics and public policy, including *Eco-Nomics: What Everyone Should Know About Economics* and the *Environment, Economics: Private and Public Choice* (with James Gwartney), and co-editor (with Roger Meiners) of *Cutting Green Tape: Toxic Pollutants, Environmental Regulation and the Law*. His articles have appeared in such journals as *Public Choice*, *Journal of Forestry*, *American Economic Review*, and *Harvard Journal of Law and Public Policy.*

BRUCE YANDLE is Alumni Professor of Economics at Clemson University, a senior associate of PERC—the Property and Environment Research Center, and Distinguished Adjunct Professor of Economics for the Mercatus Center's Capitol Hill Campus program. He is the author or co-author of numerous books, including *Taking the Environment Seriously*, *The Political Limits of Environmental Regulation*, *Environmental Use and the Market*, *Land Rights*, *The Economics of Environmental Quality*, and most recently, *Common Sense and Common Law for the Environment*. He is a member of the South Carolina State Board of Economic Advisors and a former executive director of the Federal Trade Commission.

Index

Page ranges in **bold** indicate an essay on that topic.

INDEPENDENT STUDIES IN POLITICAL ECONOMY

THE ACADEMY IN CRISIS: The Political Economy of Higher Education | *Ed. by John W. Sommer*

AGAINST LEVIATHAN: Government Power and a Free Society | *Robert Higgs*

AGRICULTURE AND THE STATE: Market Processes and Bureaucracy | *E. C. Pasour, Jr.*

ALIENATION AND THE SOVIET ECONOMY: The Collapse of the Socialist Era | *Paul Craig Roberts*

AMERICAN HEALTH CARE: Government, Market Processes and the Public Interest | *Ed. by Roger D. Feldman*

ANTITRUST AND MONOPOLY: Anatomy of a Policy Failure | *D. T. Armentano*

ARMS, POLITICS, AND THE ECONOMY: Historical and Contemporary Perspectives | *Ed. by Robert Higgs*

BEYOND POLITICS: Markets, Welfare and the Failure of Bureaucracy | *William C. Mitchell & Randy T. Simmons*

THE CAPITALIST REVOLUTION IN LATIN AMERICA | *Paul Craig Roberts & Karen LaFollette Araujo*

CHANGING THE GUARD: Private Prisons and the Control of Crime | *Ed. by Alexander Tabarrok*

CUTTING GREEN TAPE: Toxic Pollutants, Environmental Regulation and the Law | *Ed. by Richard Stroup & Roger E. Meiners*

THE DIVERSITY MYTH: Multiculturalism and Political Intolerance on Campus | *David O. Sacks & Peter A. Thiel*

DRUG WAR CRIMES: The Consequences of Prohibition | *Jeffrey A. Miron*

THE EMPIRE HAS NO CLOTHES: U.S. Foreign Policy Exposed | *Ivan Eland*

ENTREPRENEURIAL ECONOMICS: Bright Ideas from the Dismal Science | *Ed. by Alexander Tabarrok*

FAULTY TOWERS: Tenure and the Structure of Higher Education | *Ryan C. Amacher & Roger E. Meiners*

FREEDOM, FEMINISM, AND THE STATE: An Overview of Individualist Feminism | *Ed. by Wendy McElroy*

HAZARDOUS TO OUR HEALTH?: FDA Regulation of Health Care Products | *Ed. by Robert Higgs*

HOT TALK, COLD SCIENCE: Global Warming's Unfinished Debate | *S. Fred Singer*

LIBERTY FOR WOMEN: Freedom and Feminism in the Twenty-First Century | *Ed. by Wendy McElroy*

MARKET FAILURE OR SUCCESS: The New Debate | *Ed. by Tyler Cowen & Eric Crampton*

MONEY AND THE NATION STATE: The Financial Revolution, Government and the World Monetary System | *Ed. by Kevin Dowd & Richard H. Timberlake, Jr.*

OUT OF WORK: Unemployment and Government in Twentieth-Century America | *Richard K. Vedder & Lowell E. Gallaway*

A POVERTY OF REASON: Sustainable Development and Economic Growth | *Wilfred Beckerman*

PRIVATE RIGHTS & PUBLIC ILLUSIONS | *Tibor R. Machan*

RECLAIMING THE AMERICAN REVOLUTION: The Kentucky & Virginia Resolutions and Their Legacy | *William J. Watkins, Jr.*

REGULATION AND THE REAGAN ERA: Politics, Bureaucracy and the Public Interest | *Ed. by Roger Meiners & Bruce Yandle*

RESTORING FREE SPEECH AND LIBERTY ON CAMPUS | *Donald A. Downs*

RE-THINKING GREEN: Alternatives to Environmental Bureaucracy | *Ed. by Robert Higgs and Carl P. Close*

SCHOOL CHOICES: True and False | *John D. Merrifield*

STRANGE BREW: Alcohol and Government Monopoly | *Douglas Glen Whitman*

TAXING CHOICE: The Predatory Politics of Fiscal Discrimination | *Ed. by William F. Shughart, II*

TAXING ENERGY: Oil Severance Taxation and the Economy | *Robert Deacon, Stephen DeCanio, H. E. Frech, III, & M. Bruce Johnson*

THAT EVERY MAN BE ARMED: The Evolution of a Constitutional Right | *Stephen P. Halbrook*

TO SERVE AND PROTECT: Privatization and Community in Criminal Justice | *Bruce L. Benson*

THE VOLUNTARY CITY: Choice, Community and Civil Society | *Ed. by David T. Beito, Peter Gordon & Alexander Tabarrok*

WINNERS, LOSERS & MICROSOFT: Competition and Antitrust in High Technology | *Stan J. Liebowitz & Stephen E. Margolis*

WRITING OFF IDEAS: Taxation, Foundations, & Philanthropy in America | *Randall G. Holcombe*